TAKING SIDES

Clashing Views on Controversial

Issues in World Politics

ELEVENTH EDITION, EXPANDED

Selected, Edited, and with Introductions by

John T. Rourke
University of Connecticut

McGraw-Hill/Dushkin
A Division of The McGraw-Hill Companies

For my son and friend—John Michael

Cover image: © 2004 by PhotoDisc, Inc.

Cover Art Acknowledgment
Charles Vitelli

Manufactured in the United States of America

Eleventh Edition, expanded

123456789DOCDOC7654

Library of Congress Cataloging-in-Publication Data
Main entry under title:
Taking sides: clashing views on controversial issues in world politics/selected, edited, and with introductions by John T. Rourke.—11th ed.
Includes bibliographical references and index.
1. World Poltiics—1989–. I. Rourke, John T., *comp.*
909.82
0-07-304395-8
ISSN: 1094-754X

Printed on Recycled Paper

Preface

In the first edition of *Taking Sides: Clashing Views on Controversial Issues in World Politics,* I wrote of my belief in informed argument:

> [A] book that debates vital issues is valuable and necessary. . . . [It is important] to recognize that world politics is usually not a subject of absolute rights and absolute wrongs and of easy policy choices. We all have a responsibility to study the issues thoughtfully, and we should be careful to understand all sides of the debates.

It is gratifying to discover, as indicated by the success of *Taking Sides* over 10 editions, that so many of my colleagues share this belief in the value of a debate-format text.

The format of this edition follows a formula that has proved successful in acquainting students with the global issues that we face and in generating discussion of those issues and the policy choices that address them. This book addresses 19 issues on a wide range of topics in international relations. Each issue has two readings, one pro and one con. Each is accompanied by an issue *introduction,* which sets the stage for the debate, provides some background information on each author, and generally puts the issue into its political context. Each issue concludes with a *postscript* that summarizes the debate, gives the reader paths for further investigation, and suggests additional readings that might be helpful. I have also provided relevant Internet site addresses (URLs) in each postscript and on the *On the Internet* page that accompanies each part opener. At the back of the book is a listing of all the *contributors to this volume,* which will give you information on the political scientists and commentators whose views are debated here.

I have continued to emphasize issues that are currently being debated in the policy sphere. The authors of the selections are a mix of practitioners, scholars, and noted political commentators.

Changes to this edition   The dynamic, constantly changing nature of the world political system and the many helpful comments from reviewers have brought about significant changes to this edition. Of the 38 readings in this edition, 30, or 79 percent, are new, with only 8 readings being carried over from the previous edition.

The kaleidoscopic dynamism of the international system is also evident in the high turnover in issues from one edition to the next of this reader. Only 4 (21 percent) of the 19 issues are carried over directly from the previous edition. In contrast, 10 issues (53 percent) and their readings are completely new. They are: *Is Globalization Likely to Create a Better World?* (Issue 1); *Will the "Bush Doctrine" Promote a More Secure World?* (Issue 3); *Should the United States Continue to Encourage a United Europe?* (Issue 4); *Was the War With Iraq Unjustified?* (Issue 7); *Should*

North Korea's Arms Program Evoke a Hard-Line Response? (Issue 8); *Should the Rich Countries Forgive All the Debt Owed by the Poor Countries?* (Issue 10); *Are Patents on HIV/AIDS Drugs Unfair to Poor Countries?* (Issue 11); *Does the Moscow Treaty Advance Nuclear Arms Reductions?* (Issue 12); *Are Military Means the Best Way to Defeat Terrorism?* (Issue 13); and *Is Government-Ordered Assassination Sometimes Acceptable?* (Issue 14).

Another 5 issues have been recast to reflect changes in the specific concerns related to general topics that were included in the last edition. These "semi-new" debates, all with new readings, are: *Do China's Armaments and Intentions Pose a Long-Term Threat?* (Issue 5); *Would It Be an Error to Establish a Palestinian State?* (Issue 6); *Is Free Economic Interchange Beneficial?* (Issue 9); *Do International Financial Organizations Require Radical Reform?* (Issue 17); and *Do Environmentalists Overstate Their Case?* (Issue 19).

It is important to note that the changes to this edition from the last should not disguise the fact that most of the issues address enduring human concerns, such as global political organization, arms and arms control, justice, development, and the environment. Also important is the fact that many of the issues have both a specific and a larger topic. For instance, Issue 16 is about the specific topic of strengthening the UN's peacekeeping (or peacemaking) ability, but it is also about more general topics. These include whether or not international organizations should be given supranational powers; the propriety of interventionism using UN, NATO, or international forces; and the argument by some small countries that there is a growing neocolonialism in the world today.

A word to the instructor An *Instructor's Manual With Test Questions* (multiple-choice and essay) is available through the publisher for instructors using *Taking Sides* in the classroom. A general guidebook, *Using Taking Sides in the Classroom,* which discusses methods and techniques for integrating the pro-con approach into any classroom setting, is also available. An online version of *Using Taking Sides in the Classroom* and a correspondence service for *Taking Sides* adopters can be found at http://www.dushkin.com/usingts/.

Taking Sides: Clashing Views on Controversial Issues in World Politics is only one title in the Taking Sides series. If you are interested in seeing the table of contents for any of the other titles, please visit the Taking Sides Web site at http://www.dushkin.com/takingsides/.

A note especially for the student reader You will find that the debates in this book are not one-sided. Each author strongly believes in his or her position. And if you read the debates without prejudging them, you will see that each author makes cogent points. An author may not be "right," but the arguments made in an essay should not be dismissed out of hand, and you should work to remain tolerant of those who hold beliefs that are different from your own.

There is an additional consideration to keep in mind as you pursue this debate approach to world politics. To consider divergent views objectively does not mean that you have to remain forever neutral. In fact, once you are informed, you ought to form convictions. More important, you should try to influence international policy to conform better with your beliefs. Write letters to policymakers; donate to causes you support; work for candidates who agree with your views; join an activist organization. *Do* something, whichever side of an issue you are on!

Acknowledgments I received many helpful comments and suggestions from colleagues and readers across the United States and Canada. Their suggestions have markedly enhanced the quality of this edition of *Taking Sides.* If as you read this book you are reminded of a selection or an issue that could be included in a future edition, please write to me in care of McGraw-Hill/Dushkin with your recommendations or e-mail them to me at john.rourke@uconn.edu.

My thanks go to those who responded with suggestions for the 11th edition. I would also like to thank Ted Knight, managing editor, and David Brackley, senior developmental editor for the Taking Sides series, for their help in refining this edition.

John T. Rourke
University of Connecticut

Contents in Brief

Contents

Thomas Friedman, a columnist for the *New York Times,* contends that with good leadership, globalization will create more openness in government and business, a strong rule of law, and greater opportunities for people to experience personal freedom and to challenge government authority. Robert Kaplan, a correspondent for *The Atlantic Monthly,* argues that although globalization might someday bring the benefits that Friedman foresees, that is uncertain. He maintains that in the next few decades globalization is likely to cause more and more turbulence.

Professor of international relations Stephen D. Krasner contends that the nation-state has a keen instinct for survival and will adapt to globalization and other challenges to sovereignty. Kimberly Weir, an assistant professor of political science, maintains that the tide of history is running against the sovereign state as a governing principle, which will soon go the way of earlier, now-discarded forms of governance, such as empire.

George W. Bush, president of the United States, tells Americans that the United States possesses unrivaled power but is threatened by terrorists, the acquisition of weapons of mass destruction by rogue countries, and other perils. He argues that the United States should take strong action, preferably with other countries but alone if necessary, to ensure that enemies cannot attack or intimidate Americans and others. John Steinbruner, director of the Center for International and Security Studies at Maryland, contends that what has become called the "Bush doctrine" is unnecessarily provocative and that its threat to use preemptive military power against whomever the United States perceives to be a threat will increase global violence and create such hostility toward the United States that its security will be decreased.

A. Elizabeth Jones, assistant secretary of state for European and Eurasian affairs, maintains that the United States looks forward to working cooperatively with such exclusively or mostly European institutions as the European Union, the Organization for Cooperation and Security in Europe, and the North Atlantic Treaty Organization. John C. Hulsman, a research fellow for European affairs in the Kathryn and Shelby Cullom Davis Institute for International Studies at the Heritage Foundation, contends that mutual exchanges of pleasantries and vague rhetoric about the value of a strong and united Europe obscure the reality that U.S.-European relations are increasingly strained. He argues that the United States should support European countries on a selective basis but not be closely tied to Europe as a whole.

Richard D. Fisher, Jr., a senior fellow with the Jamestown Foundation, characterizes China's military as growing in sophistication and strength, and he argues that China's buildup has worrisome implications for U.S. national interests. Ivan Eland, director of defense policy studies at the Cato Institute, contends that China's military modernization is less of a threat to U.S. interests than recent studies by the Pentagon and the U.S.–China Security Review Commission indicate.

P. J. Berlyn, an author of studies on Israel, primarily its ancient history and
culture, refutes 12 arguments supporting the creation of an independent
state of Palestine, maintaining that such a state would not be wise, just, or
desirable. Rosemary E. Shinko, who teaches in the Department of Political
Science at the University of Connecticut, contends that a lasting peace
between Israelis and Palestinians must be founded on a secure and sovereign homeland for both nations.

Issue 7.   Was the War With Iraq Unjustified?   108

Professor of political science John Mueller characterizes U.S. policy toward Iraq as an overreaction and an unhealthy reflection of the U.S. sense
that the United States is an indispensable country. Brink Lindsey, a senior
fellow at the Cato Institute, asserts that Iraq's potential nuclear capability
combined with Saddam Hussein's malevolence provided sufficient justification for a U.S.-led invasion.

Issue 8.   Should North Korea's Arms Program Evoke a Hard-Line Response?   122

William Norman Grigg, senior editor of *The New American*, argues that
North Korea is a dangerous country with an untrustworthy regime and that
it is an error for the United States to react to North Korea's nuclear arms
program and other provocations by offering it diplomatic and economic incentives to be less confrontational. Fred Kaplan, author of the "War Stories" column in *Slate*, agrees that North Korea is dangerous and unstable,
but he maintains that the horrendous costs of a conflict with North Korea
mean that the United States and others would be better off trying to assuage North Korea than confronting it.

PART 3   ECONOMIC ISSUES   135

Issue 9.   Is Free Economic Interchange Beneficial?   136

Anne O. Krueger, first deputy managing director of the International Monetary Fund, asserts that the growth of economic globalization is unstoppable and that supporting it is one of the best ways to improve global conditions. José Bové, a French farmer and anti-globalization activist, contends that multinational corporations, government leaders, and others are engaged in a propaganda campaign to sell the world on the false promise of economic globalization.

Romilly Greenhill, an economist with Jubilee Research at the New Economics Foundation, contends that if the world community is going to achieve its goal of eliminating world poverty by 2015, as stated in the UN's Millennium Declaration, then there is an urgent need to forgive the massive debt owed by the heavily indebted poor countries. William Easterly, a senior adviser in the Development Research Group at the World Bank, maintains that while debt relief is a popular cause and seems good at first glance, the reality is that debt relief is a bad deal for the world's poor.

The international nongovernmental organization Médecins Sans Frontières (Doctors Without Borders), which is based in France, argues that the prevalence of HIV/AIDS is a crisis and that as part of the effort to meet that crisis, the high cost of medicines protected by patents needs to be reduced. Alan F. Holmer, president and chief executive officer of the industry trade group Pharmaceutical Research and Manufacturers of America, contends that the cost of research to develop new medicines is very high and that selling medicines at artificially low prices will harm the development of new pharmaceutical products.

PART 4 ISSUES ABOUT VIOLENCE 183

Donald H. Rumsfeld, U.S. secretary of defense, argues that the Moscow
Treaty represents a decision by the United States and Russia to move to-
ward historic reductions in their deployed offensive nuclear arsenals.
Christopher E. Paine, codirector of the Nuclear Warhead Elimination and
Nonproliferation Project, Natural Resources Defense Council, argues that
the Moscow Treaty is too vague to be legally binding on either the United
States or Russia and, worse, that the treaty is actually a step backward in
the long-standing effort to reduce, even eliminate, nuclear weapons.

Benjamin Netanyahu, former prime minister of Israel, argues that there
would be little or no organized terrorism if it were not supported by the
governments of various countries and that the only way to reduce or eliminate terrorism is to make it clear to countries that support it that they face
military retaliation. Bill Christison, a former member of the Central Intelligence Agency, identifies what he describes as the six root causes of terrorism and argues that using force to address them may not only be
ineffective but also counterproductive.

Bruce Berkowitz, a research fellow at the Hoover Institution at Stanford
University, argues that while government-directed political assassinations
are hard to accomplish and are not a reliably effective political tool, there
are instances where targeting and killing an individual is both prudent and
legitimate. Margot Patterson, a senior writer for *National Catholic Reporter*, contends that assassinations are morally troubling, often counterproductive, and have a range of other drawbacks.

Francis Fukuyama, the Hirst Professor of Public Policy at George Mason University, contends that a truly matriarchal world would be less prone to conflict and more conciliatory and cooperative than the largely male-dominated world that we live in now. Assistant professor of political science Mary Caprioli contends that Fukuyama's argument is based on a number of unproven assumptions and that when women assume more political power and have a chance to act aggressively, they are as apt to do so as men are.

PART 5 INTERNATIONAL LAW AND ORGANIZATION ISSUES 253

Lionel Rosenblatt and Larry Thompson, president and senior associate, respectively, of Refugees International in Washington, D.C., advocate the creation of a permanent UN peacekeeping force on the grounds that the present system of peacekeeping is too slow, too cumbersome, too inefficient, and too prone to failure. John Hillen, a policy analyst for defense and national security issues at the Heritage Foundation in Washington, D.C., contends that the United Nations was never intended to have, nor should it be augmented to have, the authority and capability to handle significant military operations in dangerous environments.

In an interview conducted by Lucy Komisar, Joseph Stiglitz, former chief economist of the World Bank, argues that the policies of the World Bank and the International Monetary Fund (IMF) are driven by the economic model favored by the United States and other powerful and prosperous countries rather than the interests of the poor countries that the World Bank and the IMF are supposed to be helping. Kenneth Rogoff, economic counselor and director of research for the International Monetary Fund, concedes that the World Bank and the IMF, like all human organizations, fall short of perfection. He contends, however, that Stiglitz's unbridled criticisms of the two organizations are often factually faulty and that the reforms he favors are unwise.

The Lawyers Committee for Human Rights, in a statement submitted to the U.S. Congress, contends that the International Criminal Court (ICC) is an expression, in institutional form, of a global aspiration for justice. John R. Bolton, senior vice president of the American Enterprise Institute in Washington, D.C., contends that support for an international criminal court is based largely on naive emotion and that adhering to its provisions is not wise.

Professor of statistics Bjørn Lomborg argues that it is a myth that the world is in deep trouble on a range of environmental issues and that drastic action must be taken immediately to avoid an ecological catastrophe. Fred Krupp, executive director of Environmental Defense, asserts that although Lomborg's message is alluring because it says we can relax, the reality is that there are serious problems that, if not addressed, will have a deleterious effect on the global environment.

Secretary of Defense Donald H. Rumsfeld argues that national security policy of the Bush administration should be lauded for its remarkable achievements, and that the president has a sound policy. The Task Force on a Unified Security Budget for the United States, established by the Foreign Policy in Focus Project of the Institute for Policy Studies for the Center for Defense Information, two private think tanks in Washington, D.C., contends that despite significant increases in the U.S. national security budget, the Bush administration is not spending the money wisely to increase U.S. security.

U.S. Senator Byron Dorgan, a Democrat representing North Dakota, contends that under the administration of President George W. Bush the United States has a trade strategy that is in total chaos and that is undermining the economic health of the country. U.S. Trade Representative Robert B. Zoellick, President Bush's chief trade official, maintains that open international markets create new jobs for Americans and build U.S. economic strength and that a retreat to protectionism would be destructive.

Introduction

World Politics and the Voice of Justice

John T. Rourke

Some years ago, the Rolling Stones recorded "Sympathy With the Devil." If you have never heard it, go find a copy. It is worth listening to. The theme of the song is echoed in a wonderful essay by Marshall Berman, "Have Sympathy for the Devil" (*New American Review,* 1973). The common theme of the Stones' and Berman's works is based on Johann Goethe's *Faust.* In that classic drama, the protagonist, Dr. Faust, trades his soul to gain great power. He attempts to do good, but in the end he commits evil by, in contemporary paraphrase, "doing the wrong things for the right reasons." Does that make Faust evil, the personification of the devil Mephistopheles among us? Or is the good doctor merely misguided in his effort to make the world better as he saw it and imagined it might be? The point that the Stones and Berman make is that it is important to avoid falling prey to the trap of many zealots who are so convinced of the truth of their own views that they feel righteously at liberty to condemn those who disagree with them as stupid or even diabolical.

It is to the principle of rational discourse, of tolerant debate, that this reader is dedicated. There are many issues in this volume that appropriately excite passion—for example, Issue 6 on whether or not Israel should agree to an independent Palestinian state, or Issue 13, which examines the degree to which military action is the best approach to lessening the danger of terrorism. Few would deny, for example, that a danger from terrorism exists. But what is not clear is which is the best approach to reducing that threat: putting an emphasis on attacking terrorists and the states that support them or trying to resolve the social and economic issues that many believe foster terrorism.

In other cases, the debates you will read do diverge on goals. In Issue 16, Lionel Rosenblatt and Larry Thompson argue that the world will be better off if the United Nations is given much stronger military capabilities to keep and, if necessary, forcefully make peace. John Hillen disagrees. Behind the specifics of the debate, the authors of the two readings differ on the overall goal of strengthening the UN, with Rosenblatt and Thompson favoring and Hillen being wary of the goal of greater global governance.

As you will see, each of the authors in all the debates strongly believes in his or her position. If you read these debates objectively, you will find that each side makes cogent points. They may or may not be right, but they should not be dismissed out of hand. It is important to repeat that the debate format does not imply that you should remain forever neutral. In fact, once you are informed, you *ought* to form convictions, and you should try to act on those convictions and try to influence international policy to conform better with your beliefs. Ponder the similarities in the views of two very different leaders, a very young president in a relatively young democracy and a very old emperor in a very old country: In 1963 President John F. Kennedy, in recalling the words of the author of the epic poem *The Divine Comedy* (1321), told a West German audience, "Dante once said that the hottest places in hell are reserved for those who in a period of moral crisis maintain their neutrality." That very same year, while speaking to the United Nations, Ethiopia's emperor Haile Selassie (1892–1975) said, "Throughout history it has been the inaction of those who could have acted, the indifference of those who should have known better, the silence of the voice of justice when it mattered most that made it possible for evil to triumph."

The point is: Become Informed. Then *do* something! Write letters to policymakers, donate money to causes you support, work for candidates with whom you agree, join an activist organization, or any of the many other things that you can do to make a difference. What you do is less important than that you do it.

Approaches to Studying International Politics

As will become evident as you read this volume, there are many approaches to the study of international politics. Some political scientists and most practitioners specialize in *substantive topics,* and this reader is organized along topical lines. Part 1 (Issues 1 and 2) features debates on the evolution of the international system in the direction of greater globalization. In Issue 1, Thomas Friedman and Robert Kaplan debate the general topic of globalization. Friedman is optimistic, maintaining that globalization is a positive trend; Kaplan is pessimistic. Issue 2 takes up the relatively less advanced but still important aspect of globalization represented by the growth in importance of international organizations and international law. In it, Stephen D. Krasner and Kimberly Weir engage in a debate over whether countries will continue to maintain their sovereignty in the future or whether they will become at least partially subordinate to regional and global organizations.

Part 2 (Issues 3 through 8) focuses on country-specific issues, including the role of the United States in the international system, the future of U.S.-European relations, the diplomatic posture of China, the possibility of a Palestinian state, the 2003 U.S.-led war against Iraq, and the crisis over North Korea's nuclear weapons program. Part 3 (Issues 9 through 11) deals with specific concerns of the international economy, a topic introduced more generally in Issue 1. Issue 9 takes up the economic aspect of globalization. Anne O. Krueger argues that free economic interchange is beneficial. José Bové condemns globalization for enriching the few at the expense of the many and other deleterious impacts. In Issue 10, the debate turns to the huge financial debt owed by poor countries to international financial organizations such as the International Monetary Fund (IMF) and to the governments, banks, and other investors of the wealthiest countries. Romilly Greenhill argues that the debt should be forgiven;

William Easterly disagrees. Issue 11 focuses on the difficult issue of supplying medicine to treat HIV and AIDS in the world's poor countries. One view is that pharmaceutical companies should sell the medicine cheaply or give it outright to the people in these countries. Another view is that however emotionally appealing that argument might be, the research and development cost of new drugs is immense, and giving them away will leave no funds for the research that is necessary to develop new drugs.

Part 4 (Issues 12 through 15) examines violence in the international system. The past few decades have witnessed a number of treaties that have helped slow, then reverse, the buildup of nuclear weapons. In Issue 12, Donald H. Rumsfeld argues that the Moscow Treaty signed in 2002 by President George W. Bush of the United States and President Vladimir Putin of Russia is a further step along the road to nuclear weapons reduction and safety. Christopher E. Paine argues that the Moscow Treaty accomplishes little or nothing and leaves the door open for an actual increase in nuclear weapons. Issue 13 turns to what has become a high-profile threat: terrorism. Benjamin Netanyahu argues that the best strategy for combating terrorism is to fight fire with fire by launching strong military attacks against terrorists and terrorist states. Bill Christison counters that such strikes do nothing to address the root causes of terrorism. Indeed, he views an emphasis on military retribution as counterproductive. Issue 14 debates whether or not government-orchestrated assassinations are an effective and moral way of conducting foreign policy. Although it was generally shunned in the past, in the current era of increased terrorism, assassination has become more acceptable to some people. This debate is followed by Issue 15, which focuses on the potential impact of the growing political roles of women on the level of violence in the world. There can be little doubt that national and international politics have historically been dominated by males. There can also be little doubt that women are more frequently playing larger roles in the world by, among other things, holding such positions as president, prime minister, and foreign secretary. Certainly, equity demands that women have the same opportunities as men to achieve leadership positions. But will that equity also promote a decrease in world violence? Therein lies the debate.

Part 5 (Issues 16 through 18) addresses controversies related to international law and organizations. The ability of the United Nations to deploy effective peacekeeping forces is severely constrained by a number of factors. In Issue 16, Lionel Rosenblatt and Larry Thompson contend that enhancing the strength of UN peacekeepers would represent progress. John Hillen disputes this view, arguing that the UN was never intended to have a powerful military arm and that it would be an error to create such a capability. There are also a number of important controversies surrounding the application of international law in world politics. Issue 17 takes up the issue of whether or not the policies and decision-making structures of the major international financial institutions, especially the IMF, need radical reform. The two authors, one a former ranking official in the World Bank, the other a current ranking official in the IMF, disagree over whether or not the IMF's rules and structure are destructive and undemocratic. The issue on the law of war flows into the debate in Issue 18, which evaluates the wisdom of establishing a permanent international criminal court to punish those who violate the law of war. It is easy to advocate such a court as long as it is trying and sometimes punishing alleged war criminals from other countries. But one has to understand that one day a citizen of one's own country could be put on trial.

Part 6, which consists of Issue 19, addresses the environment issue. Over the past few decades there has been a growing concern that population growth; the discharge of liquid, gaseous, and solid waste into the air, water, and ground; the overconsumption of natural resources; and other human activities are severely, even irreparably, damaging the Earth's ecosphere. From this perspective, significant changes have to be made to avert further damage and an environmental disaster. Bjørn Lomborg argues that the cries of alarm are vastly overstated and that whatever ills do exist can be addressed without radical change. Fred Krupp takes the concerned view, arguing that action needs to be taken to ensure a protected environment.

Political scientists also approach their subject from differing *methodological perspectives*. You will see, for example, that world politics can be studied from different *levels of analysis*. The question is, What is the basic source of the forces that shape the conduct of politics? Possible answers are world forces, the individual political processes of the specific countries, or the personal attributes of a country's leaders and decision makers. Various readings will illustrate all three levels.

Another way for students and practitioners of world politics to approach their subject is to focus on what is called the realist versus the idealist (or liberal) debate. Realists tend to assume that the world is permanently flawed and therefore advocate following policies in their country's narrow self-interests. Idealists take the approach that the world condition can be improved substantially by following policies that, at least in the short term, call for some risk or self-sacrifice. This divergence is an element of many of the debates in this book. Issue 3 is a particular example. In the first reading, President Bush advocates a very realpolitik approach of unilateral, preemptive military action when it is deemed necessary to preserve U.S. interests. John Steinbruner argues that coercive preemption is dangerous. He favors a more liberal approach involving multilateral diplomacy and action through the UN and other international organizations.

Dynamics of World Politics

The action on the global stage today is vastly different from what it was a few decades ago, or even a few years ago. *Technology* is one of the causes of this change. Technology has changed communications, manufacturing, health care, and many other aspects of the human condition. Technology has given humans the ability to create biological, chemical, and nuclear compounds and other material that in relatively small amounts have the ability to kill and injure huge numbers of people. Another negative byproduct of technology may be the vastly increased consumption of petroleum and other natural resources and the global environmental degradation that has been caused by discharges of waste products, deforestation, and a host of other technology-enhanced human activities.

Another dynamic aspect of world politics involves the *changing axes* of the world system. For about 40 years after World War II ended in 1945, a bipolar system existed, the primary axis of which was the *East-West* conflict, which pitted the United States and its allies against the Soviet Union and its allies. Now that the cold war is over, one broad debate is over what role the United States should play. A related issue is whether or not there are potential enemies to the United States and its allies and, if so, who they are. The advocates on either side of Issue 3 disagree about the degree to which the United States should try to act unilaterally or work cooperatively. As for potential rivals to U.S. hegemony, Issue 4 deals with Europe and its increasing uneasiness with U.S. power. Could Europe or some European countries become rivals instead of allies? More conventionally, Issue 5 deals with China, a cold war antagonist of the United States. Some people believe that China is the next superpower and that it will pose a threat to U.S. security and interests. As such, the debate, beyond the specific issues involved, also deals with how to interact with former and potential enemies.

Technological changes and the shifting axes of international politics also highlight the *increased role of economics* in world politics. Economics have always played a role, but traditionally the main focus has been on strategic-political questions—especially military power. This concern still strongly exists, but now it shares the international spotlight with economic issues. One important change in recent decades has been the rapid growth of regional and global markets and the promotion of free trade and other forms of international economic interchange. As Issue 9 on economic interdependence indicates, many people support these efforts and see them as the wave of the future. But there are others who believe that free economic interchange undermines sovereignty and the ability of governments to control their destinies. One topic related to control, which is taken up in Issue 17, is whether or not the developed countries should continue their control of the major international financial organizations and whether those organizations—particularly the IMF—are making a contribution or using the desperation of needy countries to impose alien and often destructive standards on those countries.

Another change in the world system has to do with the main *international* actors. At one time states (countries) were practically the only international actors on the world stage. Now, and increasingly so, there are other actors. Some actors are regional. Others, such as the United Nations, are global actors. Issue 2 discusses the future of countries as principal and sovereign actors in the international system. Turning to the most notable international organization, Issue 16 examines the call for strengthening the peacekeeping and peacemaking capabilities of the United Nations by debating the wisdom of establishing a permanent UN military force. Issue 18 focuses on whether or not a supranational criminal court should be established to take over the prosecution and punishment of war criminals from the domestic courts and ad hoc tribunals that have sometimes dealt with such cases in the past. And Issue 17 addresses some of the issues about voting formulas and other processes within international organizations.

Perceptions Versus Reality

In addition to addressing the general changes in the world system outlined above, the debates in this reader explore the controversies that exist over many of the fundamental issues that face the world.

One key to these debates is the differing *perceptions* that protagonists bring to them. There may be a reality in world politics, but very often that reality is obscured. Many observers, for example, are alarmed by the seeming rise in radical actions by Islamic fundamentalists. However, the image of Islamic radicalism is not a fact but a perception; perhaps correct, perhaps not. In cases such as this, though, it is often the perception, not the reality, that is more important because policy is formulated on what decision makers *think,* not necessarily on what *is.* Thus, perception becomes the operating guide, or *operational reality,* whether it is true or not. Perceptions result from many factors. One factor is the information that decision makers receive. For a variety of reasons, the facts and analyses that are given to leaders are often inaccurate or represent only part of the picture. The conflicting perceptions of Israelis and Palestinians, for example, make the achievement of peace in Israel very difficult. Many Israelis and Palestinians fervently believe that the conflict that has occurred in the region over the past 50 years is the responsibility of the other. Both sides also believe in the righteousness of their own policies. Even if both sides are well-meaning, the perceptions of hostility that each holds means that the operational reality often has to be violence. These differing perceptions are a key element in the debate in Issue 6.

A related aspect of perception is the tendency to see oneself differently than some others do. Specifically, the tendency is to see oneself as benevolent and to perceive rivals as sinister. This reverse image is partly at issue in the debate on China's future (Issue 5). Most Americans, especially those who favor increased defense expenditures, see U.S. policy as benign and the U.S. military as purely defensive. Americans are apt to see the recent changes in China, which include a more active regional role and increased military spending, as threatening. Most analysts in China see a reverse image, picturing themselves as arming to defend China against a United States with hegemonic intentions. Perceptions, then, are crucial to understanding international politics. It is important to understand objective reality, but it is also necessary to comprehend subjective reality in order to be able to predict and analyze another country's actions.

Levels of Analysis

Political scientists approach the study of international politics from different levels of analysis. The most macroscopic view is *system-level analysis.* This is a top-down approach that maintains that world factors virtually compel countries to follow certain foreign policies. Governing factors include the number of powerful actors, geographic relationships, economic needs, and technology. System analysts hold that a country's internal political system and its leaders do not have a major impact on policy. As such, political scientists who work from this perspective are interested in exploring the governing factors, how they cause policy, and how and why systems change.

After the end of World War II, the world was structured as a *bipolar* system, dominated by the United States and the Soviet Union. Furthermore, each superpower was supported by a tightly organized and dependent group of allies. For a variety of reasons, including changing economics and the nuclear standoff, the bipolar system has faded. Some political scientists argue that the bipolar system is being replaced by a *multipolar* system. In such a configuration, those who favor *balance-of-power* politics maintain that it is unwise to ignore power considerations.

State-level analysis is the middle and most common level of analysis. Social scientists who study world politics from this perspective focus on how countries, singly or comparatively, make foreign policy. In other words, this perspective is concerned with internal political dynamics, such as the roles of and interactions between the executive and legislative branches of government, the impact of bureaucracy, the role of interest groups, and the effect of public opinion. The dangers to the global environment, which are debated in Issue 19, extend beyond rarified scientific controversy to important issues of public policy. For example, should the United States and other industrialized countries adopt policies that are costly in terms of economics and lifestyle to significantly reduce the emission of carbon dioxide and other harmful gases? This debate pits interest groups against one another as they try to get the governments of their respective countries to support or reject the steps necessary to reduce the consumption of resources and the emission of waste products. To a large degree, it is the environmentalists versus the business groups.

A third level of analysis, which is the most microscopic, is *human-level analysis.* This approach focuses, in part, on the role of individual decision makers. This technique is applied under the assumption that individuals make decisions and that the nature of those decisions is determined by the decision makers' perceptions, predilections, and strengths and weaknesses. Issue 15, for example, explores whether or not increasing the number of women in positions of political authority around the world will make a difference in policy, especially in the use of violence as a political tool.

The Political and Ecological Future

Future *world alternatives* are discussed in many of the issues in this volume. Abraham Lincoln once said, "A house divided against itself cannot stand." One suspects that the 16th president might say something similar about the world today if he were with us. Issue 1, for example, debates whether growing globalization is a positive or negative trend. The world has responded to globalization by creating and strengthening the UN, the IMF, the World Bank, the World Trade Organization, and many other international organizations to try to regulate the increasing number of international interactions. There can be little doubt that the role of global governance is growing, and this reality is the spark behind the debate in Issue 2 over whether or not the traditional sovereignty of states will persist in a time of increasing globalization. More specific debates about the future are taken up in many of the selections that follow the two debates in Part 1. Far-reaching alternatives to a state-centric system based on sovereign countries include international organizations' taking over some (or all) of the sovereign responsibilities of national governments, such as peacekeeping and peacemaking (Issue 16) and the prosecution of international war criminals (Issue 18). The global future also involves the ability of the world to prosper economically while not denuding itself of its natural resources or destroying the environment. This is the focus of Issue 19 on the environment.

The Axes of World Division

The world is politically dynamic, and the nature of the political system is undergoing profound change. As noted, the once-primary axis of world politics, the East-West confrontation, has broken down. Yet a few vestiges of the conflict on that axis remain. These can be seen in Issue 5 about future relations with still-communist China. In what could be an ironic turn, it is even arguable that the growing tensions between the United States and Europe, as discussed in Issue 4, could lead to a newly configured East-West divide, with the Atlantic Ocean rather than the Iron Curtain separating the antagonists.

In contrast to the moribund East-West axis, the *North-South axis* has increased in importance and tension. The wealthy, industrialized countries (North) are at one end, and the poor, less developed countries (LDCs, South) are at the other extreme. Economic differences and disputes are the primary dimension of this axis, in contrast to the military nature of the East-West axis. Issues 10 and 11 explore these differences and debate the terms under which the North should give economic aid to the South.

Something of military relations between the North and the South are present in the war with Iraq and the crisis with North Korea, which are taken up in Issues 7 and 8, respectively. There are some, especially in the South, who believe that the negative U.S. reactions to Iraqi and North Korean power, especially their possible possession of weapons of mass destruction, constitute evidence that the powerful countries are determined to maintain their power by keeping weak countries from obtaining the same types of weapons that the United States and other big powers have. There is also some suspicion in the South that an enhanced UN military controlled by the North-dominated UN Security Council, as discussed in Issue 16, might be used as an intervention force to promote the interests of the North over the South.

Increased Role of Economics

As the growing importance of the North-South axis indicates, economics are playing an increased role in world politics. The economic reasons behind the decline of the East-West axis is further evidence. Economics have always played a part in international relations, but the traditional focus has been on strategic-political affairs, especially questions of military power.

Political scientists, however, are increasingly focusing on the international political economy, or the economic dimensions of world politics. International trade, for instance, has increased dramatically, expanding from an annual world export total of $20 billion in 1933 to $7.5 trillion in 2001. The impact has been profound. The domestic economic health of most countries is heavily affected by trade and other aspects of international economics. Since World War II there has been an emphasis on expanding free trade by decreasing tariffs and other barriers to international commerce. In recent years, however, a downturn in the economies of many of the industrialized countries has increased calls for more protectionism. Yet restrictions on trade and other economic activity can also be used as diplomatic weapons. The intertwining of economies and the creation of organizations to regulate them, such as the World Trade Organization, is raising issues of sovereignty and other concerns. This is a central matter in the debate in Issue 9 over whether or not the trend toward global economic integration is desirable.

Conclusion

Having discussed many of the various dimensions and approaches to the study of world politics, it is incumbent on this editor to advise against your becoming too structured by them. Issues of focus and methodology are important both to studying international relations and to understanding how others are analyzing global conduct. However, they are also partially pedagogical. In the final analysis, world politics is a highly interrelated, perhaps seamless, subject. No one level of analysis, for instance, can fully explain the events on the world stage. Instead, using each of the levels to analyze events and trends will bring the greatest understanding.

Similarly, the realist-idealist division is less precise in practice than it may appear. As some of the debates indicate, each side often stresses its own standards of morality. Which is more moral: defeating a dictatorship or sparing the sword and saving lives that would almost inevitably be lost in the dictator's overthrow? Furthermore, realists usually do not reject moral considerations. Rather, they contend that morality is but one of the factors that a country's decision makers must consider. Realists are also apt to argue that standards of morality differ when dealing with a country as opposed to an individual. By the same token, most idealists do not completely ignore the often dangerous nature of the world. Nor do they argue that a country must totally sacrifice its short-term interests to promote the betterment of the current and future world. Thus, realism and idealism can be seen most accurately as the ends of a continuum—with most political scientists and practitioners falling somewhere between, rather than at, the extremes. The best advice, then, is this: think broadly about international politics. The subject is very complex, and the more creative and expansive you are in selecting your foci and methodologies, the more insight you will gain. To end where we began, with Dr. Faust, I offer his last words in Goethe's drama, "*Mehr licht,*" . . . More light! That is the goal of this book.

The Ultimate Political Science Links Page

Under the editorship of Professor P. S. Ruckman, Jr., at Rock Valley College in Rockford, Illinois, this site provides a gateway to the academic study of not just world politics but all of political science. It includes links to journals, news, publishers, and other relevant resources.

http://www.rvc.cc.il.us/faclink/pruckman/PSLinks.htm

Poly-Cy: Internet Resources for Political Science

This is a worthwhile gateway to a broad range of political science resources, including some on international relations. It is maintained by Robert D. Duval, director of graduate studies at West Virginia University.

http://www.polsci.wvu.edu/polycy/

The WWW Virtual Library: International Affairs Resources

Maintained by Wayne A. Selcher, professor of international studies at Elizabethtown College in Elizabethtown, Pennsylvania, this site contains approximately 2,000 annotated links relating to a broad spectrum of international affairs. The sites listed are those that the Webmaster believes have long-term value and that are cost-free, and many have further links to help in extended research.

http://www.etown.edu/vl/

The Globalization Website

The goals of this site are to shed light on the process of globalization and contribute to discussions of its consequences, to clarify the meaning of globalization and the debates that surround it, and to serve as a guide to available sources on globalization.

http://www.emory.edu/SOC/globalization/

PART 1

Globalization

*T*he most significant change that the international system is experiencing is the trend toward globalization. Countries are becoming interdependent, the number of international organizations and their power are increasing, and global communications have become widespread and almost instantaneous. As reflected in the issues that make up this part, these changes and others have led to considerable debate about the value of globalization and what it will mean with regard to human governance.

- Is Globalization Likely to Create a Better World?

- Will State Sovereignty Survive Globalism?

ISSUE 1

Is Globalization Likely to Create a Better World?

YES: Thomas Friedman, from "States of Discord," *Foreign Policy* (March/April 2002)

NO: Robert Kaplan, from "States of Discord," *Foreign Policy* (March/April 2002)

ISSUE SUMMARY

YES: Thomas Friedman, a columnist for the *New York Times*, contends that with good leadership, globalization will create more openness in government and business, a strong rule of law, and greater opportunities for people to experience personal freedom and to challenge government authority.

NO: Robert Kaplan, a correspondent for *The Atlantic Monthly*, argues that in the next few decades globalization is likely to cause more and more turbulence.

What is globalization? Many commentators have offered various definitions, but they all contain the same kernel of reality. That is, globalization is a process that is diminishing—some would say eventually erasing—many of the factors that divide the world. It is important to note that globalization is not new. Many precursors of modern globalization date far back into history, even into antiquity. This lineage should not distract you, however, from seeing that most of the impetus behind modern globalization has occurred in a relatively short time and, in many ways, is continuing to accelerate and strengthen.

Factors of traditional division and how they are being changed by globalization are as follows:

Geography. Distances and topographical features, such as oceans and mountain ranges, "shrank" as humans learned to ride horses and to use them to pull wagons; created sailing ships, then steamships, that could travel further and further at ever increasing speed; and invented railroads, cars, and other motorized land vehicles and aircraft.

Communications. The invention of the printing press in 1454 and many other innovations has increased global communications, but that process has greatly accelerated with the creation of telegrams, radio, telephones, television, satellite communications, and the Internet.

Economics. Tariffs and other barriers to trade have decreased significantly since the end of World War II, and investment across national borders and monetary exchange have grown very rapidly. For example, global trade, measured in the value of exported goods and services, grew over 1,200 percent during the last half of the twentieth century. Huge multinational corporations (MNCs) now dominate global commerce, and the economic prosperity of almost all countries is heavily dependent on what they import and export, the flow of investment in and out of each country, and the exchange rates of the currency of each country against the currencies of other countries. The European Union (EU) has even largely abandoned traditional national currencies, such as the French franc, in favor of the common EU currency, the euro.

Culture. People around the world are increasingly dressing alike, listening to the same kind of music, and availing themselves of the cuisines at fast-food restaurants such as McDonald's and KFC. Also, for good or ill, English is increasingly the global lingua franca.

Governance. The rapid growth of international interchange has engendered a broad range of global and regional international governmental organizations (IGOs), such as the United Nations, the International Monetary Fund, the World Bank, and the European Union, to manage the process of globalization. These organizations require at least some level of authority, and many analysts believe that they might, and perhaps should, diminish the authority of the traditional central unit of governance, the state (country).

Where will all this lead? Many observers believe that globalization is both unstoppable and desirable. They argue that it leads to increased prosperity for all, that greater interdependence and common acculturation will make for a more peaceful world, and that many other benefits will accrue. Critics of globalization charge that it is enriching the few at the expense of the many; that it threatens to wipe away cultural diversity; that powerful MNCs are damaging the environment; that many of the IGOs are not democratic; that violence occurs as traditional cultures, social structures, and units of governance are undermined; and that globalization is responsible for a host of other ills.

The topic is huge, and no single pair of articles, or even books, would sufficiently explore all the pros and cons of globalization. A good place to start, however, is the following writings of Thomas Friedman and Robert Kaplan. Neither thinks the future course of globalization is certain. On the whole, though, Friedman is optimistic. Kaplan is considerably more pessimistic.

 **YES**

States of Discord

Techno Logic

What is globalization? The short answer is that globalization is the integration of everything with everything else. A more complete definition is that globalization is the integration of markets, finance, and technology in a way that shrinks the world from a size medium to a size small. Globalization enables each of us, wherever we live, to reach around the world farther, faster, deeper, and cheaper than ever before and at the same time allows the world to reach into each of us farther, faster, deeper, and cheaper than ever before.

I believe this process is almost entirely driven by technology. There's a concept in strategic theory—the sort of things Bob Kaplan has written about—stating that capabilities create intentions. In other words, if you give people B-52s [bombers], they will find ways to use them. This concept is quite useful when thinking about globalization, too. If I have a cell phone that can call around the world at zero marginal cost to 180 different countries, I will indeed call around the world to 180 different countries. If I have Internet access and can do business online, a business in which my suppliers, customers, and competitors are all global, then I will be global, too. And I will be global whether there is a World Trade Organization agreement or not.

Since September 11, 2001, many people have asked me if terrorism will stop the process of globalization. I had often wondered about this sort of situation: What would happen if we did reach a crisis moment, a crisis like terrorism, or a major financial crisis, and things started to go in reverse? People would say, "Bring back the walls!" But I knew that was going to be a particularly defining moment for us, because that's when we were all going to wake up and finally realize that technology had destroyed the walls already—that the September 11 terrorists made their reservations on Travelocity.com.

State of Progress

Bob brought up the role of the state and of governance; these are absolutely crucial issues. Some people believe that the state will wither away and matter less in an era of globalization. I believe exactly the opposite: The state matters much more in a globalized world.

Why? Well, the first thing we have to understand about globalization is that, oddly enough, it's not global. It affects different regions in different ways, and it links different countries in different ways. Yet every part of the world is directly or indirectly being globalized in some way. In this context, the state matters more, not less. If I could just use one image to describe the state—including political institutions, courts, oversight agencies, the entire system of governance—I would say it's like a plug, and it's the plug that your country uses to connect with globalization.

If that plug is corroded, corrupted, or the wires aren't connected, the flow between you and that global system—what I call the "electronic herd"—is going to be very distorted, and you are going to feel the effects of that distortion. But if the plug works well, the flow between you and the global system will be much more enriching.

The dirty little secret about globalization—and it takes a lot of countries a long time to figure it out—is that the way to succeed in globalization is to focus on the fundamentals. It's not about the wires or about bandwidth or about modems. It's about reading, writing, and arithmetic. It's about churches, synagogues, temples, and mosques. It's about rule of law, good governance, institution building, free press, and a process of democratization. If you get these fundamentals right, then the wires will find you, and the wires will basically work. But if you get them wrong, nothing will save you.

Consider Botswana and Zimbabwe. Both countries have problems, but Botswana was probably in the top 20 percent of countries last year in per capita income growth. Zimbabwe was, I dare say, certainly in the bottom 20 percent. These two countries are right next to each other. Botswana has its problems, to be sure; it's not some ideal paradise. But it has decent democratization, decent institutions left over by the British, decent free press, and decent oversight and regulatory bodies.

And Zimbabwe? Zimbabwe has President Robert Mugabe. Now, if you told me right now that in five years I'll be able to get a fair trial in Zimbabwe, I'd say that Zimbabwe is going to be fine. But if you tell me I won't get a fair trial in Zimbabwe in five years, then it doesn't matter if everyone in Zimbabwe has an Internet address, a personal computer, a Palm Pilot, and a cell phone. If the institutions through which these people have to operate to generate growth and interact with the global system are corrupted and corroded, then all the gadgetry in the world won't make a dime's worth of difference.

Or just compare Egypt and the East Asian countries. They started out with about the same per capita income in 1953, but now there's a huge disparity between them. People who have studied these parts of the world point to two fundamental differences: One is the value placed on education. The other is how leaders justify their rule.

In Asia, which had autocratic regimes for several decades, leaders tended to justify their rule with a simple trade-off: Give me your democracy and I will give you prosperity. And people gave up many democratic rights and they got prosperity. The more prosperous they became, the more the relationship between them and the regime changed, until ultimately, you had a tip-over point, and these became democratic countries in almost every case. But what happened in Egypt? The leader said, don't judge me on whether I brought you a better standard of living; judge me on how I confronted the British, how I confronted the Americans, and how I confronted the Israelis. Give me your rights, and I'll give you the Arab-Israeli conflict. That was a bad trade. As a result, we see a huge gap between the two.

That's why with globalization, leadership matters more, not less. If you have the calcified Brezhnevite management that Egypt has, then it's no wonder that the Cairo skyline has barely changed in 50 years. But if your management "gets it"—as a corporation or a country—then you'll benefit from globalization.

Let Them Eat Pizza

Bob mentioned the divides in Europe between local and national and regional identities. That discussion raises a larger question: Will globalization turn out to be simply Americanization? Some people believe globalization will homogenize us only on the surface. Japanese kids may wear jeans and my own kids may eat sushi. But underneath those jeans, those kids may still remain completely Japanese, while my kids are in reality still just hot dog–loving American teenagers. Others argue that globalization is going to homogenize us to our very roots, in which case I would agree that it really will become culturally lethal.

But I think that is an unanswerable question right now. Early 20th-century U.S. novelist Thomas Wolfe wrote that you can't go home again. With globalization, I worry whether we won't be able to leave home again, that everywhere will start to look like everywhere else. There are two ways to make people homeless: One is to take away their home, and the other is to make their home look like everybody else's home. Yet I take some succor from the fact that the most popular food in the world today is not the Big Mac, but pizza. And what is pizza? It's a flat piece of bread that every society has, on which every society and every community throws its own local ingredients and culture. In India, you can get tandoori pizza; in Japan, you can get sushi pizza; and in Mexico, you can get salsa pizza.

In the same way, my hope for the Internet is that it will become an instrument for sharing cultures, not for spreading some kind of American cultural dominance to the world. And while I have great sympathy for people who fear the latter outcome, I think it's really still an open question as to which will prevail. For example, people have proclaimed the death of regionalism in America for a long time. But I tell you, when I go back to my home state of Minnesota, I speak Minnesotan. I slip right back in there. Anyone who thinks that there's no place like home anymore, and there's no Minnesota, has really never come from a region with a strong identity.

Bikinis in a Nudist Colony

I'm a bit more optimistic than Bob when thinking about the impact of globalization on personal freedoms and democracy. Consider China. I've been visiting China for the last 15 years. And every time I go, it's a more open place, with more personal freedoms, more rule of law, and more people empowered to challenge the government. And I don't know where it ends or where the tip-over point is. All I know is that this transformation in China is largely due to the forces of globalization.

Of course, we also saw the forces of globalization on September 11, 2001. Again, globalization goes both ways. It can threaten democracies as well as strengthen them. But on net, I do believe that with the right leadership, globalization will be a force for more openness, more rule of law, and more opportunities for people to enjoy personal freedoms and challenge authorities.

My own view is that China is on the verge of experiencing the biggest leveraged buyout in history. The business community has now been invited into the Communist Party—I was there when that happened. Think about that: Capitalists in the Communist Party!

I can deal with a lot of contradictions in my life, but wearing bikinis in a nudist colony or serving steak at a vegetarian restaurant strikes me as no longer manageable. The presence of capitalists in the Communist Party represents a very deep change; it tells us that the system is really moving somewhere else. I believe that over time we will basically see the business community—the capitalist side of China—buying out the Communist Party.

States of Discord

Bad News Is Next

Let me try to give a slightly richer definition of globalization [than that offered by Tom Friedman]. The best historical metaphor I can come up with is China in the third century B.C., when the Han overlordship replaced the period of the warring states (following the short interlude of the Qin dynasty). Think of it: You had this massive mainland China, thousands of miles across, with little states constantly coalescing into bigger states over the centuries. And then, for a long period, you had six or so major states fighting each other. Finally, they were unified by a series of balance-of-power agreements, by an embryonic bureaucracy developing in all of them, and by the Chinese language. What the Han dynasty represented was not a single state; it was a serious reduction of conflict among the warring states, so that the highest morality was the morality of order, with everyone giving up a share of their independence for the sake of greater order. I'm not talking about some sort of "world government" over China. It was just a loose form of governance, where everything affected or constrained everything else.

Today, too, everything affects everything else—we're affected by disease pandemics in Africa, by *madrassas* (seminaries) in Pakistan—but there is still nothing like a global leviathan or a centralizing force. The world is coming together, but the international bureaucracy atop it is so infantile and underdeveloped that it cannot cope with growing instability.

And more complexity does lead to more instability. Today, we have several factors driving this relationship. First, we are seeing youth bulges in many of the most unstable countries. Big deal if the world population is aging; that doesn't interest me for the next five or ten years. I care about the many countries or areas like the West Bank, Gaza, Nigeria, Zambia, and Kenya where over the next 20 years the population of young, unemployed males between the ages of 14 and 29 is going to grow. And as we all know from television, one thing that unites political unrest everywhere is that it's carried out by young males. Another factor is resource scarcity—the amount of potable water available throughout the Middle East, for instance, is going to decrease substantially over the next 25 years. When you put them together, these driving forces lead to sideswipes, such as the September 11 attacks. Another sideswipe could be an environmental event like an earthquake in an intensely settled area, like Egypt or China, that could lead to the removal of a strategic regime.

In *The Lexus and the Olive Tree*, Tom wrote that globalization doesn't end geopolitics. That's the key. Globalization is not necessarily *good* news; it's just *the* news. And the news could get scarier and scarier, because more interconnections will lead to complexity before they'll lead to stability.

State of War

I agree that good things are going to happen in a more global world (and humanists will duly celebrate them), but foreign policy crises are about what goes wrong. In the short run, I'm pessimistic. Remember that poverty does not lead to revolutions—development does. The revolutions in Mexico and France were preceded by years of dramatic and dynamic economic development, as well as urbanization and population movements. And what have we seen for the last 10 years, which Tom has described so well in his columns? Incredible dynamism, with middle classes emerging in China, Indonesia, and Brazil. But this development will not automatically lead to West German–style democracies. They may eventually, but for the next 20 or 30 years, they will experience more and more turbulence.

Of course, part of the trouble with some states is that they are states in name only. North Africa exemplifies this problem. In North Africa, you have three age-old civilization clusters: Egypt, Tunisia, and Morocco. They all have their problems, but all three are far healthier as states than, say, Algeria or Libya. Why is that? Why did Algeria and Libya get so radical and suffer through civil wars? Because they were never states to begin with.

Tunisia has been a state since Roman times. There's a state mentality there, even without democracy. Citizens argue about the budget and about education. The leader does not have to be oppressive, because a state community already exists. But Algeria and Libya are geographic expressions that were not cobbled together as states until relatively recently. The only institutionalizing force there is radical ideology.

Certainly, you could argue that Africa is in a class by itself. But consider the European Union. Despite its fits and starts, if you look at the European Union emerging out of a coal and steel consortium in the early 1950s, which included France, Germany, the Benelux countries, and Italy, and then expanded to England, etc., you see a gradual superstructure growing. At the same time, however, localism emerges. You see Catalonia and the German *länder* (states) reasserting themselves. And yet, I must ask, who today would fight for Belgium? Who would fight for Germany? So, on balance, I think the state is weakening in Europe. But if the European community develops into a vapid, insipid bureaucratic despotism that only excites the upper-middle classes and the Brussels [headquarters of the European Union] Eurocrats, there will be a backlash. That's why you have retrograde nationalist backlashes.

But here's where Tom is right: States make war. And they make war because they have political accountability to people who are stuck in geographic space and because they must defend their citizenry and are therefore willing to take big risks on military strikes. The United Nations would never do that. To have the guts to make war, you have to have your own citizenry on your back; you must be physically responsible for them. I believe that, in the next 10 years or so, major wars will be very state-driven.

Say Your Prayers

I'd like to take a more specific angle on the question of culture and values. One of the biggest elements of globalization has been urbanization. Fifty years ago, the Middle East was rural; cities like Tunis, Casablanca, and Damascus had 200,000 or 300,000 or 400,000 people. Karachi, for instance, had 400,000 people in 1947. Today, Karachi has 9 million inhabitants. Even Tunis has 2 million.

And there has been a value change as well, with a more stark, more abstract, more ideological form of Islam emerging to cope with urbanization. When people lived in villages, religion was part and parcel of the daily routine of life; it wasn't conscious. In the hill villages of Afghanistan, women did not wear veils, because virtually every man they saw was a relative of some kind. But when they migrated into cities, suddenly they were among strangers, so the veil came on.

When they migrated into pseudo-Western cities such as Cairo, they were suddenly confronted with anonymity. Do you know what is amazing about cities like Cairo, Damascus, and Tunis? Yes, they're poor: Services are horrible, street lights don't work, and the police forces are fairly useless. Yet crime is actually quite low. Except for the odd pickpocket, who's in an area where tourists tend to go, it's perfectly safe to walk around with a lot of cash in your wallet.

How could that be? How can you minimize petty, random crime despite the kind of urbanization that sent crime through the roof in New York City in the mid-19th century, when Walt Whitman was writing about it? The answer: intensification of religion. Indeed, any time in history when there was tremendous economic change—with dynamism, development, and disruption—religion adapted. For a more contemporary example, just look at the megachurches in the midwestern United States. Sure, East Coast sophisticates laugh at them, but that is a classic form of adaptation to economic change. And in the Middle East, what we call fundamentalism was just a Darwinian way of coping with urbanization. Unfortunately, it also provided a fertile petri dish for the emergence of disease germs like terrorists.

Freedom From Language

. . . I think there are a lot of unnecessary arguments about democracy and globalization, because we have become trapped by language. Unfortunately, we've defined democracy as the holding of parliamentary elections now or in six months. This definition is simply wrong. At the end of World War I, the novelist

Joseph Conrad sent a letter to a friend in which he wrote that we don't really fight for elections or "democracy" but for openness and freedom and human rights—in whatever form they may take in any particular country. I think that's the right way to look at it.

Countries that already have sizable middle classes and decent political institutions may be ripe for the icing on the cake: democratic elections. We've seen that in Taiwan and South Korea, and in the southern cone of Latin America, despite Argentina's troubles. But there are other places where holding elections too soon could lead to the opposite result.

For instance, Tunisia has increased the size of its middle class from 6 percent to roughly 50 percent of the population. Tunisia has one of the most open societies in the Arab world, with cybercafes everywhere—yet it has all been done through benign despotism. Had they held elections eight or so years ago, I believe there would be less freedom today. Similarly, Egypt is in a terrible situation, but if you demanded elections there tomorrow, there is a good likelihood that more oppression and a worse human rights situation could result. So what we have to do on this question of democracy is look at each individual country and place as it comes.

POSTSCRIPT

Is Globalization Likely to Create a Better World?

Globalization is both old and new. It is old in that the efforts of humans to overcome distance and other barriers to increased interchange have long existed. The first canoes and signal fires are part of the history of globalization. The first event in true globalization occurred between 1519 and 1522, when Ferdinand Magellan circumnavigated the globe.

However, globalization is mostly a modern phenomenon. The progress of globalization until the latter half of the 1800s might be termed "creeping globalization." There were changes, but they occurred very slowly. Since then, the pace of globalization has increased exponentially. Friedman is correct when he points to technology as a prime mover of globalization. The ability of modern transportation to move masses of goods and people quickly over long distances, the virtually instant access to information from satellite communications, and many of the other aspects of globalization technology are brand new relative to the scope of human history. But globalization is not just technology-driven. It has been pushed since World War II by governments, especially that of the United States, that believe that globalization is a prescription for greater prosperity and peace. Those policy beliefs have led to policy decisions promoting greater international interchange and interdependence.

Although voices have long been raised against globalization, they have been relatively few, little heard, and even less heeded until recently. Indeed, as a robust phenomenon, the antiglobalization movement is only about a decade old. Because globalization is so multifaceted, the antiglobalization movement is fragmented, with various groups focusing on economics, culture, governance, the environment, or some other concern with the globalization process.

While the governments of the world's most powerful countries and the leading international organizations remain committed to even greater globalization, the voices of the antiglobalization movement have increasingly been acknowledged by the leaders of the world. As Secretary-General Kofi Annan of the United Nations told world political and economic leaders gathered at the World Economic Forum in February 2002, "The perception, among many, is that this is the fault of globalization, and that globalization is driven by a global elite, composed of—or at least, represented by—the people who attend this gathering." While the secretary-general went on to say, "I believe that perception is wrong," he also cautioned, "[I]t is up to you to prove it wrong, with actions that translate into concrete results for the downtrodden, exploited and excluded. . . . You must show that economics, properly applied, and profits, wisely invested, can bring social benefits within reach not only for the few but for the many, and eventually for all." Annan's message is that although technology makes globalization possible, the course of globalization and whether it is a force for good or ill will be based on the policy decisions that humans make.

Two publications that can help introduce the topic of globalization are David Held, Anthony McGrew, and Maurice G. Cotterell, eds., *Global Transformations Reader: An Introduction to the Globalization Debate* (Policy Press, 2003) and Tony Schirato and Jennifer Webb, *Understanding Globalization* (Sage, 2003). The Internet provides further resources that examine globalization. A good site is The Globalization Website, http://www.emory.edu/SOC/globalization/, hosted by Emory University. It is even-handed and is particularly meant to support undergraduates.

ISSUE 2

Will State Sovereignty Survive Globalism?

YES: Stephen D. Krasner, from "Sovereignty," *Foreign Policy* (January/February 2001)

NO: Kimberly Weir, from "The Waning State of Sovereignty," An Original Essay Written for This Volume (2002)

ISSUE SUMMARY

YES: Professor of international relations Stephen D. Krasner contends that the nation-state has a keen instinct for survival and will adapt to globalization and other challenges to sovereignty.

NO: Kimberly Weir, an assistant professor of political science, maintains that the tide of history is running against the sovereign state as a governing principle, which will soon go the way of earlier, now-discarded forms of governance, such as empire.

There are political, economic, and social forces that are working to break down the importance and authority of states and that are creating pressures to move the world toward a much higher degree of political, economic, and social integration. Whatever one may think of international organizations and their roles, the increasing number and importance of them provide evidence of the trend toward globalization.

Many have questioned whether or not the growth of international law and norms, international organizations, and other transnational phenomena are lessening the sovereignty of states. The debate concerning this issue takes the discussion over the international political system yet another step by asking what the future holds for states. Will the state persist as the principal actor in the international system or be eclipsed?

To grasp this debate it is crucial to understand that countries (states) are not a natural political order. Instead, there are two things to bear in mind. One is that states as we know them have not always existed as a form of political organization. Logically, this means that states need not always exist. Second, states, or any form of governance, are best regarded as tools. They are vehicles to serve the interests of their citizens and arguably only deserve loyalty as long as they provide benefits.

The modern state is largely a Western creation. For almost a millennium after the universalistic Roman Empire fell in 476, political power rested at two levels of authority—one universal, the other local. On the universal level authority existed in the form of the Roman Catholic Church in an era when kings and other secular leaders were subordinate in theory, and often in practice, to popes. Later, the broad authority of the Catholic Church was supplemented and, in some cases, supplanted by the Holy Roman Empire and other multiethnic empires that exercised control over many different peoples. Most of the people within these empires had little if any sense of loyalty to the empire.

The second level was local feudal authority in principalities, dukedoms, baronies, and other such fiefdoms in political units that were smaller than most modern states. Here, too, the common people were not citizens, as we know the concept today. Instead, the minor royalty came close to owning the peasantry.

For a variety of economic, military, and other reasons, this old system failed, and a new system based on territorially defined sovereign states slowly evolved.

States came to dominate the international system not because of any ideological reason but simply because they worked better than the models of governance that had failed or any of the other models (such as a rebirth of city-states like Venice) that were tried. The Treaty of Westphalia (1648), which divided Europe into Catholic and Protestant states and marked the end of the pan-European dominance of the Holy Roman Empire, is often used to symbolize the establishment of the state. The treaty was certainly important, but, in fact, that establishment of the state and the growth of the concept of citizenry and patriotism evolved over centuries.

Once this occurred, all of the basic parameters for the modern state that exists today were in place. Yet in the ceaseless ebb and flow of world forces, pressures were beginning to build that would work to undermine the modern states. Those economic, military, and other forces have, in the view of some, built up rapidly since the mid-twentieth century and cast doubt on the future role, even existence, of states as sovereign entities at the center of the international system.

There are some observers who contend that globalization of trade, communications, and other processes do not and should not threaten the existence of the state as a fundamental political unit. Analysts of this persuasion believe that countries can, will, and should continue to exist as the sovereign entities they are today. Stephen D. Krasner takes this point of view in the following selection.

Other analysts believe that states have run their course as a model for governance and are becoming outmoded in a world that is increasingly interdependent. They note that the world has changed greatly since the rise of the state to political dominance centuries ago. The analysts ask whether or not it is reasonable to assume that a model of governance that worked hundreds of years ago is the best model for the future. Their answer is no. Kimberly Weir represents this view in the second of the following selections.

Stephen D. Krasner

 **YES**

Sovereignty

The idea of states as autonomous, independent entities is collapsing under the combined onslaught of monetary unions, CNN, the Internet, and nongovernmental organizations [NGOs]. But those who proclaim the death of sovereignty misread history. The nation-state has a keen instinct for survival and has so far adapted to new challenges—even the challenge of globalization.

The Sovereign State Is Just About Dead

Very wrong Sovereignty was never quite as vibrant as many contemporary observers suggest. The conventional norms of sovereignty have always been challenged. A few states, most notably the United States, have had autonomy, control, and recognition for most of their existence, but most others have not. The polities of many weaker states have been persistently penetrated, and stronger nations have not been immune to external influence. China was occupied. The constitutional arrangements of Japan and Germany were directed by the United States after World War II. The United Kingdom, despite its rejection of the euro, is part of the European Union [EU].

Even for weaker states—whose domestic structures have been influenced by outside actors, and whose leaders have very little control over transborder movements or even activities within their own country—sovereignty remains attractive. Although sovereignty might provide little more than international recognition, that recognition guarantees access to international organizations and sometimes to international finance. It offers status to individual leaders. While the great powers of Europe have eschewed many elements of sovereignty, the United States, China, and Japan have neither the interest nor the inclination to abandon their usually effective claims to domestic autonomy.

In various parts of the world, national borders still represent the fault lines of conflict, whether it is Israelis and Palestinians fighting over the status of Jerusalem, Indians and Pakistanis threatening to go nuclear over Kashmir, or Ethiopia and Eritrea clashing over disputed territories. Yet commentators nowadays are mostly concerned about the erosion of national borders as a consequence of globalization. Governments and activists alike complain that multilateral institutions such as the United Nations, the World Trade Organization, and the International Monetary Fund overstep their authority by promoting universal standards for everything from human rights and the environment to monetary policy and immigration. However, the most important impact of economic globalization and transnational norms will be to alter the scope of state authority rather than to generate some fundamentally new way to organize political life.

Sovereignty Means Final Authority

Not anymore, if ever When philosophers Jean Bodin and Thomas Hobbes first elaborated the notion of sovereignty in the 16th and 17th centuries, they were concerned with establishing the legitimacy of a single hierarchy of domestic authority. Although Bodin and Hobbes accepted the existence of divine and natural law, they both (especially Hobbes) believed the word of the sovereign was law. Subjects had no right to revolt. Bodin and Hobbes realized that imbuing the sovereign with such overweening power invited tyranny, but they were predominately concerned with maintaining domestic order, without which they believed there could be no justice. Both were writing in a world riven by sectarian strife. Bodin was almost killed in religious riots in France in 1572. Hobbes published his seminal work, *Leviathan,* only a few years after parliament (composed of Britain's emerging wealthy middle class) had executed Charles I in a civil war that had sought to wrest state control from the monarchy.

This idea of supreme power was compelling, but irrelevant in practice. By the end of the 17th century, political authority in Britain was divided between king and parliament. In the United States, the Founding Fathers established a constitutional structure of checks and balances and multiple sovereignties distributed among local and national interests that were inconsistent with hierarchy and supremacy. The principles of justice, and especially order, so valued by Bodin and Hobbes, have best been provided by modern democratic states whose organizing principles are antithetical to the idea that sovereignty means uncontrolled domestic power.

If sovereignty does not mean a domestic order with a single hierarchy of authority, what does it mean? In the contemporary world, sovereignty primarily has been linked with the idea that states are autonomous and independent from each other. Within their own boundaries, the members of a polity are free to choose their own form of government. A necessary corollary of this claim is the principle of nonintervention: One state does not have a right to intervene in the internal affairs of another.

More recently, sovereignty has come to be associated with the idea of control over transborder movements. When contemporary observers assert that the sovereign state is just about dead, they do not mean that constitutional structures are about to disappear. Instead, they mean that technological change has made it very difficult, or perhaps impossible, for states to control movements across their borders of all kinds of material things (from coffee to cocaine) and not-so-material things (from Hollywood movies to capital flows). Finally, sovereignty has meant that political authorities can enter into international agreements. They are free to endorse any contract they find attractive. Any treaty among states is legitimate provided that it has not been coerced. . . .

Universal Human Rights Are an Unprecedented Challenge to Sovereignty

Wrong   The struggle to establish international rules that compel leaders to treat their subjects in a certain way has been going on for a long time. Over the centuries the emphasis has shifted from religious toleration, to minority rights (often focusing on specific ethnic groups in specific countries), to human rights (emphasizing rights enjoyed by all or broad classes of individuals). In a few instances states have voluntarily embraced international supervision, but generally the weak have acceded to the preferences of the strong: The Vienna settlement following the Napoleonic wars guaranteed religious toleration for Catholics in the Netherlands. All of the successor states of the Ottoman Empire, beginning with Greece in 1832 and ending with Albania in 1913, had to accept provisions for civic and political equality for religious minorities as a condition for international recognition. The peace settlements following World War I included extensive provisions for the protection of minorities. Poland, for instance, agreed to refrain from holding elections on Saturday because such balloting would have violated the Jewish Sabbath. Individuals could bring complaints against governments through a minority rights bureau established within the League of Nations.

But as the Holocaust tragically demonstrated, interwar efforts at international constraints on domestic practices failed dismally. After World War II, human, rather than minority, rights became the focus of attention. The United Nations Charter endorsed both human rights and the classic sovereignty principle of nonintervention. The 20-plus human rights accords that have been signed during the last half century cover a wide range of issues including genocide, torture, slavery, refugees, stateless persons, women's rights, racial discrimination, children's rights, and forced labor. These U.N. agreements, however, have few enforcement mechanisms, and even their provisions for reporting violations are often ineffective.

The tragic and bloody disintegration of Yugoslavia in the 1990s revived earlier concerns with ethnic rights. International recognition of the Yugoslav successor states was conditional upon their acceptance of constitutional provisions guaranteeing minority rights. The Dayton accords established externally controlled authority structures in Bosnia, including a Human Rights Commission (a majority of whose members were appointed by the Western European states). NATO [North Atlantic Treaty Organization] created a de facto protectorate in Kosovo.

The motivations for such interventions—humanitarianism and security— have hardly changed. Indeed, the considerations that brought the great powers into the Balkans following the wars of the 1870s were hardly different from those that engaged NATO and Russia in the 1990s.

Globalization Undermines State Control

No   State control could never be taken for granted. Technological changes over the last 200 years have increased the flow of people, goods, capital, and ideas—but the problems posed by such movements are not new. In many ways, states are better able to respond now than they were in the past.

The impact of the global media on political authority (the so-called CNN effect) pales in comparison to the havoc that followed the invention of the printing press. Within a decade after Martin Luther purportedly nailed his 95 theses to the Wittenberg church door, his ideas had circulated throughout Europe. Some political leaders seized upon the principles of the Protestant Reformation as a way to legitimize secular political authority. No sovereign monarch could contain the spread of these concepts, and some lost not only their lands but also their heads. The sectarian controversies of the 16th and 17th centuries were perhaps more politically consequential than any subsequent transnational flow of ideas.

In some ways, international capital movements were more significant in earlier periods than they are now. During the 19th century, Latin American states (and to a lesser extent Canada, the United States, and Europe) were beset by boom-and-bust cycles associated with global financial crises. The Great Depression, which had a powerful effect on the domestic politics of all major states, was precipitated by an international collapse of credit. The Asian financial crisis of the late 1990s was not nearly as devastating. Indeed, the speed with which countries recovered from the Asian flu reflects how a better working knowledge of economic theories and more effective central banks have made it easier for states to secure the advantages (while at the same time minimizing the risks) of being enmeshed in global financial markets.

In addition to attempting to control the flows of capital and ideas, states have long struggled to manage the impact of international trade. The opening of long-distance trade for bulk commodities in the 19th century created fundamental cleavages in all of the major states. Depression and plummeting grain prices made it possible for German Chancellor Otto von Bismarck to prod the landholding aristocracy into a protectionist alliance with urban heavy industry (this coalition of "iron and rye" dominated German politics for decades). The tariff question was a basic divide in U.S. politics for much of the last half of the 19th and first half of the 20th centuries. But, despite growing levels of imports and exports since 1950, the political salience of trade has receded because national governments have developed social welfare strategies that cushion the impact of international competition, and workers with higher skill levels are better able to adjust to changing international conditions. It has become easier, not harder, for states to manage the flow of goods and services.

Globalization Is Changing the Scope of State Control

Yes The reach of the state has increased in some areas but contracted in others. Rulers have recognized that their effective control can be enhanced by walking away from issues they cannot resolve. For instance, beginning with the Peace of Westphalia [1648 treaty often cited as the political big bang that created the modern system of autonomous states], leaders chose to surrender their control over religion because it proved too volatile. Keeping religion within the scope of state authority undermined, rather than strengthened, political stability. Monetary policy is an area where state control expanded and then ultimately contracted. Before the 20th century, states had neither the administrative competence nor the inclination to conduct independent monetary policies. The mid-20th-century effort to control monetary affairs, which was associated with Keynesian economics, has now been reversed due to the magnitude of short-term capital flows and the inability of some states to control inflation. With the exception of Great Britain, the major European states have established a single monetary authority. Confronting recurrent hyperinflation, Ecuador adopted the U.S. dollar as its currency in 2000.

Along with the erosion of national currencies, we now see the erosion of national citizenship—the notion that an individual should be a citizen of one and only one country, and that the state has exclusive claims to that person's loyalty. For many states, there is no longer a sharp distinction between citizens and noncitizens. Permanent residents, guest workers, refugees, and undocumented immigrants are entitled to some bundle of rights even if they cannot vote. The ease of travel and the desire of many countries to attract either capital or skilled workers have increased incentives to make citizenship a more flexible category.

Although government involvement in religion, monetary affairs, and claims to loyalty has declined, overall government activity, as reflected in taxation and government expenditures, has increased as a percentage of national income since the 1950s among the most economically advanced states. The extent of a country's social welfare programs tends to go hand in hand with its level of integration within the global economy. Crises of authority and control have been most pronounced in the states that have been the most isolated, with sub-Saharan Africa offering the largest number of unhappy examples.

NGOs Are Nibbling at National Sovereignty

To some extent Transnational nongovernmental organizations (NGOs) have been around for quite awhile, especially if you include corporations. In the 18th century, the East India Company possessed political power (and even an expeditionary military force) that rivaled many national governments. Throughout the 19th century, there were transnational movements to abolish slavery, promote the rights of women, and improve conditions for workers. The number of transnational NGOs, however, has grown tremendously, from around 200 in

1909 to over 17,000 today. The availability of inexpensive and very fast communications technology has made it easier for such groups to organize and make an impact on public policy and international law—the international agreement banning land mines being a recent case in point. Such groups prompt questions about sovereignty because they appear to threaten the integrity of domestic decision making. Activists who lose on their home territory can pressure foreign governments, which may in turn influence decision makers in the activists' own nation.

But for all of the talk of growing NGO influence, their power to affect a country's domestic affairs has been limited when compared to governments, international organizations, and multinational corporations. The United Fruit Company had more influence in Central America in the early part of the 20th century than any NGO could hope to have anywhere in the contemporary world. The International Monetary Fund and other multilateral financial institutions now routinely negotiate conditionality agreements that involve not only specific economic targets but also domestic institutional changes, such as pledges to crack down on corruption and break up cartels.

Smaller, weaker states are the most frequent targets of external efforts to alter domestic institutions, but more powerful states are not immune. The openness of the U.S. political system means that not only NGOs, but also foreign governments, can play some role in political decisions. (The Mexican government, for instance, lobbied heavily for the passage of the North American Free Trade Agreement [NAFTA].) In fact, the permeability of the American polity makes the United States a less threatening partner; nations are more willing to sign on to U.S.-sponsored international arrangements because they have some confidence that they can play a role in U.S. decision making.

Sovereignty Blocks Conflict Resolution

Yes, sometimes   Rulers as well as their constituents have some reasonably clear notion of what sovereignty means—exclusive control within a given territory— even if this norm has been challenged frequently by inconsistent principles (such as universal human rights) and violated in practice (the U.S.- and British-enforced no-fly zones over Iraq). In fact, the political importance of conventional sovereignty rules has made it harder to solve some problems. There is, for instance, no conventional sovereignty solution for Jerusalem, but it doesn't require much imagination to think of alternatives: Divide the city into small pieces; divide the Temple Mount vertically with the Palestinians controlling the top and the Israelis the bottom; establish some kind of international authority; divide control over different issues (religious practices versus taxation, for instance) among different authorities. Any one of these solutions would be better for most Israelis and Palestinians than an ongoing stalemate, but political leaders on both sides have had trouble delivering a settlement because they are subject to attacks by counterelites who can wave the sovereignty flag.

Conventional rules have also been problematic for Tibet. Both the Chinese and the Tibetans might be better off if Tibet could regain some of the autonomy it had as a tributary state within the traditional Chinese empire. Tibet had extensive local control, but symbolically (and sometimes through tribute payments) recognized the supremacy of the emperor. Today, few on either side would even know what a tributary state is, and even if the leaders of Tibet worked out some kind of settlement that would give their country more self-government, there would be no guarantee that they could gain the support of their own constituents.

If, however, leaders can reach mutual agreements, bring along their constituents, or are willing to use coercion, sovereignty rules can be violated in inventive ways. The Chinese, for instance, made Hong Kong a special administrative region after the transfer from British rule, allowed a foreign judge to sit on the Court of Final Appeal, and secured acceptance by other states not only for Hong Kong's participation in a number of international organizations but also for separate visa agreements and recognition of a distinct Hong Kong passport. All of these measures violate conventional sovereignty rules since Hong Kong does not have juridical independence. Only by inventing a unique status for Hong Kong, which involved the acquiescence of other states, could China claim sovereignty while simultaneously preserving the confidence of the business community.

The European Union Is a New Model for Supranational Governance

Yes, but only for the Europeans   The European Union (EU) really is a new thing, far more interesting in terms of sovereignty than Hong Kong. It is not a conventional international organization because its member states are now so intimately linked with one another that withdrawal is not a viable option. It is not likely to become a "United States of Europe"—a large federal state that might look something like the United States of America—because the interests, cultures, economies, and domestic institutional arrangements of its members are too diverse. Widening the EU to include the former communist states of Central Europe would further complicate any efforts to move toward a political organization that looks like a conventional sovereign state.

The EU is inconsistent with conventional sovereignty rules. Its member states have created supranational institutions (the European Court of Justice, the European Commission, and the Council of Ministers) that can make decisions opposed by some member states. The rulings of the court have direct effect and supremacy within national judicial systems, even though these doctrines were never explicitly endorsed in any treaty. The European Monetary Union created a central bank that now controls monetary affairs for three of the union's four largest states. The Single European Act and the Maastricht Treaty provide for majority or qualified majority, but not unanimous, voting in some issue areas. In one sense, the European Union is a product of state sovereignty because it has been created through voluntary agreements among its member states. But, in another sense, it fundamentally contradicts conventional understandings of sovereignty because these same agreements have undermined the juridical autonomy of its individual members.

The European Union, however, is not a model that other parts of the world can imitate. The initial moves toward integration could not have taken place without the political and economic support of the United States, which was, in the early years of the Cold War, much more interested in creating a strong alliance that could effectively oppose the Soviet Union than it was in any potential European challenge to U.S. leadership. Germany, one of the largest states in the European Union, has been the most consistent supporter of an institutional structure that would limit Berlin's own freedom of action, a reflection of the lessons of two devastating wars and the attractiveness of a European identity for a country still grappling with the sins of the Nazi era. It is hard to imagine that other regional powers such as China, Japan, or Brazil, much less the United States, would have any interest in tying their own hands in similar ways. (Regional trading agreements such as Mercosur and NAFTA have very limited supranational provisions and show few signs of evolving into broader monetary or political unions.) The EU is a new and unique institutional structure, but it will coexist with, not displace, the sovereign-state model.

The Waning State of Sovereignty

T hose who think that the sovereign state will reign supreme forever in the international system have been fooled. Like a torrential downpour, a multitude of things—ranging from the industrial revolution to the inception of the Internet, from the end of colonialism to the formation of the European Union [EU]—are at work eroding state sovereignty. Although they will try to weather the storm, states will eventually go the way of the Holy Roman Empire.

The Sovereign State Is Just About Dead

No, but they are dying.   The state as a political unit is not on its immediate deathbed, but states are outdated institutions that have difficulty meeting the needs of their citizens. One argument is that states are too big to do the small things. The size of states and the scope of their governments leave many people complaining about "big government," where the multiple layers of bureaucracy complicate accomplishing any task. Securing a passport can take months, and implementing a new social welfare program can take years to get through the system.

It can also be said that states are too small to do the big things. We live in an era of global problems. Just a few of the concerns that affect the Earth and which are difficult or impossible for any state to address alone include: global warming; the depletion of the ozone layer; the increase of nuclear, biological, and chemical weapons of mass destruction which can be delivered intercontinentally by missiles or terrorists; the movement of trade, investment capital, and money across borders; transnational communications and travel; and the global spread of AIDS and other diseases. Can states any longer truly protect the health, wealth, and very lives of their citizens? The answer in many cases is, "not very well."

Then there is the problem of small states. Microstates are on the increase as more and more people seek national self-determination. Too often what that means, however, is the creation of sovereign entities that more resemble a small city than a country. Nauru is a good example. With a territory that is but eight square miles, a population of barely 10,000, and an economy that depends on the export of guano (seabird droppings) for fertlizer, Nauru is a sham state. Yes, it may take only a day to obtain a Nauruan passport, but no, Nauru cannot provide a secure future for its population.

In 1648, the Treaty of Westphalia marked the birth of the modern state. The states that formed and survived did so because at that time they were the most available means to provide people with security and resources. A competition ensued between the states that formed in Europe. They sought to secure trade relations and to stake claims on territories throughout the world to create empires. This process continued until World War II. The atrocities brought about by the fascist ideology driving the conflict pushed the victors of the war towards working to preserve human dignity and identity. To facilitate this goal of self-determinism, the colonial empires were broken up, and the former colonies were granted their independence and recognized as states. As a result, the number of states increased exponentially during the mid-twentieth century. It was expected that through a process of decolonization, the former empires would endow the colonies with the institutions they needed to become successful and independent states. Instead, a division has emerged between the wealthy North (most of the former imperial countries) and poor South (most of the former colonies) that structures the international system. The geographical description of these countries does not fully illustrate the economic, political, social, and ecological disparities that separate these states.

Indeed many factors have contributed to the developmental problems encountered by the former colonies. Not the least of these is rather abstractly-drawn political borders that failed to take into consideration the indigenous settlements established long before colonization. The intrastate (internal) ethnic conflict that now plagues the majority of all developing states—particularly in Africa, the region most randomly dissected—constitutes the majority of all conflict in the post–cold war era.

One of the main functions of the state is that it is supposed to provide security for its citizens. Yet the millions of people who have been killed as a result of conflict since the formation of states raises questions as to the ability of states to provide security. Considering that weapons are increasingly powerful, and that biological, chemical, and weapons of mass destruction know no political boundaries, the state does not appear to be a very effective means of protection. In most cases, it is the state that incites war and violence. Furthermore, there are countless examples where states even do this against their own citizens who [are] the very ones the state is supposed to protect.

Those who argue that states are still strong sovereign units are missing the reality of state existence. In theory, all states are equal. Any state recognized as sovereign by the international community earns the same status as other states. This is true regardless of the state's resources, its level of development, or its ability to sustain itself and its population. International recognition also almost automatically earns the state a seat in the United Nations (UN) General Assembly, gives the state the opportunity to make treaties with other states, and otherwise gives the state legal standing as one among equals with other states. This theory is largely fiction, however. According all states the same status masks the reality of the situation. It appears that states possess all of the qualities necessary to act as sovereign units, when, indeed, many states fail to provide economic security for their own citizens. Many people in such states would suffer without international assistance offered by nongovernmental organizations (NGOs) and intergovernmental organizations (IGOs).

So, those who argue that state sovereignty is not in decline ignore the fact that a majority of the states have very limited sovereignty to begin with. Recognizing the inconsistencies between the entities collectively called states prompted scholar Robert Jackson to coin the term 'quasi-state' to refer to states that are recognized as sovereign, yet do not meet the criteria for actual statehood. Even more discouraging is the fact that Jackson was also moved to coin the term 'failed-state' to describe states like Somalia, Chad, Liberia, and Afghanistan, whose infrastructures have disintegrated almost beyond repair. Despite their inability to function, however, these states continue to be recognized as legitimate actors because the international system does not permit states to "quit" or decide they are no longer going to be states.

Sovereignty Means Final Authority

Not anymore, if ever.   Many, even most, states have never been free to [do] whatever they want, and some small states with powerful neighbors have been on very short leashes. That continues to occur. Furthermore, most developing states still struggle to lay the foundation necessary to hold free and fair elections, keep government corruption in check, and maintain a state of order sufficient enough to withstand meddling by other states and the international community.

Though having sovereign recognition gives states the ability to enter into agreements and treaties, enables them to become members of IGOs, makes them eligible for international aid, and is supposed to protect states from unwanted foreign intervention, these "benefits" also compromise state authority. A closer look at sovereignty reveals that it is not necessarily what it appears to be.

While no government wants to relinquish any of its sovereignty and most claim that no other authority supersedes its own, the international system is pushing states more and more to abide by decisions made by IGOs. Membership in an international organization necessarily requires a state to relinquish at least some of its authority for the organization to function effectively. Changes in the international system (the rising importance, number, and influence of NGOs, emergent transnational issues, and the increasing number of overall actors) have required states and other actors to cooperate through international organizations. The primary incentive for states to cooperate is to minimize the effects of the changes and the challenges to state sovereignty. China took considerable heat during talks over its application for admission to the World Trade Organization (WTO) from the United States and the European Union (EU) because of its human rights record. And the WTO members, including both the powerful and less powerful members alike, have been subject to rulings by the WTO that require these states to change their domestic policies to be consistent with those at the international level.

Most states attempting to meet the basic needs of their citizens require international assistance. However, that aid usually has stipulations attached, ranging from revamping state budgets to downsizing bureaucracy size to requiring that states adopt a democratic form of government. States are then evaluated on a regular basis and may have their aid withdrawn if they are noncompliant.

While the notion of nonintervention is considered sacred in the international community, the reality of the situation is that intervention, both overt and covert, happens regularly. Instances abound, from the justifiable drive to end apartheid in South Africa to the equally understandable wish to end ethnic cleansing in Kosovo, where the international community, as well as individual states, have interfered in one or another state's domestic affairs.

Universal Human Rights Are an Unprecedented Challenge to Sovereignty

Indeed they are.   The Universal Declaration of Human Rights, as negotiated by the UN member-states in 1948 following the atrocities of World War II, invites interference in states' domestic affairs in the name of humanitarian intervention. In the post–cold war era, the number of conflicts throughout the world has exploded, and the Somalias, Bosnias, Rwandas, Yugoslavias, and Haitis of the world have prompted the international community to intervene for humanitarian reasons. However, the sovereignty of these states is depleted as foreigners intervene in the internal affairs of these countries. This is not to say that the massacre and violence taking place is justified, but rather that events are indicators of how states have become obsolete.

Not only is the sovereignty of domestic states diminished by the Universal Declaration of Human Rights, but the authority of Eastern states is undermined by the Western-oriented values that are the foundation of this so-called universal declaration. The difference is that the Western idea of rights is individually-based while the Eastern concept of rights is communally-based. Many Eastern cultures feel that Western values predominated in defining what a "universal" set of human rights should encompass. Thus, states like China are constantly criticized for violating the Universal Declaration of Human Rights, even though their practices coincide with their own values of the good of the community taking precedence over that of the individual.

Although some argue that UN agreements regarding the various human rights issues are often ineffective, more people are now conscious of these issues. Oppression, hate, slavery, abuse, and violence have not been wiped out by these agreements, but international agreements have spread awareness about these issues. Furthermore, the establishment of international war crime tribunals now provides a way to evaluate and determine whether or not actions warrant punishment. For those found guilty of unnecessarily inhumane practices during a time of war, the international system now has a means for seeking retribution. NGOs such as Amnesty International and WorldWatch monitor state action as well as lobby, protest, and boycott states that violate human rights. At least, their efforts have put states' human rights violations in the spotlight. Their perseverance has helped to push countries to amend their policies, with South Africa as the most stark example of a state's authority being undermined because of its human rights violations.

Globalization Undermines State Control

Yes.   While states undeniably remain significant actors in the international system, the effects of globalization are steadily eroding away at state authority. Trade, communications, technology, and travel all serve to undermine state control.

The increase in international trade has had devastating effects throughout various sectors of economically advanced economies. One does not have to see [the film] *The Full Monty* to be reminded of the power that MNCs [multinational corporations] wield over states. A visit to Bethlehem, Pennsylvania, or Birmingham, England, finds huge factories abandoned, houses vacant due to bank foreclosures, and unemployment rates through the sky. Whole cities have lost their main source of jobs when MNCs decided to utilize cheap foreign labor and have moved abroad, leaving both blue-collar workers and middle-managers jobless. Consider that U.S. labor union members—fighting to keep jobs at home—constituted the largest number of protestors at the Seattle Millennium Round of the WTO talks in 1999.

States have been forced into devising social welfare strategies in attempts to salvage their economies from the effects of globalization. Indeed, it would appear that developing countries benefit from the industry losses of the industrially advanced countries. Instead, however, studies indicate that in most cases the high trade-offs, such as land, tax incentives, and capital, for developing countries offer only a few small returns. In the end, it is the MNCs that both contribute to and benefit from waning state sovereignty.

Communications have been chipping away at the state since the printing press was invented. Since that time, countless communications innovations, from Radio Free Europe to televised satellite broadcasts, have whittled away at government control. Undoubtedly the most recent of these is the Internet. Despite the many opportunities and convenience it brings, surfing the web diminishes state control. Citizens are moving from interacting with one another in civil society to virtual reality. The more time citizens spend alone at home in front of their computers, the less likely they are to take part in community or civic activities, play on a softball team, or sing in the church choir. Furthermore, the Internet provides the freedom to move beyond political boundaries. People have endless opportunities to voice their points of view or network globally to challenge states' authority on issues like environmental practices or human rights records. States, including China and Singapore, have attempted to regulate Internet access in order to preserve their authority. But, it has been an uphill battle as more people gain access to information outside of what their governments report.

Just as traditional forms of terrorism challenge a state's authority over its territory and/or nationals, cyberterrorism also undermines state control. The number of cyberterrorist attacks increases daily, with infiltrators planting viruses that destroy entire databanks, as has happened in the United States, or posting slanderous information, as occurred in Japan. As governments increasingly rely on the Internet to disseminate information and to process passports, driver's licenses, and welfare benefits, states open themselves to cyberterrorist attacks.

There is yet another way that technology presents problems for governments, especially those that do not have it. The increasing technology gap between the North and South only serves to undermine the authority of developing states as they struggle to advance. Falling farther behind the cutting edge of technology decreases job opportunities, chances for development, and hopes for improving living standards. As people become frustrated with the lack of services provided by the state, the possibility for social unrest increases tremendously.

The technological advances of the agricultural industry also affect state sovereignty. Though genetically modified organisms (GMOs) are touted for the fortified grains and vegetables that can be produced and the plants genetically encoded with built-in pesticides, there is a downside to altering genes. Biotech companies can also produce seeds that do not reproduce, forcing poor countries to buy new seeds every season, rather than storing seeds from the last crop for the next year. GMOs undermine state development programs by keeping poor countries dependent upon international aid to feed their people.

More efficient travel presents a plethora of problems for the state. To begin, travel facilitates terrorist activities. It provides terrorists with an easy way to enter a foreign country or to take hostage a country's nationals who are traveling abroad. Transportation also moves migrants. Consider the effect of the break-up of Yugoslavia and the consequent conflict. Tens of thousands of people abandoned their homes, seeking refuge. They spread across Europe, only to be faced with neofascist movements protesting their arrival. More affordable travel also undermines state sovereignty because it carries people abroad where, for any number of reasons, they choose to remain illegally. In many cases, these people rely on the social welfare of their host country to help to meet their needs, thus depleting the state's resources. Finally, more efficient travel means more efficient spread of disease. Thanks to mobility, AIDS has spread like wildfire across Africa, now threatening into Asia. The situation has become so horrific that many states now consider AIDS to be a threat to their national security because states cannot guarantee their citizens' protection from this threat.

Globalization Is Changing the Scope of State Control

Yes.   Globalization is forcing states to alter their authority. States have not voluntarily changed the amount or level of control they hold. Rather, the nature of the Global Village requires states to change if they want to remain players in the international system. This argument, however, diminishes the fact that the world is not a static thing and that, as Darwin concluded, all things must evolve if they are to survive. It is not enough to argue that state sovereignty is not waning but just trying to survive by transforming. Instead, what needs to be considered is that the significance of the state to its people is in decline, regardless of the changes that happen to a state as an evolving entity.

States are being challenged by international organizations. These include both the intergovernmental bodies that the states created and are members [of] and the myriad private membership nongovernmental organizations (NGOs) that have sprung up. If states did not concede at least some of their authority to the international organizations they formed, it would defeat the purpose of creating them, i.e., to facilitate relations between them.

NGOs Are Nibbling at National Sovereignty

To a large extent. This is especially if, as is appropriate, one considers corporations to be NGOs.

MNCs have power, mobility, and resources. They are very mobile; if a state is not willing to grant incentives to MNCs to set up or to stay, corporations will move. In a global business environment where states compete for industries that provide revenue and jobs, MNCs have the upper hand over many small developing economies. MNCs seek out places like the *maquiladoras* (factories near the U.S. border) in Mexico that draws in businesses because of the low cost of labor, or havens in the Caribbean that attract banks with lax financial regulations. And it is not just developing states that are willing to bargain away their control over tax and environmental regulations to entice businesses to build new factories in their backyards; indeed developed states are just as tenacious in offering enticements and incentives with hopes of boosting their economies.

Big businesses are not the only NGOs that nibble on states. The recent growth spurt of private citizen organizations can be attributed to a phenomenon called "post-materialism" by Professor Ronald Inglehart. Enough citizens of the North have achieved sufficient economic security that they are willing to spend time and money on substantive issues like the environment and human rights. These post-materialist trends have brought people with similar concerns together to create, join, and support NGOs. The result is people uniting across political boundaries to form transnational networks that challenge state authority. Many of the groups that protested the Seattle Millennium Round of the WTO or the United Nations Council on Trade and Development (UNCTAD) in Bangkok are NGOs that demonstrated to press governments to make responsible decisions regarding enforcement of stricter environmental codes or labor regulations. Their efforts at the WTO talks have recently been rewarded, as the major IGO that determines and governs trade policy between states is now trying to work with NGOs rather than exclude them, as had previously been the case.

As the overall number of NGOs increases, so does the number of NGOs moving into developing countries in need of assistance. In many instances, these groups provide essential social welfare services that struggling economies and fractured governments cannot supply. Increasingly, industrially advanced states and IGOs are using NGOs to distribute aid and evaluate progress because they have a local knowledge of the situation. Though countless people benefit from these services, the consequence is that the legitimacy of these developing states is even further undermined because people come to depend on the NGOs, rather than the states. The net effect is to put the NGOs in a position of power over those governments.

Sovereignty Blocks Conflict Resolution

Increasingly less. It appears that conflict resolution is based more on issues of self-identity and expression than on preserving state sovereignty. Before Lebanon crumbled to pieces, the Jews, Christians, and Arabs lived peaceably together under a well-functioning government. But Lebanon was eventually affected by neighboring religious conflict. As long as each of these ethnically diverse groups was recognized, the state prospered. When groups attempted to oppress one another, conflict broke out.

A majority of conflict occurring in the post–cold war era is a result of ethnic intrastate clashes in developing countries. Oppressing minorities without fair representation chips away at state sovereignty. Rwanda's Tutsi population was massacred as a result of ethnic hatred. Somalia is still divided by clan rivalries. Kurds throughout Turkey, Iran, and Iraq are persecuted daily. Though they express their desires in very different ways, the Zapatista rebels in the southern Mexican state of Chiapas and many of the French-heritage Quebecois in Canada continue to fight for separation seeking representation.

When the international community does intervene in these conflicts, most efforts are channeled through regional organizations and IGOs. The North Atlantic Treaty Organization (NATO) member-states as well as the UN forces intervened in the Balkans in attempts to save Yugoslavia from itself because of ethnic intrastate differences.

The European Union Is a New Model for Supranational Governance

Yes. It provides a model of what is likely to come.

Given the success of the EU, particularly in the global market, it is not only desirable, but also necessary for other countries to emulate its success. Though the laws of capitalism are based on competition, cooperation has proven to be successful for the member-countries of the EU. The strengthening of the EU has even prompted other industrially advanced countries like the United States to join trade groups like the North American Free Trade Agreement (NAFTA) and seek even larger multilateral trade organizations, such as the Free Trade Agreements of the Americas that, if it comes into being as planned, will include virtually all the countries in the Western Hemisphere.

Globalization is pressuring individual states into regional blocs in order to compete with other regionally based economic organizations that have formed. Across the globe, few states do not belong to at least one regional economic organization. Despite resistance to external political and economic forces, even previously closed countries such as Burma/Myanmar have realized the benefits of regional cooperation.

Stephen Krasner argues that expanding the EU to include more Central and Eastern European countries would make it difficult to "move toward a political organization that looks like a conventional sovereign state." But, that's the point—the conventional states are becoming obsolete as regional and international organizations supplant them by providing more of what citizens need. The formation and relative success of the EU provided the inspiration for other regional economic organizations, such as MERCOSUR (the Southern Cone Common Market in South America), the Association of Southeast Asian Nations (ASEAN), and the Southern Africa Development Community (SADC), that hope to emulate the EU's successes. By joining together within regions, member-states have already increased their potential in terms of a larger market for their products and possibilities for joint ventures, such as the hydroelectric plants being built in the Southern Cone region. Rather than competing against international powerhouses in a highly heterogeneous international market, Latin American, Asian, and African states open their markets regionally, thus exponentially increasing the size of their markets for their domestic businesses by targeting similarly situated consumers and economies. Just as the EU started out with joint economic ventures, these regional economic organizations have begun to lay the same foundations that have proven successful for the advancement of the EU.

What Can We Conclude About the Future of the Sovereign State?

Its prospects are poor.   Stephen Krasner is whistling past the proverbial graveyard of states. The growth of separatist movements and the general rising tide of frustration that people feel about the national governments are evidence that states have gotten too big to tend efficiently to the needs of their people. States also cannot hope to deal with global pollution, the spread of weapons of mass destruction, the rocketing level of economic interdependence, and a host of other problems that ignore national borders.

Those who cannot see the end of the state coming also seem oblivious to the impact that rapid international travel, almost instantaneous international communications, the Internet, the increasing homogenization of culture, and a host of other transnational trends are having on obliterating national distinctions. "What is now must always be," may seem comforting, but it is self-delusion.

POSTSCRIPT

Will State Sovereignty Survive Globalism?

Many people find it difficult to debate the issue of the survival of the sovereign state. For a general view of the role of states in the contemporary international system, see Walter C. Opello, Jr., and Stephen Rosow, *The Nation-State and Global Order: A Historical Introduction to Contemporary Politics* (Lynne Rienner, 1999). Nationalism, the identity link between people and their country, is such a powerful force that some find it almost treasonous to suggest that their country—whether the United States, Canada, Zimbabwe, or any other country—is destined to be eclipsed and perhaps to cease to exist as the dominant unit of the international political system.

Yet it could happen. Since states as we know them did not always exist as a political unit, there is little reason to believe that they will necessarily persist. Instead, as economic, military, and other factors change, it is reasonable to suspect that the form of governance best suited to address these new realities will also change. It may be that just as feudal units once proved to be economically nonviable, states may also fail to meet the tests of a global economy. So, too, just as small units could not provide adequate security amid new weapons and tactics, critics say that states provide little protection from weapons of mass destruction (WMDs) and terrorism. Maryann Cusimano Love, ed., *Beyond Sovereignty: Issues for a Global Agenda* (Wadsworth, 2002) discusses the problems that are beyond the ability of any single state to solve.

Krasner and Weir both make thoughtful arguments, but it is unclear which scholar the verdict of history will ultimately uphold on the question of whether or not states should continue to dominate the political system and to be the principal focus of political identity. Clearly, states continue to exercise great political strength and remain most people's main focus of political identification. But it is also the case that states exist in a rapidly changing political environment and that, ultimately, they need to adapt and serve the needs of their people. A good Web site to begin further exploration on the future of sovereignty is that of the Global Policy Forum at http://www.globalpolicy.org. Click the box that says "Nations & States."

More reading from Krasner's perspective is available in his edited volume *Problematic Sovereignty* (Columbia University Press, 2001) and in Frederick C. Turner and Alejandro L. Corbacho, "New Roles for the State," *International Social Science Journal* (Winter 2000). For a different perspective, see Gerard Kreijen, ed., *State, Sovereignty, and International Governance* (Oxford University Press, 2002), which investigates the response of the global community to states that collapse and that fail to observe the basic principles of international law.

On the Internet . . .

Country Indicators for Foreign Policy (CIFP)

Hosted by Carlton University in Canada, the Country Indicators for Foreign Policy project represents an ongoing effort to identify and assemble statistical information conveying the key features of the economic, political, social, and cultural environments of countries around the world.

http://www.carleton.ca/cifp/

U.S. Department of State

The information on this site is organized into categories based on countries, topics, and other criteria. "Background Notes," which provide information on regions and specific countries, can be accessed through this site.

http://www.state.gov/index.cfm

PART 2

Regional Issues

*T*he issues in this section deal with countries that are major regional powers. In this era of interdependence among nations, it is important to understand the concerns that these issues address and the actors involved because they will shape the world and will affect the lives of all people.

- Will the "Bush Doctrine" Promote a More Secure World?

- Should the United States Continue to Encourage a United Europe?

- Do China's Armaments and Intentions Pose a Long-Term Threat?

- Would It Be an Error to Establish a Palestinian State?

- Was the War With Iraq Unjustified?

- Should North Korea's Arms Program Evoke a Hard-Line Response?

ISSUE 3

Will the "Bush Doctrine" Promote a More Secure World?

YES: George W. Bush, from "The National Security Strategy of the United States of America" (September 2002)

NO: John Steinbruner, from "Confusing Ends and Means: The Doctrine of Coercive Pre-emption," *Arms Control Today* (January/ February 2003)

ISSUE SUMMARY

YES: George W. Bush, president of the United States, argues that the United States should take strong action, preferably with other countries but alone if necessary, to ensure that enemies cannot attack or intimidate Americans and others.

NO: John Steinbruner, director of the Center for International and Security Studies at Maryland, contends that what has become called the "Bush doctrine" is unnecessarily provocative and that its threat to use preemptive military power against whomever the United States perceives to be a threat will increase global violence and create such hostility toward the United States that its security will be decreased.

In September 2002 President George W. Bush issued a document entitled "The National Security Strategy of the United States of America," an annual report to Congress required by the National Security Act of 1947. The yearly documents have usually been of interest to those following U.S. defense planning and foreign policy, but seldom has one of the reports sparked the sort of intense debate that the 2002 document began. The report begins by characterizing the United States as possessing "unprecedented—and unequaled—strength and influence in the world" and goes on to argue that "this position comes with unparalleled responsibilities, obligations, and opportunity. The great strength of this nation must be used to promote a balance of power that favors freedom." A number of the report's sections pledge that the United States will use its strength globally to promote human dignity, democracy, and free economic interchange, and the document also declares that the United States intends to develop cooperation with "other main centers of global power." These sections drew relatively little criticism, although some commentary did object that, if taken literally, the pledges would lead the country into an unending series of draining efforts around the world.

It was the sections on threats to the United States and its allies, especially by terrorism and weapons of mass destruction (biological, chemical, nuclear, and radiological weapons) that caused the greatest response. As you will see in the first of the following readings, President Bush not only promises that attacks on the United States and its allies will bring swift, sure, and devastating consequences to perpetrators. He also indicates that the United States will not wait to be attacked but might act preemptively to destroy perceived enemies and their capabilities before an attack could occur.

The president's position is almost certainly the product of both events and his foreign policy orientation. The relevant events include the September 11, 2001, terrorist attacks and the ongoing terrorist threat against Americans. Also important are the increasing number of countries that are able to build an arsenal of weapons of mass destruction and to use them against increasingly distant targets either by passing the weapons on to terrorists or by using increasingly available and effective missile technology.

The second impetus behind the president's national security approach stems from his general foreign policy orientation. Two factors are relevant here. One is that he is more apt than was his predecessor, President Bill Clinton, to see the world in stark terms. That was evident in his labeling of Iraq, Iran, and North Korea an "axis of evil" in his January 2002 State of the Union Address, a reference that not only cast them as immoral but also equated them with Nazi Germany and other Axis powers from World War II.

The second factor related to Bush's foreign policy approach is his greater willingness (than Clinton, among others) to take unilateral action and to take positions that are vastly unpopular with much of the world community, including many U.S. allies. For example, to the dismay of many, the Bush administration has rejected U.S. adherence to the Kyoto treaty to reduce the emissions of carbon dioxide and other so-called greenhouse gases that most scientists believe promote global warming. Bush also abrogated the Anti-Ballistic Missile Treaty with Russia in order to try to build a national missile defense system, refused to join the countries ratifying the treaty establishing the International Criminal Court, defied a great deal of global sentiment on the road to war with Iraq, and has taken many other unilateral actions.

The selections that follow lay out the essence of the debate about Bush's foreign policy approach. In the first, the president himself presents his national security doctrine. He provides a great deal of strong rhetoric aimed at establishing "moral clarity," and he pledges the United States to take decisive strong action, unilaterally and preemptively when necessary, to defeat terrorists and rogue nations. These steps are necessary and justified, according to Bush. In the second selection, John Steinbruner takes strong exception to Bush's policy stance, arguing that unilateralism and preemption are both unrealistic and dangerous.

George W. Bush

 **YES**

The National Security Strategy of the United States of America

Today, the United States enjoys a position of unparalleled military strength and great economic and political influence. In keeping with our heritage and principles, we do not use our strength to press for unilateral advantage. We seek instead to create a balance of power that favors human freedom: conditions in which all nations and all societies can choose for themselves the rewards and challenges of political and economic liberty. In a world that is safe, people will be able to make their own lives better. We will defend the peace by fighting terrorists and tyrants. We will preserve the peace by building good relations among the great powers. We will extend the peace by encouraging free and open societies on every continent.

Defending our Nation against its enemies is the first and fundamental commitment of the Federal Government. Today, that task has changed dramatically. Enemies in the past needed great armies and great industrial capabilities to endanger America. Now, shadowy networks of individuals can bring great chaos and suffering to our shores for less than it costs to purchase a single tank. Terrorists are organized to penetrate open societies and to turn the power of modern technologies against us.

To defeat this threat we must make use of every tool in our arsenal—military power, better homeland defenses, law enforcement, intelligence, and vigorous efforts to cut off terrorist financing. The war against terrorists of global reach is a global enterprise of uncertain duration. America will help nations that need our assistance in combating terror. And America will hold to account nations that are compromised by terror, including those who harbor terrorists—because the allies of terror are the enemies of civilization. The United States and countries cooperating with us must not allow the terrorists to develop new home bases. Together, we will seek to deny them sanctuary at every turn.

The gravest danger our Nation faces lies at the crossroads of radicalism and technology. Our enemies have openly declared that they are seeking weapons of mass destruction, and evidence indicates that they are doing so with determination. The United States will not allow these efforts to succeed. We will build defenses against ballistic missiles and other means of delivery. We

From George W. Bush, "The National Security Strategy of the United States of America," Washington, D.C. (September 2002).

will cooperate with other nations to deny, contain, and curtail our enemies' efforts to acquire dangerous technologies. And, as a matter of common sense and self-defense, America will act against such emerging threats before they are fully formed. We cannot defend America and our friends by hoping for the best. So we must be prepared to defeat our enemies' plans, using the best intelligence and proceeding with deliberation. History will judge harshly those who saw this coming danger but failed to act. In the new world we have entered, the only path to peace and security is the path of action. . . .

Overview of America's International Strategy

. . . The United States possesses unprecedented—and unequaled—strength and influence in the world. Sustained by faith in the principles of liberty, and the value of a free society, this position comes with unparalleled responsibilities, obligations, and opportunity. The great strength of this nation must be used to promote a balance of power that favors freedom.

For most of the twentieth century, the world was divided by a great struggle over ideas: destructive totalitarian visions versus freedom and equality.

That great struggle is over. The militant visions of class, nation, and race which promised utopia and delivered misery have been defeated and discredited. America is now threatened less by conquering states than we are by failing ones. We are menaced less by fleets and armies than by catastrophic technologies in the hands of the embittered few. We must defeat these threats to our Nation, allies, and friends.

This is also a time of opportunity for America. We will work to translate this moment of influence into decades of peace, prosperity, and liberty. The U.S. national security strategy will be based on a distinctly American internationalism that reflects the union of our values and our national interests. The aim of this strategy is to help make the world not just safer but better. Our goals on the path to progress are clear: political and economic freedom, peaceful relations with other states, and respect for human dignity. . . .

[Several long sections relating to the goals of the United States have been omitted entirely because they do not directly relate to U.S. military national security policy. These sections pledge the United States to "champion aspirations for human dignity," to "ignite a new era of global economic growth through free markets and free trade," to "develop agendas for cooperative action with other main centers of global power," and to "expand the circle of development by opening societies and building the infrastructure of democracy." The entire document can be found on the White House Web site at http://www.whitehouse.gov/nsc/nss.html.*—Ed.]*

Strengthen Alliances to Defeat Global Terrorism and Work to Prevent Attacks Against Us and Our Friends

... The United States of America is fighting a war against terrorists of global reach. The enemy is not a single political regime or person or religion or ideology. The enemy is terrorism—premeditated, politically motivated violence perpetrated against innocents.

In many regions, legitimate grievances prevent the emergence of a lasting peace. Such grievances deserve to be, and must be, addressed within a political process. But no cause justifies terror. The United States will make no concessions to terrorist demands and strike no deals with them. We make no distinction between terrorists and those who knowingly harbor or provide aid to them.

The struggle against global terrorism is different from any other war in our history. It will be fought on many fronts against a particularly elusive enemy over an extended period of time. Progress will come through the persistent accumulation of successes—some seen, some unseen. . . .

Our priority will be first to disrupt and destroy terrorist organizations of global reach and attack their leadership; command, control, and communications; material support; and finances. This will have a disabling effect upon the terrorists' ability to plan and operate.

We will continue to encourage our regional partners to take up a coordinated effort that isolates the terrorists. Once the regional campaign localizes the threat to a particular state, we will help ensure the state has the military, law enforcement, political, and financial tools necessary to finish the task.

The United States will continue to work with our allies to disrupt the financing of terrorism. We will identify and block the sources of funding for terrorism, freeze the assets of terrorists and those who support them, deny terrorists access to the international financial system, protect legitimate charities from being abused by terrorists, and prevent the movement of terrorists' assets through alternative financial networks. . . .

While our focus is protecting America, we know that to defeat terrorism in today's globalized world we need support from our allies and friends. Wherever possible, the United States will rely on regional organizations and state powers to meet their obligations to fight terrorism. Where governments find the fight against terrorism beyond their capacities, we will match their willpower and their resources with whatever help we and our allies can provide.

As we pursue the terrorists in Afghanistan, we will continue to work with international organizations such as the United Nations, as well as non-governmental organizations, and other countries to provide the humanitarian, political, economic, and security assistance necessary to rebuild Afghanistan so that it will never again abuse its people, threaten its neighbors, and provide a haven for terrorists. . . .

Work With Others to Defuse Regional Conflicts

. . . Concerned nations must remain actively engaged in critical regional disputes to avoid explosive escalation and minimize human suffering. In an increasingly interconnected world, regional crisis can strain our alliances, rekindle rivalries among the major powers, and create horrifying affronts to human dignity. When violence erupts and states falter, the United States will work with friends and partners to alleviate suffering and restore stability.

No doctrine can anticipate every circumstance in which U.S. action—direct or indirect—is warranted. We have finite political, economic, and military resources to meet our global priorities. The United States will approach each case with these strategic principles in mind:

- The United States should invest time and resources into building international relationships and institutions that can help manage local crises when they emerge.

- The United States should be realistic about its ability to help those who are unwilling or unready to help themselves. Where and when people are ready to do their part, we will be willing to move decisively. . . .

Prevent Our Enemies From Threatening Us, Our Allies, and Our Friends With Weapons of Mass Destruction

. . . The nature of the Cold War threat required the United States—with our allies and friends—to emphasize deterrence of the enemy's use of force, producing a grim strategy of mutual assured destruction. With the collapse of the Soviet Union and the end of the Cold War, our security environment has undergone profound transformation.

Having moved from confrontation to cooperation as the hallmark of our relationship with Russia, the dividends are evident: an end to the balance of terror that divided us; an historic reduction in the nuclear arsenals on both sides; and cooperation in areas such as counterterrorism and missile defense that until recently were inconceivable.

But new deadly challenges have emerged from rogue states and terrorists. None of these contemporary threats rival the sheer destructive power that was arrayed against us by the Soviet Union. However, the nature and motivations of these new adversaries, their determination to obtain destructive powers hitherto available only to the world's strongest states, and the greater likelihood that they will use weapons of mass destruction against us, make today's security environment more complex and dangerous.

In the 1990s we witnessed the emergence of a small number of rogue states that, while different in important ways, share a number of attributes. These states:

- brutalize their own people and squander their national resources for the personal gain of the rulers;
- display no regard for international law, threaten their neighbors, and callously violate international treaties to which they are party;
- are determined to acquire weapons of mass destruction, along with other advanced military technology, to be used as threats or offensively to achieve the aggressive designs of these regimes;
- sponsor terrorism around the globe; and
- reject basic human values and hate the United States and everything for which it stands.

. . . We must be prepared to stop rogue states and their terrorist clients before they are able to threaten or use weapons of mass destruction against the United States and our allies and friends. Our response must take full advantage of strengthened alliances, the establishment of new partnerships with former adversaries, innovation in the use of military forces, modern technologies, including the development of an effective missile defense system, and increased emphasis on intelligence collection and analysis. . . .

For centuries, international law recognized that nations need not suffer an attack before they can lawfully take action to defend themselves against forces that present an imminent danger of attack. Legal scholars and international jurists often conditioned the legitimacy of preemption on the existence of an imminent threat—most often a visible mobilization of armies, navies, and air forces preparing to attack.

We must adapt the concept of imminent threat to the capabilities and objectives of today's adversaries. Rogue states and terrorists do not seek to attack us using conventional means. They know such attacks would fail. Instead, they rely on acts of terror and, potentially, the use of weapons of mass destruction—weapons that can be easily concealed, delivered covertly, and used without warning. . . .

The United States has long maintained the option of preemptive actions to counter a sufficient threat to our national security. The greater the threat, the greater is the risk of inaction—and the more compelling the case for taking anticipatory action to defend ourselves, even if uncertainty remains as to the time and place of the enemy's attack. To forestall or prevent such hostile acts by our adversaries, the United States will, if necessary, act preemptively.

The United States will not use force in all cases to preempt emerging threats, nor should nations use preemption as a pretext for aggression. Yet in an age where the enemies of civilization openly and actively seek the world's most destructive technologies, the United States cannot remain idle while dangers gather. We will always proceed deliberately, weighing the consequences of our actions. . .

The purpose of our actions will always be to eliminate a specific threat to the United States or our allies and friends. The reasons for our actions will be clear, the force measured, and the cause just. . . .

Transform America's National Security Institutions to Meet the Challenges and Opportunities of the Twenty-First Century

. . . The unparalleled strength of the United States armed forces, and their forward presence, have maintained the peace in some of the world's most strategically vital regions. However, the threats and enemies we must confront have changed, and so must our forces. A military structured to deter massive Cold War–era armies must be transformed to focus more on how an adversary might fight rather than where and when a war might occur. We will channel our energies to overcome a host of operational challenges.

The presence of American forces overseas is one of the most profound symbols of the U.S. commitments to allies and friends. Through our willingness to use force in our own defense and in defense of others, the United States demonstrates its resolve to maintain a balance of power that favors freedom. To contend with uncertainty and to meet the many security challenges we face, the United States will require bases and stations within and beyond Western Europe and Northeast Asia, as well as temporary access arrangements for the long-distance deployment of U.S. forces. . . .

Innovation within the armed forces will rest on experimentation with new approaches to warfare, strengthening joint operations, exploiting U.S. intelligence advantages, and taking full advantage of science and technology. We must also transform the way the Department of Defense is run, especially in financial management and recruitment and retention. Finally, while maintaining near-term readiness and the ability to fight the war on terrorism, the goal must be to provide the President with a wider range of military options to discourage aggression or any form of coercion against the United States, our allies, and our friends.

We know from history that deterrence can fail; and we know from experience that some enemies cannot be deterred. The United States must and will maintain the capability to defeat any attempt by an enemy—whether a state or non-state actor—to impose its will on the United States, our allies, or our friends. We will maintain the forces sufficient to support our obligations, and to defend freedom. Our forces will be strong enough to dissuade potential adversaries from pursuing a military build-up in hopes of surpassing, or equaling, the power of the United States. . . .

In exercising our leadership, we will respect the values, judgment, and interests of our friends and partners. Still, we will be prepared to act apart when our interests and unique responsibilities require. When we disagree on particulars, we will explain forthrightly the grounds for our concerns and strive to forge viable alternatives. We will not allow such disagreements to obscure our determination to secure together, with our allies and our friends, our shared fundamental interests and values.

Ultimately, the foundation of American strength is at home. It is in the skills of our people, the dynamism of our economy, and the resilience of our institutions. A diverse, modern society has inherent, ambitious, entrepreneurial energy. Our strength comes from what we do with that energy. That is where our national security begins.

NO

John Steinbruner

Confusing Ends and Means:
The Doctrine of Coercive Pre-emption

In a speech at West Point last June [2002], in a more formal statement of national security strategy submitted to Congress in September, and in a White House document published in December, President George W. Bush has proclaimed what appears to be a new security doctrine. Reduced to its essentials, the doctrine suggests that the United States will henceforth attack adversaries to prevent them not only from using but also from acquiring the technologies associated with weapons of mass destruction. If it were systematically implemented, this doctrine would represent a major redirection of policy and a radical revision of established international security rules.

The Bush administration evidently intends to make Iraq the first test case, but the doctrine also has direct implications for the two other countries—North Korea and Iran—that the president has named as members of an "axis of evil." The doctrine is backed by the unprecedented degree of military superiority the United States has acquired. It has also been accompanied by repudiation of prominent agreements that have long been pillars of international regulation—most notably the Anti-Ballistic Missile (ABM) Treaty and the Comprehensive Test Ban Treaty. In that context, the announced doctrine projects an assertive form of American nationalism that is sure to inspire considerable animosity—and not just among potential adversaries. Signs of an international backlash are already evident to those who are willing to look for them—in the recent elections in South Korea and in Germany, for example.

In attempting to understand the significance of this development, it is important to remember that blunt talk and practical accomplishment are not the same thing. The president's inherently provocative pronouncements will force deliberation and reaction throughout the world. The eventual consequences of Bush's declared doctrine will be shaped by compelling interests and competing principles that his pronouncements only dimly acknowledged. When those are considered, as eventually they must be, the importance of cooperatively establishing greater control over mass destruction technologies will overpower the impulse to attack alleged rogues pre-emptively. The idea of using decisive force against implacable evil may be emotionally satisfying, but it is hardly the basis for responsible policy against today's most likely threats. Pre-emptive actions are the result of policy failures, not the triumph of superior virtue or strategic reason.

The Vital Importance of Legitimacy

The central problem with the Bush administration's doctrine is that it fundamentally confuses ends and means. Obviously, the aspiration to prevent warfare is intrinsically legitimate and increasingly important. It is also much better to pre-empt the conditions that generate violence than to prevail in a process of countervailing destruction. The question has to do with the methods that are used to accomplish these purposes. The Bush doctrine of pre-emption apparently proposes to rely primarily on coercive power, that is, to initiate violence in order to prevent it, and it appears to neglect and indeed to disdain international legal restraint. In the judgment of much of the world, that formula is more likely to generate violence than to contain it. Civilized security policy is primarily a matter of establishing and preserving a viable rule of law, and the use of coercive power is subordinate to that objective for very practical reasons. Coercion alone is too inefficient and too ineffective to provide adequate protection. Most of normal life depends on consensual rules, so they are necessarily the foundation of security.

A related problem with the Bush administration's doctrine concerns the scale and character of threat. Before and during the Cold War, security policy was primarily concerned with territorial aggression on a continental scale and with massive destruction by remote bombardment. Preparations for missions of that magnitude would have to be very extensive, readily observable, and centrally organized. Now, threats of primary concern are smaller in scale; much more readily concealed; and, potentially at least, more widely distributed and more diffusely organized. The legitimacy and effectiveness of pre-emptive action depends a great deal on the type of threat to which it is applied.

The most broadly accepted form of pre-emption would be directed against an observably imminent threat of conventional invasion. The prohibition on territorial aggression and the right to defend against it are the most solidly established international legal standards. It is plausible to believe that World War II and the 1991 Persian Gulf War could both have been prevented had timely pre-emption been undertaken. In October 1994, the United States and the United Kingdom successfully reversed a second Iraqi mobilization against Kuwait by credibly threatening a pre-emptive attack, and their actions were backed by a UN Security Council resolution. In any currently foreseeable situation of that sort in which the United States is seriously engaged, the doctrine is likely to be successfully applied. The Bush administration documents cite this established application but attempt to extend it to circumstances where the perceived threat is neither large nor imminent. They do not explain how they will determine aggressive intent before it is demonstrated in deeds or how they will prevent errors of judgment that would make the enacted punishment outweigh the anticipated threat.

Pre-emption against the threat of massive nuclear attack was seriously considered when U.S. and Russian forces were first being formed. Indeed, today U.S. and Russian nuclear forces remain configured to attack enemy nuclear forces in the hopes of destroying them before they can be launched—an operational inclination that could be extremely dangerous during a crisis. It is so difficult, however, to execute a first strike that destroys all enemy weapons—and to be certain that you have that capability—that pre-emption has never been a responsible option for nuclear self-defense. It has also been recognized that a systematic effort to acquire that level of military ability and psychological confidence would lead to destructive competition between potential adversaries (i.e., an arms race). The ABM Treaty and associated offensive force limitation treaties were devised to prevent that from happening.

Bush's new strategic pronouncements reopen this issue with a new twist. They assert the right to use coercive force against the acquisition of mass destruction weapons and imply that mass destruction weapons might themselves be used for this purpose. That form of pre-emption, traditionally termed preventive war, might well succeed if practiced against a smaller adversary early enough in the cycle of weapons development. It sets an inherently discriminatory and implicitly imperial standard, however, that has no chance of ever being broadly accepted, and in forfeiting legitimacy it promises to incite an interminable process of clandestine retribution. When resistance is widely considered justified, even socially mandatory, coercive pre-emption against all forms of clandestine retribution becomes infeasible, as is evident in the many current instances of active civil conflict.

Over the past decade, the United States and the international community have been repeatedly entangled in instances of civil conflict that could not be resolved by the direct combatants and the nominally responsible sovereign authority. There is as yet no settled interpretation of this experience, but the outlines of an intervention doctrine with pre-emptive implications are nonetheless visible. Interventions in Bosnia and Kosovo generated a reluctant and belated but ultimately acknowledged understanding that sustained violence in those areas would pose an intolerably dangerous threat to the surrounding region. UN Security Council Resolution 1244, which authorizes an indefinite international occupation of Kosovo, asserts an international interest in basic standards of legal order that overrides the traditional prerogatives of sovereignty. In retrospect, it is apparent that these interventions could have been more successful and less costly had they been undertaken earlier than they were. Similarly, it is now widely believed that a forceful intervention could have and should have halted the 1994 genocide in Rwanda well before more than half a million people had been slaughtered and millions more driven from their homes.

In the aftermath of the September 11 terrorist attacks, it has been widely recognized that a sustained breakdown of legal order anywhere in the world would provide an organizational base for global terrorism and that forceful intervention to establish basic civil order is justified. The U.S. assault on Afghanistan was generally accepted under that understanding. The implication

is that situations of that sort demand pre-emptive correction, but repeated instances would have to be authorized by the international community as a
whole for reasons of general interest. Despite September 11, the United States
will not be conceded exclusive responsibility for determining the circumstances under which pre-emptive intervention is required to restore civil order,
and it does not have the capacity to assert that prerogative against widespread
resistance.

Coercive pre-emption against terrorists and terrorist organizations is pre-
sumed to be legitimate, as dramatically demonstrated by an incident in Yemen
on November 3 [2002]. On that day, the CIA used an unmanned aerial vehicle
to fire a missile at a car traveling in a remote area of the country, killing all five
of the vehicle's occupants. One of them was said to be a key al Qaeda figure, and
that assertion was generally accepted as valid justification for the attack. There
was no public protest from the Yemeni government, which was apparently consulted in advance but not otherwise involved in the operation. The precedent is
nonetheless inherently contentious. The Yemeni operation was in effect a summary execution with no semblance of legal due process—no disputable presen-
tation of evidence, no equivalent of an impartial judge or jury. If repeated often
enough, that type of action will assuredly generate incidents that exceed the
bounds of accepted justification and will incite recrimination. One cannot defend legal order by violating its central principles. One cannot fight terrorism
by actions that are themselves terrorist in character. In fact, terrorism's strategic
purpose is to exploit the target's natural impulse to respond in kind—to provoke
a decisively stronger opponent into reactions that damage and discredit it.

Practical Judgments

Whether ultimately wise or not, coercive pre-emption against Iraq is obviously
an imminent possibility. Saddam Hussein's regime has so indicted itself that
due process concerns are not likely to be a significant restraint. The legitimacy
of denying Iraq access to mass destruction technology is established in UN resolutions, and a substantial part of the world would apparently acquiesce to a U.S.
military campaign dedicated to that purpose. No one doubts the United States'
ability to undertake such a campaign. The major question is whether an attack
perceived to be designed for the broader notion of "regime change" would trigger a cascading political reaction sufficiently adverse to discredit pre-emption
as a doctrine. If so, that might take some time to recognize.

Even a decisive and enduring success in Iraq would not establish coercive
pre-emption against programs to build weapons of mass destruction as a general
principle. The international community cannot categorically deny the right of
North Korea, Iran, or any other country to nuclear weapons. As non-nuclear-
weapon states under the nuclear Nonproliferation Treaty, both North Korea and
Iran committed themselves not to exercise their inherent right, but that did not
take away the right itself or the associated right to acquire fissile material. North
Korea can legally withdraw from the treaty as it stated it did January 10 [2003].
In the current diplomatic crisis concerning its uranium-enrichment program,

North Korea has cited that right, and in its evident reluctance to apply the doctrine of coercive pre-emption the Bush administration has so far implicitly conceded it.

The categorical prohibition on the offensive application of biotechnology formulated in the 1925 Geneva Protocol and in the Biological Weapons Convention is more plausibly considered a inviolable standard. Although the United States has accused some 13 countries, including North Korea and Iran, of having illegal offensive programs, none of them admits to that allegation, and no country currently claims either the right to biological weapons or the possession of them. All countries, however, assertively and legitimately proclaim the right to conduct biomedical research, and most of them actively do it. The United States cannot restrict that right by simplistically labeling a country "evil."

In addition to having to concede that states have the right to pursue nuclear and biological technology—an admission that undermines the justification for coercive pre-emption—the United States will have to acknowledge that its capability to pre-emptively attack North Korea and Iran is more questionable than its ability to attack Iraq. Since the U.S. military operates within a network of foreign basing rights and access agreements that require consent from the host governments, it would be difficult to organize a pre-emptive attack that did not enjoy general approval. The bottom line is that the United States needs not merely permissive acquiescence but active collaboration of most major countries in order to deal with emerging security problems that cannot be addressed by military force of any sort.

The most urgent of these problems is the management of biotechnology. Fundamental understanding of basic life processes emerging out of a global biomedical research community is enabling extremely powerful applications, both therapeutic and destructive. The eradication of some devastating diseases is becoming feasible as is the deliberate creation of yet more devastating ones. In general, the prospective benefits and the potential dangers are both greater than is the case for any of the other technologies that carry the "mass destruction" label.

The pattern of development is also distinctive. Biotechnology is the product of a worldwide research enterprise operating through open literature primarily for public health purposes. Dedicated weapons projects are a small part of the whole picture and are not the major source of scientific development. The momentum and diffusion of the research base makes it infeasible for any country to appropriate this technology for its exclusive use or to control the flow of information. Current attempts to impose such controls in the hopes of frustrating bioterrorists are unlikely to succeed. Moreover, since biomedical facilities need not be large and do not have inherently identifying features—unlike nuclear facilities—it is more difficult to fathom their activities through satellite imagery and other means.

The relentless implication is that the deliberately destructive use of biotechnology is a threat to all human societies of a scope and magnitude greater than any other. That threat could be developed and delivered by clandestine means, and current national security methods cannot provide adequate protection no matter how they might be elaborated. Under prevailing circumstances of access, it would be impossible to identify and disable all dedicated terrorists and rogues before they have accomplished nefarious deeds, and it would be foolish to attempt to do so by national military operations. A campaign of that sort conducted by the United States under the doctrine of coercive pre-emption is more likely to stimulate the destructive application of biotechnology than to prevent it.

The only reasonable hope is to establish comprehensive oversight procedures within the scientific community robust enough to make dangerous research far more difficult to conceal and simultaneously to organize the research process so that protective applications of biotechnology outpace any destructive ones that might evade oversight. An arrangement of that sort would require intimate, equitable collaboration on a global basis without exception. Impossible as that kind of cooperation might seem given current attitudes, it will be considered, and probably attempted, as the nature of the threat from biotechnology is absorbed. The process of deliberation will impose a major amendment on the doctrine of coercive pre-emption.

The management of fissile material presents a similar imperative in somewhat weaker form. It is technically feasible for terrorists or rogues to use nuclear explosives to wreck devastating havoc with small operations that could be successfully concealed and would therefore evade coercive pre-emption. Doing so is inherently more difficult than using biotechnology to cause damage because the scale of activity required to produce fissile material is far more difficult to conceal and access controls over material already produced are far more developed. Current national standards of accounting and physical security for fissile material are not impermeable, however, and they could be substantially improved by establishing a common international arrangement. The problems involved are more political than technical in character. If the confrontational policies forged during the Cold War were transcended in fact as well as in rhetoric, more robust protection of fissile material could be achieved, but that would assuredly require very convincing restriction on the doctrine of coercive pre-emption. No country will subject its fissile material to international accounting if it believes that coercive pre-emption is a serious possibility.

Conclusion

In the end, the Bush administration's doctrinal pronouncements may prove to be a transient political exercise of little enduring significance or possibly a useful threat with exclusive application to the Iraq situation. They also might spark major international disputes and eventual adjustment. However it turns out, the central contention—that pre-emptive attack can prevent the acquisition of mass destruction technology—is not realistic and does not provide a responsible basis for protecting the United States or anyone else. Preventive action against potentially unmanageable threats is indeed an increasingly vital security interest, but that cannot be accomplished by coercive methods. It will require the systematic exchange of sensitive monitoring information for mutual protection, and arrangements of that sort cannot be established while one party is wielding a confrontational threat against the others. If coercive preemption is to be done at all, it must be done by the international community as a whole for common benefit, not by the United States alone for its own exclusive purposes. The confusion of ends and means presented in the Bush administration's documents will have to be corrected. That is a direct responsibility of the U.S. political system in which the rest of the world has a very substantial stake.

POSTSCRIPT

Will the "Bush Doctrine" Promote a More Secure World?

The debate over the Bush doctrine is multifaceted, and it addresses many of the most pressing issues of the day. One issue is that of danger. How great are the dangers that the United States and other countries face from terrorism and the spread of weapons of mass destruction? One good source of reading on terrorism from many perspectives is Thomas J. Badey, ed., *Annual Editions: Violence and Terrorism 03/04* (McGraw-Hill/Dushkin, 2003). Then there is the matter of how to best deter and defeat terrorism.

As for weapons of mass destruction, one good comprehensive source is Robert J. Einhorn, ed., *Protecting Against the Spread of Nuclear, Biological, and Chemical Weapons: An Action Agenda for the Global Partnership* (Center for Strategic and International Studies, 2003). Also valuable is Joseph Cirincione et al., *Deadly Arsenals: Tracking Weapons of Mass Destruction* (Carnegie Endowment for International Peace, 2002). A fine Web site of the Federation of American Scientists can be found at http://www.fas.org/irp/threat/wmd.htm.

A second important issue inherent in the debate over the Bush doctrine is that of unilateralism versus multilateralism. As the international system has become more interdependent and as organizations such as the United Nations have grown in importance, the norms of the global community have increasingly emphasized the multilateral approach of operating in cooperation with others. Unilateralism is a more traditional approach that emphasizes common (compared to national) interests and independent (compared to collective) decisions and action. It would be inaccurate to see the issue of unilateralism versus multilateralism as a simple "yes-no" debate. Certainly, very few people would argue that the United States (or any other sovereign state) always takes action regardless of what the world thinks or never takes action if it does not have broad international support. Rather, the debate is over the degree to which policy tends in one direction or the other. One study that specifically addresses the intersection of the U.S. superpower status and its role in international organizations is Rosemary Foot, S. Neil Macfarlane, and Michael Mastanduno, eds., *U.S. Hegemony in an Organized World: The United States and Multilateral Institutions* (Oxford University Press, 2002). For a defense of President Bush's unilateralism, read Charles Krauthammer, "The Bush Doctrine: ABM, Kyoto, and the New American Unilateralism," *The Weekly Standard* (June 4, 2001). For more on the often critical international view of Bush's foreign policy, read David M. Malone and Yuen Foong Khong, eds., *Unilateralism and U.S. Foreign Policy: International Perspectives* (Lynne Rienner, 2003).

A third major focus of the debate is the wisdom and, to a degree, the morality of preemptive military action. Here again, there are not many who would contend that a country has to always wait until after an attack has occurred to take action. That goes to the old adage that an ounce of prevention is worth a pound of cure. However, many analysts worry that once preemption becomes part of a country's military doctrine, that country might strike on the basis of a misperceived threat or use a claimed threat to justify an otherwise unsupportable attack. An article detailing the downside of preemptive war is William Galston, "Perils of Preemptive War," *The American Prospect* (September 23, 2002). An article that supports the Bush doctrine, including preemption, is Thomas Donnelly, "The Underpinnings of the Bush Doctrine," http://www.aei.org/publications/pubID.15845/pub_detail.asp (February 1, 2003).

ISSUE 4

Should the United States Continue to Encourage a United Europe?

YES: A. Elizabeth Jones, from Testimony Before the Subcommittee on Europe, Committee on International Relations, U.S. House of Representatives (March 13, 2002)

NO: John C. Hulsman, from "European Arrogance and Weakness Dictate Coalitions of the Willing," *Heritage Lecture No. 777* (February 10, 2003)

ISSUE SUMMARY

YES: A. Elizabeth Jones, assistant secretary of state for European and Eurasian affairs, maintains that the United States looks forward to working cooperatively with such exclusively or mostly European institutions as the European Union, the Organization for Cooperation and Security in Europe, and the North Atlantic Treaty Organization.

NO: John C. Hulsman, a research fellow for European affairs in the Kathryn and Shelby Cullom Davis Institute for International Studies at the Heritage Foundation, argues that the United States should support European countries on a selective basis but not be closely tied to Europe as a whole.

At the end of World War II, Europe was physically and economically devastated. World War II resulted in the deaths of tens of millions of Europeans and the destruction of a great deal of Europe's industry and infrastructure (transportation and communication networks, buildings, etc.). Soon after, the cold war between the United States and the Soviet Union added to Europe's problems by dividing the continent.

The United States took steps to strengthen Europe economically and politically. One path was by extending aid through the multibillion-dollar European Recovery Plan (also called the Marshall Plan) beginning in 1947. The United States aided European recovery in part because American economic health required Europe to once again become a strong trading partner and partly because Washington worried that left-wing movements in France, Italy, and elsewhere might result in communist influence in those countries' governments.

One of the strings that the United States tied to its aid was the demand that Europeans plan together how to best utilize it. This forced economic cooperation and the views of such Europeans as French statesman Jean Monnet (1888–1979) combined to promote the establishment that began in the 1950s and that evolved in the decades that followed into the European Union (EU). Today the EU has 15 member states, and it will soon be joined by an additional 13 eastern and southern European countries. In terms of the size of its population and gross domestic product, the EU rivals the United States.

On the security front, the United States encouraged the formation of the North Atlantic Treaty Organization (NATO) in 1949. The alliance consists of the United States, Canada, 13 Western European countries, and Turkey. The original purpose of NATO was to counter the threat that many in the West thought the Soviet Union posed. Other unspoken factors behind the founding of NATO were the desire to keep the United States engaged in Europe and to avoid any possibility that Germany might reemerge as a powerful, hostile country.

The cooperation in Europe, marked by NATO and what would become the EU, led to other joint organizations, such as the Organization for Security and Cooperation in Europe (OSCE). The OSCE includes 55 countries that are primarily European, but also Canada, the United States, and some states in Central Asia. The organization works through diplomats, observers, and others to prevent conflicts, manage crises, and facilitate post-conflict rehabilitation.

During the decades that followed the end of World War II, the United States served as the leader of the transatlantic bloc, and there was a great deal of unity, although some would argue that it was European subservience to Washington's direction.

However, the last decade or so has seen market changes in U.S.-European relations. One factor is the end of the cold war. The collapse of the Soviet Union and Soviet bloc (with Poland and some other former communist countries now belonging to NATO) erased any immediate military threat to Europe and, therefore, lessened the need for strategic cooperation. With no common enemy to bind Europe and the United States together, old disputes that had once been suppressed in the name of allied unity came to the fore and new disagreements became public and more acrimonious.

A second factor is the full recovery of Europe's economy and confidence. In particular, the expanded political and economic integration of the European Union have made Europeans increasingly willing to challenge U.S. hegemony on a range of political, economic, and social issues. Some in Europe even believe that its countries should work together to restrain U.S. power.

There are many people on both sides of the Atlantic Ocean who continue to believe that a strong, increasingly united Europe and strong transatlantic cooperation are in the mutual interests of the United States and Europe. That is the official U.S. position, which is reflected in the following selection by A. Elizabeth Jones. In the second selection, John C. Hulsman argues that such statements are papering over the distinct and growing division between the United States and its formal allies on the other side of the Atlantic. He contends that the United States should adopt a more realistic policy.

U.S. and Europe: The Bush Administration and Transatlantic Relations

President [George W.] Bush said last August [2001] in Warsaw that the Administration seeks a Europe "whole, free, and at peace." This is even more vital to America's national security in the aftermath of September 11th. The imperative for closer coordination has opened up new opportunities to achieve our goals in Europe and Eurasia. We are cooperating more broadly to combat terrorism. We are pursuing a deeper relationship with Russia. We are advancing throughout the region respect for democracy, the rule of law, human rights, and free market economies.

We know who our friends are when the chips are down and we need help. By this measure, we have friends in Europe. Following September 11th our European partners offered critical assistance in military deployments to Afghanistan. They cracked down on terrorist activities in their territory. European and U.S. soldiers are working side-by-side in Afghanistan. . . . German and Danish troops suffered fatalities while trying to disarm abandoned ordnance in Kabul [Afghanistan]. Europe and the U.S. are partners in every sense.

Recently, a few European leaders have expressed concerns about U.S. "unilateralism." Some wonder about our long-term goals in the War on Terrorism and our intentions regarding pariah states such as Iraq. We take these concerns seriously. But we must put them in perspective. Europeans speak as our coalition partners. They are vulnerable to the same dangers that we are. As one European explained it: "September 11th was an attack on all of us. We want to be involved in the solution." As Secretary [of State Colin] Powell says constantly, the U.S. will continue to engage vigorously with our European partners. Our policies have not changed. We will remain in close touch. U.S.-European relations remain steadfast.

We are reinvigorating our partnership with the European Union [EU]. Counter-terrorism is front and center. In December Secretary Powell signed an agreement with EUROPOL [European Police Office]. We are aiming next for an agreement on judicial cooperation. There is potential for progress on nonproliferation, intelligence sharing, asset freezes, and uprooting terrorist networks. We are taking joint action against terrorist organizations.

From U.S. House of Representatives. Committee on International Relations. Subcommittee on Europe. *U.S. and Europe: The Bush Administration and Transatlantic Relations.* Hearing, March 13, 2002. Washington, D.C.: U.S. Government Printing Office, 2002.

The U.S. and EU economies are increasingly integrated. Trade and reciprocal foreign investment rise each year, doubling since 1990. The U.S. supports a fair, open international trading system. We worked with the EU on a successful launch of the new WTO [World Trade Organization] Round at Doha [Qatar]. We pursue vigorously the resolution of U.S.-EU trade disputes. We will continue to promote U.S. business and economic interests in resolving outstanding disagreements, not just on steel, but on Foreign Sales Corporation tax, biotechnology and beef hormones. Europeans have reacted strongly to the President's decision to impose temporary safeguards on steel. We will work with our European friends and other steel producing countries to address the heart of this problem: excess global capacity in steel production. Our goal is that transatlantic trade solidify all aspects of our relationship, including security.

Our European friends and allies share our concern about the need to accord recognition to surviving Holocaust victims within their lifetimes. In the past eight months, the German foundation "Remembrance, Responsibility and the Future" distributed more than $1.1 billion to 600,000 former slave and forced laborers as provided under the July 17, 2000 agreements. The payment of insurance claims is a difficult issue. We will continue to work with the International Commission on Holocaust Era Insurance Claims and other involved parties to resolve outstanding procedural problems. We are engaged on property restitution. In this regard, the International Task Force on Holocaust Education, Remembrance and Research is an important focus. The foundation's board of trustees is working on criteria for projects of the Future Fund. The interest on the endowment will be used to combat racism and hatred.

NATO [North Atlantic Treaty Organization] remains the cornerstone of transatlantic security. In the aftermath of September 11th, Allies invoked NATO's Article 5 collective defense commitment for the first time in history. Our Allies have provided invaluable support to the anti-terrorist effort. This includes force deployments, intelligence sharing, and extensive law enforcement assistance. Allies recognize that we must intensify this cooperation to address the threats of terrorism and Weapons of Mass Destruction. That is among our goals for the Prague Summit next November.

The September 11th attacks and continued terrorist threats have underscored the need for NATO to improve its ability to meet new challenges to our common security. Allies recognized this threat in the 1991 Strategic Concept. They reinforced it at the Washington Summit in 1999. When President Bush meets with Allied Leaders in Prague, NATO is expected to approve a program of action to enhance its ability to deal with these threats. It is vital that our European Allies, who have not followed through on all the commitments made in NATO's Defense Capabilities Initiative, refocus and reprioritize their efforts to address the growing capabilities gap within NATO. Thus, the development of new capabilities is one of our priorities for the Prague Summit next November.

A second key goal for Prague is the addition of new members to the Alliance. Continued NATO enlargement will reinforce the strength and cohesion of states committed to our values. It will bolster our own defense. We are looking closely at values issues among aspirant countries. We will evaluate candidates on their ability to further NATO's principles and contribute to the security of the North Atlantic area. An inter-agency team recently visited each of the nine countries participating in the Membership Action Plan for frank discussions of their progress toward these goals. As we approach these historic decisions, we look forward to a close dialogue with the Congress. Our goal is to forge a united U.S. approach to enlargement and a solid consensus within the Alliance.

We also hope to advance new relationships at the Prague Summit. Foremost among these is a constructive NATO-Russia relationship, which I will address later. NATO's continued outreach to Partnership for Peace [PfP] member states has overcome entrenched hostility and historical divisions. Through its unique Partnerships, NATO remains the only institution that can unite the continent in security cooperation. NATO remains the indispensable nexus for broadening and deepening Euro-Atlantic security, democracy, free markets, and the rule of law. At Prague, we intend to continue building closer links with Russia, Ukraine, and all of NATO's Partners.

As NATO further evolves, we will work to strengthen Alliance links between those Partners who are not yet ready or do not seek NATO membership. Many of our Partners, such as the Nordic countries and Ireland, have contributed significantly to NATO's efforts in the Balkans. They have reached out to the states of the former Soviet Union. We will continue to work closely with these Partners to improve interoperability and capabilities of all NATO's Partners.

Most recently, our Central Asian and Caucasus Partners have stepped forward to play critical roles in the anti-terrorist effort. We intend to energize all elements of the Partnership for Peace at NATO to engage Central Asian and Caucasus Partners. Working with our Allies and more advanced Partners, we hope to increase, coordinate and target assistance to the Central Asian and Caucasus states. We believe PfP programs should address issues that have the greatest appeal to these countries. These include terrorism, border security, and civil emergency planning. We will continue to support the development of democracy and market economic institutions to help ensure the viability of our security partnerships with these countries. We look to the OSCE [Organization for Security and Cooperation in Europe] to play an increasing role in this regard.

We continue to support a European Security and Defense Policy that strengthens NATO while increasing the EU's ability to act where NATO as a whole is not engaged. At the same time, the broader value of close NATO-EU cooperation is nowhere more evident than in Southeast Europe, where NATO and the EU have worked closely to prevent instability, overcome violence and begin to build a lasting peace. The Macedonia peace settlement is a model of our collective ability to draw on the unique strengths of these organizations in a common effort.

Key to a Europe "whole, free and at peace" is a more stable, democratic and prosperous Southeast Europe. Despite the region's great strides since the Dayton Peace Accords, governments still have much to do. Working in partnership with the U.S. and the Europeans, these nations must complete reform efforts and establish an environment conducive to prosperity. Corruption, insufficient border controls and weak export control regimes contribute to trafficking throughout the region—in arms, drugs and people. Work in these areas also contributes to our global counterterrorism efforts.

NATO and its partners in SFOR [Stabilization Force in Bosnia] and KFOR [Kosovo Security Force] still have a role to play, as does the German-led NATO "Task Force Fox" in Macedonia. Our vision is that the U.S. and the international community deal with this region "normally"—without troops on the ground and through trade and investment rather than aid. We are mindful that we came into this region together with our Allies and we should go out together.

Our engagement with Southeast Europe is changing. We continue to support economic reform and regional trade development, supported by a Southeast Europe Trade Preferences Act (SETPA). We are encouraging further integration of the region with Europe. We promote rule of law, cooperation with the International Criminal Tribunal, and ethnic tolerance. With success, our European partners and we have been able to reduce force levels in Bosnia. We anticipate that NATO Military Authorities will recommend further reductions in Bosnia and Kosovo. The EU will take over the UN's police mission in Bosnia at the end of [2002]. The international community recently agreed to a blueprint for streamlining and downsizing its presence in Bosnia. The creation of a government in Kosovo will allow the transfer of many responsibilities from the international community to local democratically elected authorities. In Macedonia, the close and continuing cooperation between the EU, NATO and the OSCE is a model for transatlantic cooperation in crisis management. Task Force Fox is small. It is of limited duration and made up almost entirely of Europeans.

A critical element of achieving the President's vision of a Europe "whole, free and at peace" is the resolution of regional and ethnic conflicts in Europe and neighboring Eurasia. We are pleased by progress in the Cyprus talks. We will encourage the leaders on the island to achieve a final settlement in the coming months. The Good Friday Accord is being implemented in Northern Ireland. We will work to solidify the role of the police force there. Cooperation among all factions is crucial. In Northern Europe, we will continue to work with our Nordic Allies and friends and our Baltic and other regional partners, including Russia. It is vital that we reinforce ten years of progress in a region of shared values. Opportunities for economic progress, good neighborly relations and democratic institution building are beginning to outweigh the challenges.

OSCE remains a vital element in our engagement with Europe. It is the pre-eminent multilateral institution for upholding democracy, human rights and the rule of law. It undertakes early warning measures, conflict prevention, and post-conflict rehabilitation. OSCE also implements valuable programs to counter corruption and trafficking, and strengthen the rule of law through police training and judicial reform. Its broad membership allows it to operate throughout Europe and Eurasia.

The OSCE has said it will begin to play a role in the war against terrorism. The OSCE can encourage European and Eurasian countries to adhere to the principles of UN Resolution 1373. It will continue to be central to development of pluralistic societies in the Balkans, including solidifying the Framework Agreement in Macedonia. Implementation of CFE [Conventional Armed Forces in Europe] commitments will be an ongoing OSCE oversight responsibility. The organization can offer opportunities for cooperative engagement with Russia and the European Union.

The OSCE plays a critical role in our effort to promote democracy, human rights and rule of law throughout Eurasia. It is working to restore territorial integrity in Moldova. In Belarus, we work with the OSCE and our European partners to urge the Lukashenko regime to adopt OSCE standards of behavior and come out of its self-imposed isolation. Unfortunately, the regime shows no inclination to do so thus far. In Moldova, we work through the OSCE and with key players to resolve the separatist conflict in Transnistria and reincorporate that region into Moldova. Ukrainian involvement is important on this issue and in the region generally. Ukraine's influence is a potential force for regional stability and European integration. Ukrainian success in political and economic reform will fulfill that country's European aspirations and will inspire other post-Soviet states to follow the same path.

In the Caucasus, we are working with Armenia and Azerbaijan to resolve their conflict over Nagorno-Karabakh. We seek a comprehensive settlement through the Minsk Group peace process. Georgian sovereignty is important to the Administration. We are proposing a program to develop Georgia's internal capacity to deal with terrorism now and in the future. We also are working to support the development of democracy and human rights in the Caucasus.

Bilateral U.S.-Russia cooperation is unprecedented. Counterterrorism collaboration is central to this effort, although not the sole focus. The U.S. and Russia are cooperating more closely in intelligence sharing, nuclear weapons reduction, and resolution of Eurasian regional conflicts. We are working together in the fight against HIV-AIDS and other infectious diseases, organized crime and narcotics trafficking. We hope to expand the economic and commercial component of the relationship. While we broaden this new cooperation with the Russians, we have not forgotten the difficult issues. We continue to press our concerns over issues such as the conduct of Russian forces in Chechnya and threats to media freedom in Russia as a whole.

Russia's cooperation with us and our Allies in the war on terrorism also reflects the opportunity to bring Russia closer to NATO. We are working with our Allies on arrangements for a new NATO-Russia body that would focus on concrete, practical projects of mutual benefit. Russia would participate in this "NATO-Russia Council"—which would focus on issues with potential for cooperative initiatives—as an equal. The deepening of the Russia-NATO relationship will not be allowed to undercut NATO's ability to decide and act on its own. Russia would not get a veto over the ability of NATO's 19 Allies to act on their own. The NAC [North Atlantic Council] will continue to meet and make decisions as it always has. The mechanisms and substance of such arrangements are still being worked out. I pledge to keep the Committee apprised of progress. Moreover, I want to reiterate President Bush's and [secretary-general of NATO] Lord [George] Robertson's pledges not to give Russia a veto over NATO operations. This is not a backdoor to membership. This is an opportunity for Russia to develop a new relationship with NATO that would advance not only our interests but also its.

In the spirit of new U.S.-Russia cooperation, we believe it is time to move beyond the Cold War. Russia has made significant progress on religious freedom and emigration. Therefore, the President is pursuing the removal of Russia and eight other Eurasian countries from the application of Jackson-Vanik legislation [1974 Trade Act to Russia amendment that was intended to promote free emigration from the Soviet Union]. We hope that Congress will pass legislation to "graduate" Russia from Jackson-Vanik before the President visits Moscow this spring. The President and Secretary Powell appreciate the support of many Members of this committee in this endeavor.

Success in addressing transnational problems is more important than ever in pursuing America's transatlantic agenda. Stable countries able to withstand terrorist and other threats are based on respect for the rule of law, human rights, religious freedom, and open media. Stable countries have vibrant civil societies. They are committed to the principles of free market economies. The Administration's attention of these values with our European and Eurasian friends is even more critical as we pursue the War on Terrorism with our coalition partners. Enhanced defense and security cooperation and intelligence sharing must be buttressed by societies committed to democratic principles such as those in the Final Act in the Conference on Security and Cooperation in Europe. Moreover, we are continuing efforts with our transatlantic partners to address problems that respect no borders, e.g., HIV/AIDS and infectious disease, narcotics trafficking and environmental degradation.

Critical to the promotion of our policies in Europe and Eurasia is the use of Public Diplomacy. Training programs and exchanges offer an accurate portrayal of American views, values and traditions. Such people-to-people ties will help bind the nations of Europe and Eurasia with the United States, thereby enhancing the transatlantic relationship and American security.

John C. Hulsman

 **NO**

European Arrogance and Weakness Dictate Coalitions of the Willing

As the fabulously successful 12-step program pioneered by Alcoholics Anonymous has conclusively demonstrated, one cannot tackle a crisis until acknowledging the reality of a genuine problem. Throughout the 1990s, mutual exchanges of pleasantries and vague rhetoric of a "Europe whole and free" obscured the fact that the transatlantic relationship was increasingly in crisis, with a significant portion of the European political elite viewing the United States as part of the problem in international politics, rather than as part of the solution to global problems.

Representative of this trend is the typical anodyne statement that "a stronger Europe is also more likely to be a reliable strategic partner with the U.S." Given the resurgence of a European-wide strain of Gaullism, the long-desired European effort to emerge as a global power balancing America, this platitude is increasingly open to question.

In the past several years, genuine policy differences between the U.S. and its European allies have emerged over trade issues such as the "banana war"; genetically modified foods; the American Federal Sales Corporation (FSC) tax; America's increase in steel tariffs; Europe's refusal to substantially reform the Common Agricultural Policy (CAP) and the repercussions this holds for the Doha [Qatar] global free trade round; the moral justness of the death penalty; whether Cuba, Libya, and Iran should be engaged or isolated; Iraq; the Israeli-Palestinian crisis; the role international institutions should play in the global arena; when states ought to be allowed to use military force; ideological divisions between European Wilsonians and American realists and neoconservatives; the Kyoto Accord; the jurisdiction of the International Criminal Court (ICC); National Missile Defense (NMD) and the U.S. abrogation of the ABM [Anti-Ballistic Missile] treaty; the military debate within NATO [North Atlantic Treaty Organization] regarding burden-sharing and power-sharing; American unilateralism; Turkey's ultimate role in the West; widely varying global threat assessments; the doctrine of humanitarian intervention and the efficacy of nation-building; and how to organize an economy for the best societal effect, to name a few.

From John C. Hulsman, "European Arrogance and Weakness Dictate Coalitions of the Willing," *Heritage Lecture No. 777* (February 10, 2003). Copyright © 2003 by The Heritage Foundation. Reprinted by permission. Notes omitted.

This incomplete list should make it crystal clear to the most complacent of analysts that drift in the transatlantic relationship is about far more than carping, black-leather-clad, ineffectual Europeans glowering about American dominance from the safety of a Parisian café. It is centered on fundamental philosophical and structural differences held by people with a very different view of how the world should be ordered from that of the average American; it should be evaluated far more seriously than has been the case in Washington.

Those Europeans pushing for the creation of a more centralized, federal, coherent European Union (EU) political construct do so by increasingly defining themselves through their differences from Americans. European Gaullists [after the notably independent French president Charles de Gaulle] see the emergence of a European pole of power as an effective foil to overweening American global power. Such a reality makes a lie of American Wilsonian pretensions to advance universal values. Paradoxically, these universalist pretensions are all too often seen by Europeans of many political stripes as another, more subtle form of self-centered American unilateralism.

The French position, predictably the most suspicious of American power, could not have been clearer during the [Lionel] Jospin premiership. A more united Europe was necessary to "build counterweights" to combat "the risk of hegemony." Any thought that classical balance-of-power thinking was no longer relevant in today's global environment ought to be put to rest by any vague scrutiny of the French government's rationale for a more coherent Europe. Across the continent, Gaullism was clearly on the rise at the end of the 1990s.

Reasons for the Gaullist Resurgence

The reasons for this resurgence are structural and thus are likely to endure. With the end of the Cold War, it was to be expected that America and Europe would drift: Without the unifying growl of the Soviet bear to subsume the reality that America and various European states had quite distinct international interests, there were bound to be divergences. The U.S. has emerged as the sole superpower in the post–Cold War era, while the European states, with the partial exception of France and the UK, are at best regional powers. This structural difference, unlikely to change in even the medium to long term, does much to explain the practical policy differences increasingly emerging on both sides of the Atlantic.

Not only has America gone from strength to strength in the new era, but Europe also has conspicuously failed to emerge as a coherent power in its own right. This sense of a resurgent and increasingly unfettered America, coupled with an introverted, increasingly marginalized Europe, does much to explain not only the differences in policy between the two poles, but also the increased virulence many Europeans feel toward American policies with which they disagree.

A Question of Power

In the end, such differences are less about philosophy and more about power. It is not that European Gaullists feel American international policies are merely wrong; increasingly, they feel they have no power to affect them, even at the margins. This change in political psychology does much to explain the rise of an anti-American Gaullism in Europe, as well as the increasing drift in the transatlantic relationship.

The example of European military weakness is instructive. Given anemic European defense spending, it is little wonder that many politicians in Europe are implacably opposed to the military tool being used in international relations, that they don't want strength to matter in the international community, that they want to live in a world where international law and institutions predominate, that they want to forbid unilateral military action by powerful nations, and that they advocate all nations having equal rights that are equally protected by accepted international norms of behavior. The Europeans are merely making a philosophical virtue of a very practical necessity.

While attempting to limit through diplomacy the glaring weakness in their own power portfolio, European Gaullists are attempting one thing more: to balance the United States in a non-traditional manner, to harness overwhelming American power in multilateral institutions in such a way as to have a significant say in how such power is used. This reality explains France's implacable demand that all action against Saddam Hussein proceed institutionally through the Security Council, where Paris has a veto. It is an effort by the Lilliputians to tie Gulliver up, and it is completely understandable given the present power discrepancy between Europe and the U.S.

It also structurally explains why relations are increasingly frayed between an American Gulliver that naturally wants to preserve its freedom of action and European Lilliputians that, given their strategic weakness, want to constrain the American behemoth in multilateral institutions. The rise of European Gaullism, the desire to create a countervailing pole defined by its very un-American nature, is a logical structural response to such a world.

The Reality of European Weakness

Just as all is not well in the transatlantic relationship, rhetoric should not replace reality as to Europe's capabilities to emerge as a major power, even in the medium to long term. While the desire to compete successfully with America may be ensconced in many European chanceries, the ability to do so appears to be well beyond Europe's collective means.

Military Weakness

Militarily, despite a collective market that is slightly larger than that of the United States, Europe presently spends only two-thirds of what the U.S. does on defense (with American defense budget increases, even this paltry percentage will decrease) and produces less than one-quarter of America's deployable fighting strength. German defense spending has dropped from 1.5 percent to a laughable 1.1 percent. Likewise, except for the UK and France, all other European countries are presently incapable of mounting an expeditionary force of any size anywhere in the world without borrowing American lift capabilities.

Current U.S. defense increases are greater than the entire defense budgets of any of the individual European allies. As Richard Perle bluntly put it, Europe's armed forces have already "atrophied to the point of virtual irrelevance."

Given the moribund state of the European economies and the proclivity of the European publics to eschew significant defense spending, absolutely no empirical evidence suggests that this trend of relative military decline will change in the long term. At best, the United States can expect a multi-tiered NATO where, beyond the British and the French, individual European member states will fill niche roles in the overall American strategic conception. American decision-makers used to positive spins on the Alliance must acknowledge that not all the allies are equal—that real differences exist among European capitals over how often to side militarily with the U.S. and how much capability individual countries can bring to bear.

Economic Stagnation

Economically, the latter part of the 1990s has not led Europe into the "promised land" so confidently predicted by many. Rather, massive and largely ignored structural problems—labor rigidities, a demographic-pensions time bomb, a safety net that precludes significant cuts in unemployment, too large a state role in the economy stifling growth—have led Europe into a cul-de-sac. Staggeringly, according to the Organization for Economic Cooperation and Development, the number of private-sector jobs in the euro zone has not increased since 1970.

Germany is emblematic of this Western European problem. Germany's five wise men—the government's independent economic advisers—now forecast growth of only 0.2 percent in 2002 and 1 percent in 2003. Germany's public deficit is expected to run at a rate of 3.7 percent this year and perhaps next year as well, overshooting EU Stability Pact strictures. Efforts to lower unemployment remain stalled, with over 4 million Germans remaining out of work.

This economic snapshot is also representative of Germany's recent economic performance. After an initial post-reunification surge, German GDP [gross domestic product] increased over the past 10 years by a mere 1.5 percent per year on average. The reasons for this are as simple as they are politically intractable—Germany's non-wage labor costs are among the highest in the world, well over 42 percent of gross wages.

This factor, combined with excessive labor rigidities, a virtually unfunded pensions system, and a looming demographic crisis, as well as a crucial lack of political will in either the SPD [liberal Social Democratic Party] or the CDU [conservative Christian Democratic Union] to implement the unpopular yet necessary measures to tackle these massive problems—only the Free Democrats discussed radical economic reform during the recent campaign and received a measly 7.4 percent of the vote—means that the motor of Europe will continue to sputter. Structural economic problems common to Italy, France, and Germany, as well as the accompanying lack of political will to deal with them, signify that the only question facing Europe is whether it continues to limp along or falls into a Japan-style torpor.

In some ways, the euro [the currency of the EU] has made this difficult economic situation even worse. Its one-size-fits-all macroeconomic policy has led interest rates to be set far too high for a sputtering German economy while threatening a booming Ireland with long-term inflation. The euro zone is far from an optimal currency area. It remains to be seen whether the economies of Europe are sufficiently in sync to make the project flourish in the medium term.

Dangerous Economic Rigidity

The Stability Pact is emblematic of Europe's overly rigid macroeconomic approach. Ironically enacted to quell German fears about the long-term economic soundness of countries such as Greece, Italy, and Portugal, it is hamstringing Berlin itself (as well as Lisbon) with its strictures, limiting budget deficits to 3 percent per year. Already in recession and faced with certain warnings from the EU and the possibility of massive fines amounting to 0.5 percent of the GDP if it fails to correct its budget imbalance, Germany has been forced to enact austerity measures at a time of economic decline—the worst short-term fiscal policy imaginable.

Such a rigid economic approach seems politically doomed in the long term; already, critics ranging from EU Commission President [Romano] Prodi to the French and German governments are signaling the need to fundamentally reform the process. In the short run, the Stability Pact has proved to be just another unnecessary constraint on a German economy already caught in the doldrums. There is little sign that either Germany or Europe as a whole is likely to gain economically on the U.S. in the medium to long term. Rather, the challenge is to avoid permanent economic stagnation of the continent.

As with military matters, the overall view must be qualified. Over the past five to eight years, the British, Spanish, Dutch, and Irish economies have grown at very respectable rates. Given their more open pensions systems, neither Dublin nor London face the same demographic crisis currently looming in Italy, France, and Germany. Great Britain remains the largest direct investor in the United States, as America does in the UK.

Moving geographically around the traditional motor of EU integration—France, Germany, and Italy—economic liberalism is found flourishing on the European periphery. It is hard to characterize a common European economic state of being, as the differences outweigh the economic commonalities.

Political Disunity

This is even truer in the political realm. Contrary to any number of soothing and misleading commission communiqués, the Europeans are light years away from developing a common foreign and security policy (CFSP). One has only to look at the seminal issue of war and peace today—what to do about Saddam Hussein's Iraq—to see a complete lack of coordination at the European level.

Presently, the UK stands shoulder-to-shoulder with the U.S.; Germany's militant pacifists are against any type of military involvement, be it sanctioned by the UN or not; and France holds a wary middle position, stressing that any use of military force must emanate from UN Security Council deliberations. It is hard to imagine starker and more disparate foreign policy positions being staked out by the three major powers of Europe.

Even on issues relating to trade, there are vast differences within the EU. The recent spat between President [Jacques] Chirac of France and British Prime Minister [Tony] Blair was about far more than atmospherics. It was about whether Northern European countries, such as the UK, would continue to countenance Southern EU countries' (such as France) dogged desire to protect the wasteful Common Agricultural Policy even though it may well prove to be a deal-breaker at the Doha global free trade round.

On missile defense, relations with Turkey, and, critically, the future course of the EU—with Germany advocating adding more members and greater centralization to the EU, the UK in favor of broader membership but little additional centralization, and France stressing greater centralization—one finds a cacophony of European voices rather than everyone singing from the same hymnal.

Military weakness, economic stagnation, and political disunity—this is the reality that confronts American decision-makers today when looking at Europe. Despite positive spins and European hopes, Europe is not likely (though it remains possible) to challenge American primacy in the long run. This is not due to any general, continental love of Washington or its policies. Rather, it is the result of European political, military, and economic weakness.

Cherry-Picking as the American Answer to a Weak but Gaullist Europe

In separating rhetoric from reality, there is a comforting final conclusion that needs to be drawn by American policymakers: The very lack of European unity that hamstrings European Gaullist efforts to challenge the United States presents America with a unique opportunity. If Europe is more about diversity than uniformity, if the concept of a unified "Europe" has yet to really exist, then a general American transatlantic foreign policy based on cherry-picking—engaging coalitions of willing European allies on a case-by-case basis—becomes entirely possible. Such a stance is palpably in America's interests, as it provides a method of managing transatlantic drift while remaining engaged with a continent that will rarely be wholly for, or wholly against, specific American foreign policy initiatives.

For such an approach to work, it is essential to view Europe as less than a monolithic entity. The different approaches the Bush Administration took with the Kyoto global warming treaty and missile defense are instructive. By condemning out of hand the Kyoto agreement and offering no positive policy alternatives, the Bush Administration found itself in a public relations disaster in its early days. By failing to engage the Europeans, the White House unwittingly succeeded in uniting them.

Enlisting Support for Missile Defense

Embracing the learning curve in the wake of Kyoto and refusing to believe reports that "Europe" was implacably opposed to American desires to abrogate the

ABM treaty and begin constructing a missile defense system, the White House sent its representatives to the capitals of Europe, where they found the "European" stance on missile defense far more fragmented than it had appeared at first glance. Intensive diplomatic efforts led Spain, Italy, the UK, Poland, Hungary, and ultimately Russia to embrace the Administration's initiative to one degree or another. By searching out potential European allies at the national level, Washington engaged in successful cherry-picking and avoided the kind of diplomatic and public relations disaster that had occurred in the wake of Kyoto.

Ironically, this realist policy actually calls for more diplomatic and political engagement with Europe at a national level, even if Brussels is generally taken less seriously. As the Kyoto episode made abundantly clear, in order for cherry-picking to work, the U.S. must find divisions in "European" opinion based on differing conceptions of national interest.

America has to constantly note differences within Europe in order to exploit them to form a coalition of the willing on any given policy initiative. Europe, such as it presently exists, suits general American interests—its member states are capable of assisting the U.S. when their interests coincide with America's; yet it is too feeble to easily block America over fundamental issues of national security. Cherry-picking as a general strategy ensures the endurance of this favorable status quo.

Coalitions of the Willing and NATO

Militarily, such an approach explains present efforts at NATO reform. Beyond the sacrosanct Article V commitment, the future of NATO consists of coalitions of the willing acting out-of-area. Here, a realist cherry-picking strategy confounds the impulses of both unilateralist neoconservatives and strictly multilateralist Wilsonians.

Disregarding neoconservative attitudes towards coalitions as often not worth the bother, cherry-pickers call for full NATO consultation on almost every significant military issue of the day. As is the case with Iraq, if full NATO support is not forthcoming, realist cherry-pickers would doggedly continue the diplomatic dance rather than seeing such a rebuff as the end of the process as many Wilsonians would counsel.

A Combined Joint Task Force (CJTF), a subset of the Alliance where a coalition of the willing is formed to carry out a specific mission using common NATO resources, would be a cherry-picker's second preference. If this too proved impossible due to a general veto of such an initiative, a coalition of the willing outside of NATO—composed of states around the globe committed to a specific initiative based on shared immediate interests—would be the third best option. Only if the third option failed and fundamental national interests were at stake should America then act alone.

While agreeing with neoconservatives (and disagreeing with Wilsonians) that full, unqualified approval of specific missions may prove difficult to achieve diplomatically with NATO, cherry-pickers disagree with them about continuing to engage others at the broadest level. For, as the missile defense example illustrates, there are almost always some allies who will go along with any specific American policy initiative—that is, if they are genuinely asked. By championing initiatives such as the CJTF and the new NATO rapid deployment force, the Bush Administration is fashioning NATO as a toolbox that can further American interests around the globe by constructing ad hoc coalitions of the willing that can bolster U.S. efforts in specific cases.

Coalitions of the Willing and Free Trade

Less developed than the NATO process, free trade coalitions of the willing hold out intriguing possibilities for a future that may well see the breakdown of the Doha free trade process. As with NATO, there is no doubt that a comprehensive, all-inclusive liberalizing deal built around the Doha process (involving agricultural, services, and manufacturing liberalization) would best suit both the world and the United States.

However, given the great disparities in world opinion over the efficacy and even the definition of free trade, the United States must be prepared to enact free-trading coalitions of the willing if the Doha round stalls over European failures to respond to the developing world's demand for significant agricultural liberalization. Certainly, the "free trade by any means" mantra emanating from United States Trade Representative Bob Zoellick's office is an indication that the Bush Administration is moving in this direction.

Needed: A Global Free Trade Association

Beyond efforts to make the regional Free Trade Area of the Americas (FTAA) and bilateral deals with countries such as Singapore, Chile, and Australia viable, the Bush Administration needs to embrace the idea of a Global Free Trade Association [GFTA]—a coalition of the willing determined to maximize trade liberalization throughout its member states. States around the globe that meet certain, predetermined, numerical criteria relating to trade policy, capital flows and foreign investment, property rights, and regulation would automatically qualify for the grouping. Members would, thus, select themselves based on their genuine commitment to a liberal trading order.

Given the politico-economic commonalities such a grouping would share, the GFTA would hopefully allow for the freer movement of capital within the grouping, establish common accounting standards, set very low rates of subsidies across the board, and diminish overt and hidden tariffs. If the Doha round stalls, the U.S. must not take its ball and go home; again a coalition of the willing, this time in trade, is the way forward.

Given these specific criteria, Denmark, Estonia, Finland, Iceland, Ireland, Luxembourg, and the United Kingdom could join the U.S. and five non-European countries in a GFTA in 2003. To do so, a lessening of the ties that bind these states to the EU is necessary, given that the EU functions as a customs union. Countries in the EU, such as the UK and Ireland, would have to legally recalibrate their trading regimes with Brussels—something that, given the protectionist nature of the EU, they should do anyway. They would have to relax

EU-harmonized rules in the case of goods and services when required to do so by the GFTA.

Being able to derogate from EU rules in the case of internally traded goods and services imported from non-EU countries would be similar to obtaining an opt-out. Such a policy would strengthen efforts to transform the future architecture of the continent to resemble a Europe à la carte where individual countries would be far freer to pick and choose what elements of the European experiment they wish to join.

The point is that the United States, in the guise of this new cherry-picking initiative, will not wait for Godot any longer: The fact that Europe as a whole is not ready for further trade liberalization must not stop individual states (both within and without Europe) from continuing to press toward a freer trading world.

After 50 Years: Time for a New Realism

Politically, America must stop giving generally sympathetic countries like the UK and Poland such bad geopolitical advice. By pushing the UK into "Europe," the U.S. hoped to make the project more pro-American, more pro-free market, and pro-transatlantic alliance. After 50 years, it is time to look the results squarely in the eye: The EU is simply no more pro-American, pro-free market, or pro-transatlantic alliance than it was at the time of its inception.

Only a Europe that widens, rather than deepens, a Europe à la carte where efforts at increased centralization and homogenization are kept to a minimum, suits both American national interests and the interests of individual citizens on the continent. Any hint of further significant centralization—the UK joining the euro, CFSP becoming a reality, the closer harmonization of tax or fiscal policy across the continent—must be seen by America for what it is: a Gaullist effort to construct a pole in opposition to the United States. That will be the point at which the transatlantic tie genuinely begins to break.

Such an outcome is, however, entirely avoidable. A strategy of creating coalitions of the willing will preserve a status quo where the transatlantic relationship, despite fraying a bit at the edges, continues to provide common goods to both sides of the Atlantic. Such an overall policy acknowledges an awkward current truth of the transatlantic relationship: The United States wants Europe neither to be too successful nor to fail. As such, the Europe of today suits America's long-term strategic interests.

Cherry-picking will allow the U.S. to make the appearance of a Gaullist, centralized European rival far less likely while distributing enough shared benefits that the overall transatlantic relationship will continue to provide Europeans, as well as Americans, with more benefits than problems. Such an accurate assessment, fitting the realities of the world we now live in—where the United States behaves multilaterally where possible and unilaterally where necessary—is likely to endure.

POSTSCRIPT

Should the United States Continue to Encourage a United Europe?

Jones and Hulsman debate the future of an alliance that has persisted for over a half century. For a review of that history, consult Marc Trachtenberg, ed., *Between Empire and Alliance: America and Europe During the Cold War* (Rowman & Littlefield, 2003).

By some measures, U.S. commitment to working with and strengthening Europe remains. With the support of Washington, NATO has taken up new security roles, such as peacekeeping in Macedonia and waging an air war on Yugoslavia to end its oppression of Kosovo. The alliance has also expanded. Former communist countries, the Czech Republic, Hungary, and Poland joined NATO in 1999. Four years later, seven Eastern European countries (Bulgaria, Estonia, Latvia, Lithuania, Romania, Slovakia, and Slovenia) joined NATO, with the enthusiastic support of Washington.

The European Union is also growing. Its current membership of 15 countries is likely to grow to 28 as 13 applicants for membership are admitted over the next few years. In addition, the "Convention on the Future of Europe" has been established to try to fashion an EU constitution, an eventuality that will be a giant step toward political integration for Europe. Although the United States is wary of any attempt to create a separate European military alliance that might supplant NATO, Washington is at least passively supportive of the latest EU effort to achieve greater unity. For an article that argues that NATO has continued relevance, see Christopher Layne, "Death Knell for NATO? The Bush Administration Confronts the European Security and Defense Policy," *Policy Analysis* (April 4, 2001). Taking the opposite view is Robert Kagan, *Of Paradise and Power: America and Europe in the New World Order* (Alfred A. Knopf, 2003).

Concerns about general U.S. power in the world and about perceived U.S. unilateralism were arguably part of what caused some European countries—France and Germany in particular—to oppose a U.S.-led war against Iraq in 2003. That policy divide seriously escalated tension in the "Western alliance." Despite the constraints of diplomacy, the two factions were barely able to avoid name-calling at times. At one point, for instance, U.S. defense secretary Donald Rumsfeld described the actions of France as "inexcusable," and at another the German chancellor dismissively referred to the war as President Bush's "adventure" in Iraq.

These tensions led some on both sides of the Atlantic to see the controversy as the death rattle of the Western alliance. Others believe that the alliance will recover its health and remain a key aspect of both U.S. and European foreign policy.

ISSUE 5

Do China's Armaments and Intentions Pose a Long-Term Threat?

YES: Richard D. Fisher, Jr., from Statement Before the Committee on Armed Services, U.S. House of Representatives (July 19, 2000)

NO: Ivan Eland, from "Is Chinese Military Modernization a Threat to the United States?" *Policy Analysis No. 465* (January 23, 2003)

ISSUE SUMMARY

YES: Richard D. Fisher, Jr., a senior fellow with the Jamestown Foundation, characterizes China's military as growing in sophistication and strength, and he argues that China's buildup has worrisome implications for U.S. national interests.

NO: Ivan Eland, director of defense policy studies at the Cato Institute, contends that China's military modernization is less of a threat to U.S. interests than recent studies by the Pentagon and the U.S.-China Security Review Commission indicate.

China has a history as one of the oldest, most sophisticated, and most powerful countries (and empires) in the world. Four thousand years ago, under the semi-legendary Emperor Yu of the Hsia dynasty, the Chinese built irrigation channels, domesticated animals, engaged in cultivation, and established a written language. Through 14 Chinese dynasties—from the Hsia dynasty (1994–1523 B.C.) to the Manchu dynasty (1644–1911)—China built a civilization marked by great culture, engineering feats, and other advances.

On the other side of the globe, Europe languished amid the Dark Ages. Then, as the world moved into the second Christian millennium, Europe began to revive. Even though the West and East still had scant contact, the balance of power between the two had begun to shift by the 1500s. In search of spices and other treasures, European traders increasingly sailed to Asia, and they were quickly followed by European soldiers.

By the 1800s outside powers came increasingly to dominate a decaying China. Over the next eight decades, China underwent what was to the Chinese a period of humiliation. Huge tracts of their territory were occupied by other countries. During the Boxer Rebellion (1898–1900), Chinese nationalists tried to expel the foreigners. The Chinese forces were defeated by an international coalition that included American troops. The moribund imperial dynasty fell in 1911 and was replaced by a republic headed by Sun Yat-sen.

Sun's government marked the beginning of China's change from decline to growth, but Sun died in 1916, and a struggle for power among various factions led to the establishment of a central government under Nationalist Chinese leader Chiang Kai-shek. Although Chiang's government proved corrupt and ineffective in many ways, it did largely consolidate power and moved to edge foreign influences out of China. That trend became even stronger in 1949 when Chiang's government fell to the Communists of Mao Zedong.

For two decades, many in the West were caught up in the psychology of the cold war and perceived China to be part of the communist threat. Whatever the U.S. ideological view, however, by the late 1960s China had gained enough strength and showed enough independence from the Soviet Union that even the coldest warrior had to see that China was a rising power in its own right. This led to important shifts in U.S. policy and a normalization of relations that began with President Richard Nixon's visit to China in 1972. Relations improved even more after Chairman Mao Zedong died in 1976. The waning of the cold war further persuaded both powers to seek better relations.

Other changes occurred within China. After some turmoil following the death of Mao, the Chinese government moved to moderate the impact of communist ideology on China's economic policy. Greater economic ties with the industrialized West were sought. China moved slowly to establish limited capitalism to improve economic performance. China's economy is now rapidly expanding and stands as one of the largest in the world. The view of Americans toward China steadily improved during the 1980s.

The world was shocked in May 1989 when Chinese troops opened fire on antigovernment demonstrators in the giant Tiananmen Square in Beijing. It became clear that the so-called liberalization of China extended to economic but not political matters.

During the 1990s American investment in, and trade with, China grew rapidly, and the two countries cooperated (or at least did not oppose one another) on numerous diplomatic fronts. Yet there were also many sore points. China's human rights record was one of them. On other fronts, the United States accused China of supplying nuclear weapons and missile technology to Pakistan and a number of other countries. Economic disputes led the United States to block China's admission to the World Trade Organization. Other disputes have regularly soured relations.

The rise and fall and rise of China's power and the ambiguity that the United States feels toward China set the stage for this debate. In the first of the following selections, Richard D. Fisher, Jr., paints an alarming picture of the growing size and sophistication of China's military forces and contends that the changes do not bode well for the future of U.S. foreign policy interests. In the second selection, Ivan Eland takes the position that China's military modernization does not look especially threatening.

Richard D. Fisher, Jr.  **YES**

China's Military Capabilities

Introduction

It is clear from numerous open sources that the People's Republic of China (PRC) is in the midst of an ambitious modernization effort for its People's Liberation Army (PLA), which incorporates the Second Artillery missile forces, the PLA Army, PLA Navy and PLA Air Force. While building from a low base in the late 1980s, the leadership of the PRC has devoted considerable resources to building the research and development infrastructure plus acquiring foreign technologies to sustain PLA growth. And while in the year 2000 the PLA does not pose a serious threat to U.S. forces in Asia, it is rapidly acquiring the systems necessary to shift the military balance on the Taiwan Strait in its favor. The PRC government believes that the credible threat or even the use of force is critical to recovering "lost territories" such as Taiwan or the South China Sea. This, plus the PLA's emphasis on seeking "asymmetrical" capabilities that exploit U.S. military weaknesses, its willingness to incorporate foreign technology to establish new capabilities, and its growing hostility to U.S. missile defenses and to the U.S. military presence in Asia, should cause concern in Washington. When the PRC evolves away from Communism it may become a positive force in its region and beyond. Until then, it is incumbent upon the United States to continue to exercise the political and military leadership necessary to promote peace and stability in Asia.

Varying Assessments of the PLA

Accurate open sector assessments of the PLA are difficult as the PRC government does not make available to its own people or outsiders detailed knowledge of the budgets, doctrine, research and development, current equipment or future modernization plans of its People's Liberation Army. Concealing one's strength so as to enable maximum surprise is a key element of ancient Chinese statecraft that continues to this day. In three reports requested by the Congress, the Department of Defense has added some new knowledge about the direction of PLA modernization. [Recently] the House Select Committee on Military/ Commercial Relations with the People's Republic of China provided the most

From U.S. House of Representatives. Committee on Armed Services. *Military Capabilities of the People's Republic of China.* Hearing, July 19, 2000. Washington, D.C.: U.S. Government Printing Office, 2000. Notes omitted.

comprehensive description of PRC military high-technology goals and espionage efforts, from a U.S. government source. The Select Committee demonstrated that the Administration can and should offer much more detailed analysis of the PLA to better inform the U.S. policy process.

Due to incomplete information, and the lack of a comprehensive U.S. government statement on the PLA, widely varying assessments of the PLA have been produced in recent years. In the 1980s the U.S. and other Western governments were selling arms to PRC in a tacit alliance against the former Soviet Union, and had wide access to the PLA. The general assessment at the time was that it was mired in strategic concepts of "People's War" based on the guerrilla campaign that chased the Nationalist government to Taiwan, and burdened by military technology not much improved upon from what the Soviets sold them in the 1950s. The twin shocks of the Tiananmen Massacre [of hundreds of student activists in Beijing in 1989] and subsequent Western military embargoes, plus the demonstration of overwhelming U.S. and Allied force during the Gulf War, have prompted the current PLA modernization. But there are today differing assessments of the success of the PLA's modernization.

Two respected scholars of the PLA stated [recently] that "China's military is simply not very good." Or as two others stated in 1998, the PLA "will not in the next ten years, . . . be anything more than a nuisance," and "in 20 to 30 years China will still be a regional power." This group of scholars tends to focus on the exiting old equipment of the PLA, its not-yet-completed effort to formulate modern joint-service doctrine and operations, and a lack of realistic training and maintenance problems. The PLA's acquisition of foreign weapon systems is seen as compounding training and maintenance challenges while demonstrating the inability of its domestic military industrial sector. To be sure, the PLA faces enormous challenges, not the least of which is whether the PRC government can afford a continued comprehensive modernization of the PLA. And even here, there is great variance in estimates of the PLA's budget, from the PRC government 1998 figure of about $11 billion—which few accept—up to $40 billion a year and higher.

However, this is also an emerging view that the PRC is devoting great resources into high technology areas for which there is only scant information, usually from Chinese language journals, that could result in a formidable PLA in the next decade or soon after. While the PLA has factions of leaders who stress obtaining current generation weapons and "readiness" concerns, there is another faction of officers who promote the PLA's embrace of the Revolution in Military Affairs (RMA) [an emphasis on high-tech weapons and fewer, more mobile forces] to leap into the forefront of military powers. It can also be discerned from the three Pentagon assessments, in addition to Chinese and journalistic reports, that the PLA is rapidly gaining the weapon systems necessary to pose a credible threat to Taiwan. While there is some danger in overestimating the future capabilities of the PLA, there is perhaps even an even greater threat to U.S. security interests in underestimating the ability of the PLA to modernize. A closer examination of the impact of the PLA's access to foreign military technology also produces greater cause for concern. There is a need for both government and non-government analysts to question the seeming orthodoxy that the PLA cannot build a modern military force capable of meeting near and long-term goals. The remainder of this statement will expand on the effort that the PLA is making in the areas of high technology, missiles and space, preparations for conflict over Taiwan, and laying the foundation for greater power projections.

The PLA's High Technology Focus

While one can look at the PLA of today and see immobile liquid-fueled ICBMs [intercontinental ballistic missiles], thousands of obsolete tanks, jet fighters, and scores of submarines, all based on ancient Soviet designs, it would be a tragic error to presume that the PLA will remain so backward for much longer. While the RMA faction does not appear to dominate the current PLA leadership, their influence is apparent in the PLA's emphasis on developing an extensive and competitive military-civil high-technology sector, and the PLA's high priority on future-technology weapons.

[Recently] the PRC news agency Xinhua reported on PRC President Jiang Zemin's having asked "war industry to closely link itself with other industries," when visiting a demonstration of "war industry achievements." This statement is the most recent confirmation of the close linkage between the PRC's military and commercial science and technology sectors. . . .

Critical to the success of the PLA's short and long-term modernization plans has been greater access to foreign military and dual-use civil technology. The intensive effort to obtain foreign technology includes the purchase of weapon systems that can be reverse-engineered, to increasingly, the purchase of basic military science. . . . Russia and Israel have been principle sources of both weapons and technology in recent years, though the PRC has also sought to break down the European Community's military embargo. In addition to billions of dollars worth of weapons purchased from Russia, unconfirmed reports hold that . . . Russia agreed to form a "military equipment scientific research cooperation group" that would help the PLA with "a new generation of stealth bombers, conventional strategic missiles, water-surface aircraft, a new generation of submarines, naval vessels, laser weapons, telecommunication systems. . ." From open sources it is possible to comment on the following PLA high technology efforts.

Lasers The PLA's intense interest in laser weapons exemplifies its quest for next-generation technology that also exploits the weaknesses of potential enemies. . . . Pro-RMA officers view lasers as a key weapons technology for the future. The PLA envisions using lasers for anti-air, satellite tracking, anti-satellite, and for radar functions. In 1995 the PLA company Norinco marketed its ZM-87 battlefield laser dazzler. The 1998 Pentagon PLA report noted that the PLA might already have a ground-based laser capable of damaging low-orbit reconnaissance satellites. [Recently] the Select Committee suggested that Russia might be a source of nuclear-pump laser technology for the PRC for use in space. Last October [1999] the PLA revealed for the first time its Type-98 main battle tank, which has a box on the hull that may be a low-light camera or a laser dazzler. According to the Pentagon's PLA report released in June, China "reportedly is investigating the feasibility of shipborne laser weapons for air defense."

Space   The PRC's immanent launching of its first citizens into space will herald the PRC's emergence not just as a world-class space power, but as a rapidly developing military-space power as well. As has been the experience in the U.S. and Russia, there is in the PRC a direct linkage between military and commercial missile and space activities. The PRC's DF-5 ICBM formed the basis for its family of commercial space launch vehicles and its first man-rated launch vehicle, the Long March-2F. . . . The PRC plans to launch its own space station . . . , and is known to be researching its own reusable space shuttle. The Pentagon's June report on the PLA noted China's interest in "manned reconnaissance from space." More worrisome is the prospect of Russia selling the PRC plans or technology for its many space warfare programs that died with the former Soviet Union. The Soviets had tested an anti-satellite interceptor, and had plans for military combat space stations plus space-based laser and kinetic-kill anti-ICBM systems. . . .

Information warfare   PLA literature has revealed in recent years an intense interest in information warfare, which for the PLA includes activities ranging from anti-satellite operations to computer network attack, manipulation of information for deception and psychological warfare. It is very likely that the PLA is at least preparing to be able to attack the computer and communication infrastructure of an opponent as a key element of future military campaigns.

Radio frequency weapons   The 1998 Pentagon report on the PLA notes its interest in radio frequency and high-power microwave weapons, one of the few western references regarding the PRC and this new and dangerous class of weapons. These can be configured either as blast-bombs or beams that focus energy so as to neutralize missiles, computers or even humans.

Stealth, counter-stealth   A critical technology for current and future U.S. weapons systems, the PLA is very interested in both exploiting and countering stealth. At the 1998 Zhuhai Air Show the PRC company Seek-Optics marketed stealth coating and design technology. At the same show it also marketed the "J-231" radar, advertised as having "high anti-stealth capability." China has revealed the design of a new "F-16" frigate, intended for export that uses extensive stealth shaping. The Pentagon report on the PLA released in June notes China "reportedly is developing new fighter aircraft which will incorporate LO [Low Observable] technology."

Unmanned vehicles   Like the U.S., the PLA has a strong interest in unmanned air and sea platforms for military missions. The PLA has long used unmanned reconnaissance drones based on U.S. *Firebee* drones captured during the Vietnam War, and revealed the ASN-206 small battlefield reconnaissance drone that may incorporate some Israeli technology. In 1997 the PLA revealed a new submarine-robot. . . . It incorporated technologies such as automatic control, artificial intelligence, computers, acoustics and optics, and was said to have military benefits. An important RMA technology for the U.S. Navy is underwater combat vehicles.

Emphasis on Missiles and Space

Missile and space forces are proving to be an early emphasis in the PLA's modernization. In the absence of credible air and naval forces the PLA hopes to compensate by building its missile forces, which confer an "asymmetric" advantage over the U.S. and its Asian allies. The need to retain this advantage in part explains the PRC's vicious opposition to U.S. national and theater missile defense plans. The PLA's Second Artillery missile force, which controls both strategic ICBMs and theater intermediate range ballistic missiles (IRBMs) and short range ballistic missiles (SRBMs), is now in the process of fielding new strategic and theater missiles. Current numbers of PLA ICBMs is thought to include 20–25 older liquid fueled DF-5, plus about 12 JL-1 SLBMs on one *Xia* class SSBN [ballistic missile nuclear submarine]. PRC strategic missile numbers will increase irrespective of threats to do so as part of the PRC's attempt to forestall U.S. strategic and theater missile defense plans.

New Strategic Missiles   After a long development period, the PLA revealed its new DF-31 5,000-mile range ICBM in the October 1, 1999 military parade in Beijing. The DF-31 is a modern solid-fueled and mobile ICBM that uses a new Chinese small nuclear warhead. . . . This missile is not thought to have multiple warheads now, but could in the future. Its truck-based transporter-erector-launcher (TEL) indicates that the missile will be hidden in a network of cave shelters connected by roads. The DF-31 will also form the basis of the JL-2 SLBM that will arm a new class of SSBN being built with Russian technical assistance. . . . By the middle of this decade the Second Artillery is also expected to deploy the new 8,000-mile range DF-41 ICBM, which is expected to be mobile and armed with multiple warheads.

The PRC's current small ICBM force is designed to deter both nuclear attack, and to deter U.S. intervention on behalf of Taiwan. It is reasonable to estimate that the PRC will in this decade deploy 100 to 200 new ICBMs and SLBMs, which would allow the PRC to avoid restrictive arms control agreements, but also give it a hedge to be able to overcome limited U.S. national missile defenses.

Theater information-strike complex   China's most profound challenge to the balance of power on the Taiwan Strait, or in Asia generally, is the PLA's developing "information-strike complex" of highly accurate ballistic and cruise missiles, combined with multiple layers of long-range space and airborne sensors. China is improving the 1,125-mile range DF–21 ballistic missile. . . . This new DF–21 may have a new highly accurate warhead that uses navigation satellite data from the U.S. GPS [global positioning system] or Russian GLONASS [global navigation satellite system] network, or perhaps radar guidance technology. The DF-15 SRBM, used to intimidate Taiwan in 1995 and 1996, is being improved with GPS/GLONASS guidance systems, and may be given longer range to increase its speed and survivability. Last October [1999] the PLA also revealed the M-11 Mod 1 SRBM, which has a shorter range than the DF-15, but a larger warhead, and will likely incorporate satellite guidance. The Pentagon has noted that development of land-attack cruise missiles (LACM) for theater and strategic missions has a "relatively high development priority" for China. . . . A long-range strategic cruise missile, similar in capability to early U.S. *Tomahawk* cruise missiles, will likely enter service after 2005.

Although these theater ballistic and cruise missiles could carry a new small nuclear warhead, China is placing great emphasis on developing powerful non-nuclear warheads, thus reducing the prospect of nuclear retaliation. Missiles armed with Radio Frequency (RF) weapons that simulate the electromagnetic pulse created by nuclear explosion, could disable computers, electric power grids, or even an aircraft carrier without causing great casualties. China is also interested in building cluster munitions for ballistic or cruise missiles that could disable airbase runways.

Anti-missile, anti-satellite, and space information systems   China's government loudly protests U.S. anti-missile plans, promotes bans on outerspace warfare, but says almost nothing about its own anti-missile or anti-satellite programs, or its space warfare plans. The Pentagon's PLA report released in June notes the PRC "can be expected to try to develop a viable ATBM [anti-tactical ballistic missile] and ABM [anti-ballistic missile] capability by either producing its own weapons or acquiring them from foreign sources." The PLA is marketing a new long-range surface-to-air (SAM) missile called the FT-2000, which is based on Russian and U.S. technology, and could form the basis of an ATBM-capable missile. The PLA is aware of the need to defend against opposing missiles and of the need to exploit the U.S. military's high dependence on reconnaissance and communication satellites. The PLA's interest in laser ASAT [antisatellite] systems has already been noted.

As it seeks the means to deny space to future adversaries, China is also seeking to better exploit outer space for military missions. China is developing new military satellites for high-resolution imaging, radar imaging, signal intelligence (SIGINT) collection, navigation and communication. China has recently announced it will launch eight new reconnaissance satellites: four imaging satellites and four radar satellites. When in orbit, this network will give China twice daily satellite revisits. Radar satellites can penetrate cloud cover and are very useful for finding naval formations at sea. As does the U.S. military, China will also likely seek to integrate access to commercial satellite imaging into its military operations. For navigation and targeting, the PLA uses the GPS satellite network, is negotiating with Russia to invest in its faltering GLONASS network, and is also developing its own navigation satellite network.

Focus on Taiwan

PLA modernization is being driven to a large extent by PRC government's objective to build the military force needed to either intimidate or compel the government on Taiwan into reunification dictated largely by Beijing. The goal is not necessary to invade and conquer Taiwan, but to defeat its armed forces swiftly and decisively enough to prevent American intervention—or to even attack U.S. forces in the event they do come to Taipei's assistance. Recent advances in doctrine, training, and the modernization of missile, air and naval forces stem from this objective. The PLA faces an immense challenge to net these forces together in a cohesive whole—but it is working to do just that. The danger is that absent requisite U.S. sales of defensive weaponry to Taiwan, the PRC is likely to obtain a level of military superiority later in this decade that would make the use of force tempting to Beijing under some circumstances.

Assembling information-strike combine   The PLA is currently working to modernize its missile, information and air forces to assemble an information-strike combine to form the cutting edge of any future military campaign against Taiwan. The goal will be to undertake joint SRBM, LACM, air and information-strikes in order to deliver a decisive surprise blow against the airfields, naval bases, and communications nodes of Taiwan. . . . It is possible that the U.S. estimate may be conservative. But when counting expected LACMs, it is reasonable to estimate that the PLA could have 1,000 or more ballistic and cruise missiles aimed at Taiwan by the end of this decade. And these will be made deadly accurate with targeting data from optical and radar satellites slaved to navigation satellite data.

Modern air power   While a slow process, the PLA is devoting greater resources to the purchase of modern Russian and Israeli combat systems and technology to supplement indigenous air combat programs. By 2005 to 2010 the PLA Air Force (PLAAF) [could] assemble a significant number of 4th generation fighters [the newest U.S. planes are 4th generation], supported by radar and electronic intelligence aircraft, that could wrest air superiority away from a ROCAF that has been severely damaged by massive missile strikes. It is telling that this year's Pentagon PLA report warns that by 2005, "if current trends continue, the balance of air power across the Taiwan Strait could begin to [swing] in China's favor. . ."

The PLAAF has about 50–55 Russian Suhkoi Su-27 fighters and has started co-producing up to 200 more as the J-11. Last year Russia agreed to sell between 30–40, perhaps a first installment for the Su-30MKK twin-seat strike fighter. In the Su-30 the PLAAF will have its first all-weather strike fighter comparable to the U.S. F-15E *Strike Eagle* [a 3rd generation aircraft]. The PLAAF Su-30s will be equipped with a range of precision guided bombs and missiles, and the Kh-31 anti-radar missile. Older J-8 and J-7 fighters will soon be supplemented by the J-10, China's first 4th generation fighter designed with Israeli and Russian assistance. Having just been denied the cutting-edge Israeli *Phalcon* phased-array airborne warning and control radar, the PLAAF will likely quickly purchase the Russian A-50 AWACS [Airborne Warning and Control System aircraft] with a Russian radar. There are now over 10 H-6D aerial refueling tankers in the PLAAF, which service the J-8IID fighter. Russia is trying hard to sell the PLAAF the Il-78 aerial tanker.

New subs, missiles   The PLA also hopes that massive missile attacks will diminish Taiwan's naval defenses, which will ease the PLA Navy's [PLAN's] attempt to place a naval blockade around the island. To do so it has stressed the acquisition of new submarines, anti-ship missiles and mines. China could purchase more than the current four Russian *Kilo* conventional submarines—two of which are the ultra-quiet "636" version. The indigenous *Song*-class conventional sub is experiencing developmental difficulties and may be discontinued. Also, the PLAN has about 19 older *Ming* conventional subs that could be brought to bear. The PLAN has five older *Han*-class nuclear attack subs that will soon be supplemented by the more modern Type-093.

More modern PLAN missiles are being purchased from Russia, such as the SS-N-22 supersonic missile that will equip its two, or four or more *Sovremenniy* missile destroyers, two of which will be delivered this year. Taiwan's Navy has no defense against the SS-N-22 save to sink the destroyer. Indigenous PLAN ships are improving. . . . The latest *Luhai* class destroyer incorporates elementary stealth shaping and later versions may deploy phased-array radar and vertical-launched anti-air and cruise missiles. Less often commented upon is the PLAN's large inventory of modern naval mines. These include rocket-propelled fast rising mines that are very dangerous to military and civil shipping. The PLAN Air Force is also growing with the addition of J-8IID fighters, JH-7 attack fighters, and its new Y-8J patrol aircraft equipped with the British *Skymaster* radar to perform airborne early warning missions.

Foundation for Future Power Projection

If successfully absorbed, forces being acquired for possible use against Taiwan can also lay the foundation for future PLA power projection, on the way to attaining "preeminence" in the Asian region. [Recently] the PLAN's first *Luhai* destroyer, along with the Ukrainian-build tanker *Nancang*, [were] on a naval diplomatic mission to visit Malaysia and ports in Africa. The PLAN icebreaker-intelligence ship *Haidaio* completed a circuit of the Japanese islands in May. Beyond mere diplomacy, these exercises mark a small beginning that the PLA intends to carry the PRC's influence beyond its Mainland, where its future interests will be impacted.

Today the PLA cannot begin to approach the global reach of the forces of the United States. There is also no guarantee that the PRC will survive and not suffer the fate of most other Communist regimes. But if the Chinese Communist Party remains in power, and if it can dictate the fate of Taiwan, the PLA will then have resources to devote to the space, naval and air elements of greater projection. There is also the question of whether close political and arms sales/technology relations with Russia today in the future develops in the direction of military-operational cooperation. By the next decade the U.S. should not be surprised to see the PLA Navy being able to project power into the South China Sea, Indian Ocean and Persian Gulf.

First island chain　For Japanese and Southeast Asians, PLA power projection does not have to extend very far. PLA Navy planners long ago set the goal to be able to extend its control to what it terms the "First Island Chain," which is the chain islands that extends from the Kurils, to Japan, Taiwan, the Philippines down to Indonesia. To project power in this region the PLA simply needs to purchase or build more of the systems currently being procured. More IRBMs and long-range cruise missiles, guided by new reconnaissance and navigation satellites, will provide long-range strike or deterrent capabilities unmatched in the region save by India. More Su-30 strike fighters, supported by aerial tankers and long-range AWACS, could extend PLA air control to the disputed islands in the South China Sea, or pose a real challenge to the Japanese Air Force. And with better air support, a PLA Navy with more modern submarines and warships can undertake bolder operations in support of its extensive claims to most of the South China Sea. Today the PLA is interested in the Russian Beriev Be-200 jet amphibian, which could perform both patrol and resupply missions for the PRC's far flung chain of island bases in the South China Sea. While wealthy Southeast Asian countries like Singapore might be able to afford missile defenses, it is unlikely that others will be able to do so. They will also be hard pressed to maintain the air and naval forces necessary to blunt a serious PLA move into the South China Sea.

The PRC's development of strategic military relations with Burma, Pakistan and Iran can be seen as part of an effort to contain India and to build geostrategic points of influence that ensure growing access to Persian Gulf oil. More accurate SRBMs and cruise missiles are likely to be sold to Pakistan, Iran or even Saudi Arabia, extending PLA influence by proxy. With these missiles will likely come the sale of PLA satellite intelligence, which will be of use to their political leadership and military forces.

Second island chain   Looking further into the next decade and beyond, PLA navy planners also have set their sights on the "Second Island Chain," which extends out to Guam. By this time, it is likely that PLA planners expect that the U.S. will be a declining power not in much of a position to challenge growing PLA might. It can also be expected that by this time, the PRC's embryonic civil and military space power will have developed much further, to include space stations, Moon missions and much greater military use of outer space, to possibly include space combat capabilities. These would then be used to support more advanced naval and air forces.

New long-range bombers may also be developed for the post-Taiwan period. In the mid-1990s the PLA explored purchasing the Russian Tu-22M-3 *Backfire,* but has not done so as has India. There are occasional hints that there is a domestic program underway to replace the H-6 medium bomber, a copy of the 1950s-vintage Soviet Tu-16 *Badger.* When this new long-range bomber emerges, the PLA will also likely possess supersonic LACMs and be following U.S. and Russian efforts to build hypersonic aircraft. . . .

Implications for the United States

As this statement has stressed already, the PLA is not today a match for U.S. power, save in the area of theater ballistic missile in Asia—which the U.S. has none. What this statement does convey is that the People's Republic of China is undertaking a serious and broad effort to acquire the technical and research base, the most modern technologies, and when necessary, actual weapon systems to modernize its armed forces. The long-term goal of American policy is to live in peace and prosperity with the Chinese people, and where possible, to help their evolution toward even greater freedom and prosperity. The oft-stated justification made buy the current and previous U.S. Administrations for policies of "engagement" is that they hold a far better chance for positive and pluralistic forces to emerge and to displace Communist/authoritarian rule.

The hard truth, however, is that the PRC is still controlled by a Communist leadership for whom maintaining total power is its paramount goal. . . . [The PLA] commands a lions share of the national budget and it leaders are deeply suspicious of democracy, most of their Asian neighbors and the West. Furthermore, the PLA is standard bearer of the Communist Party's nationalist credentials, which are increasingly linked to the recovery of "lost territories," to include Taiwan and the South China Sea. Furthermore, the PRC and the PLA increasingly oppose and covet America's leadership role in Asia. . . .

However, the areas where U.S. and PRC interests are likely to conflict, Taiwan, missile defense, the U.S. alliance structure in Asia, the South China Sea, and proliferation, are not going to be resolved soon. In all of these potential conflict areas the PRC views a stronger PLA as a key tool for achieving its ends. Until China changes its Communist-dictated national priorities it will remain necessary for the United States to defend its leadership position in Asia and to ensure that its ability to militarily deter the PRC remains unquestioned.

The importance of Taiwan   The most important near-term area of conflict between the U.S. and the PRC is over the future of the democratic system on Taiwan. Long-term U.S. policy enshrined in the 1979 Taiwan Relations Act holds that the U.S. supports a peaceful resolution between the PRC and Taiwan, and will maintain defensive arms sales to Taiwan to deter war. Insuring Taiwan's survival, however, is not just a matter of principle for the U.S.; it can also hasten the day when the PRC evolves from Communism. Taiwan is a Chinese culture that has proven that freedom and prosperity can co-exist, proving to Chinese on the Mainland that they do not need dictatorship for economic development. Sustaining Taiwan's survival, therefore, will in the long run strengthen chances for peace in Asia.

Today Taiwan has sufficient strength to deter PLA attack, but that may not last for long. Taiwan has some U.S. *Patriot* ATBMs and is developing a domestic version, but their small number and vulnerability to attack is insufficient to defend Taiwan from a growing number of PLA missiles. Taiwan's Air Force currently has 330 F-16, Mirage-2000 and IDF fighters and the E-2 *Hawkeye* AWACS. But its edge will erode if it does not receive advanced air-to-air missile, anti-radar missiles, and greater defense against PLA missiles. Taiwan only has two 1970s designed Dutch submarines that actually work, plus two World War II U.S. *Guppy* subs which cannot dive deeper than they are long! The Taiwan Navy lacks an effective defense against the SS-N-22 supersonic anti-ship missile, now in PLA Navy service.

Given that the PLA is shifting the military balance on the Taiwan Strait, it is incomprehensible that the Clinton Administration has decided . . . not to sell

Taiwan major weapon systems that it is reported to have requested. The White House has refused to sell Taiwan the *AEGIS* naval air defense system which is possibly the only U.S. system that can defeat the SS-N-22. The Administration has also refused to sell Taiwan the HARM [High-speed Anti-Radiation Missile] anti-radar missile, in the face of the PLA deployment of advanced S-300 SAM sites on the Taiwan Strait that threaten Taiwan's Air Force. And the Administration refused Taiwan's request for conventional submarines, which could help Taiwan thwart the PLA Navy's increasing ability to impose a blockade. . . .

Conclusion

The PRC is not today the enemy of the United States, but the PRC's territorial and political ambitions in Asia, and resources it is devoting to building its People's Liberation Army into an advanced fighting force, dictate that the U.S. exercise great caution. While there are today widely varying views on the current and future capabilities of the PLA, there is enough evidence to conclude that the PLA is rapidly shifting the balance of power on the Taiwan Strait, and laying the foundation for greater power projection in the next decade and beyond. As long as the PRC remains governed by a Communist Party there is likely to be growing friction with the U.S. over Taiwan, missile defense, Washington's leadership in Asia, and the South China Sea. However, the U.S. can best sustain peace in Asia by working to strengthen its leadership role, strongly defending its interests in all possible areas of conflict with the PRC, and sustaining its military superiority in Asia so as to deter conflict.

NO ↵

Ivan Eland

Is Chinese Military Modernization a Threat to the United States?

Introduction

Both the Pentagon and a congressionally mandated commission recently issued studies on the Chinese military that overstated the threat to the United States posed by that force. The pessimism of both studies was understandable. The Department of Defense's [DOD's] study—the *Annual Report on the Military Power of the People's Republic of China*—was issued by a federal bureaucracy that has an inherent conflict of interest in developing assessments of foreign military threats. Because the department that is creating the threat assessments is the same one that is lobbying Congress for money for weapons, personnel, fuel, and training to combat threats, its threat projections tend to be inflated. Because China, with an economy that is seemingly growing rapidly, is the rising great power on the horizon that should shape the future posture of American conventional forces (the brushfire wars needed to combat terrorism are likely to require only limited forces), the threat from China's armed forces is critical for bringing additional money into the Pentagon. The U.S.-China Security Review Commission's work—*The National Security Implications of the Economic Relationship between the United States and China*—drew at least partially on the Pentagon's effort and was written by anti-China hawks and those with a desire to restrict commerce with China.

In contrast, this paper attempts to place the modernizing Chinese military in the context of a more balanced and limited view of U.S. strategic interests in East Asia. In addition, when the distorting perspectives of both studies are removed—that is, their focus on recent improvements in Chinese military capabilities rather than on the overall state of the Chinese military—the threat from the Chinese armed forces is shown to be modest. The bone-crushing dominance of the U.S. military remains intact. In fact, the Chinese military does not look all that impressive when compared even to the Taiwanese armed forces.

From Ivan Eland, "Is Chinese Military Modernization a Threat to the United States?" *Policy Analysis No. 465* (January 23, 2003). Copyright © 2003 by The Cato Institute. Reprinted by permission. Notes omitted.

Putting the Modernizing Chinese Military in Context

Frequently, improvements in the Chinese military are reported in the world press without any attention to context. That is, those "flows" are highlighted but the "stock"—the overall state of the Chinese military—is ignored. The state of the Chinese military and how rapidly it is likely to improve will be examined in the second half of this paper. But first, additional context is needed.

Pockets of the Chinese military are now modernizing more rapidly than in the past, but compared to what? Both the modernization and the actual state of the Chinese military must be compared with those of the U.S. military and other militaries in the East Asian region (especially Taiwan's armed forces). In addition, the geopolitical and strategic environment in which Chinese military modernization is occurring needs to be examined. Western students of the Chinese military often speak abstractly about when growing Chinese military power will adversely affect "U.S. interests." It is very important to concretely define such interests because the wider the definition, the more likely even small increments of additional Chinese military power will threaten them.

U.S. Interests in East Asia

Even before President Bush's expansive new national security strategy was published, the United States perceived that it had a vital interest in maintaining in East Asia a continuous military presence that was deployed far forward. Despite the end of the Cold War, the United States has maintained Cold War–era alliances that encircle China; indeed, it has actually strengthened them. The United States has formal alliances with Japan, South Korea, Thailand, the Philippines, and Australia. In addition, the United States has an informal alliance with Taiwan—China's arch enemy—and a friendly strategic relationship with Singapore and New Zealand. In the post–Cold War era, as the military threat to East Asia decreased, the United States strengthened its alliance with Japan by garnering a Japanese commitment to provide logistical support to the United States during any war in the theater. The Bush administration came into office with an even stronger predilection to enhance security alliances (especially the one with Japan) than its predecessor.

Also, using the war on terrorism as part of its rationale, the Bush administration has expanded U.S. military presence in the areas surrounding China. Citing the need to fight the war on terrorism, the United States sent special forces to fight Abu Sayaf—a tiny group of bandits with only a tangential connection to the al-Qaeda terrorist movement—in order to strengthen the U.S. security relationship with the government of the Philippines. . . . During the war against terrorism in Afghanistan, the United States established a "temporary" military presence on bases in Central Asian nations on China's western border. Given the Bush administration's use of the war on terrorism as a cover for deploying troops to Georgia and the Philippines and the history of the U.S. military presence in Japan, Germany, and South Korea, the U.S. military presence in those Central Asian nations will likely become permanent.

Before the September 11, 2001, terrorist attacks and the ensuing war in Afghanistan slowed the process, the administration was seeking better relations with India so as to use that country as a counterweight to a rising China. Finally, the war on terrorism has fostered a newly cooperative U.S.-Russian relationship, thus completing the encirclement of China. Moreover, the Pentagon is increasing the number of U.S. warships in the Pacific region.

Of course, the U.S. government does not admit to a policy of containing China, as it did with the Soviet Union during the Cold War. But in Asia the ring of U.S.-led alliances (formal and informal), a forward U.S. military presence, and closer American relationships with great powers capable of acting to balance against a rising China constitute a de facto containment policy. Such a policy is unwarranted by the current low threat posed by China and may actually increase the threat that it is designed to contain.

Even the DoD admits that the Chinese are recognizing and reacting to U.S. policy:

> China's leaders have asserted that the United States seeks to maintain a dominant geostrategic position by containing the growth of Chinese power, ultimately "dividing" and "Westernizing" China. . . . Beijing has interpreted the strengthening U.S.-Japan security alliance, increased U.S. presence in the Asia-Pacific region, and efforts to expand NATO [North Atlantic Treaty Organization] as manifestations of Washington's strategy.

The DoD report continues:

> Chinese analyses indicate a concern that Beijing would have difficulty managing potential U.S. military intervention in crises in the Taiwan Strait or the South China Sea. There are even indications of a concern that the United States might intervene in China's internal disputes with ethnic Tibetan or Muslim minorities. . . .

And the Bush administration recently issued a national security strategy that shows that Chinese perceptions are largely correct. The new security strategy aims at ensuring U.S. primacy—that is, keeping the United States so powerful that other nations will be dissuaded from challenging it—and "preempting" (actually preventing) amorphous threats from nations that are developing or possess weapons of mass destruction. Clearly, the portion of the White House's security strategy concerned with primacy is aimed at China, the rising great power that administration officials think is most likely to challenge the United States at some time in the future. Despite the grandiose nature of the strategy, however, the administration will probably not (one hopes) apply the "preemption" part of it to China—a nation possessing 20 long-range nuclear missiles that can hit the United States. In short, the de facto containment policy will probably continue to be followed.

The extended defense perimeter that the United States continues to maintain in East Asia to carry out that containment policy shows a failure to recognize China's security concerns. Although China remains an authoritarian state . . . , conflict might be avoided if some understanding of the calculus of a potential adversary were shown. If a foreign nation had ringed the United States with alliances, friendships with potential adversaries, and an increasing military presence, the United States would feel very threatened—as was the case, for example, when the Soviets attempted to place nuclear missiles in nearby Cuba during the 1960s.

The United States fears any attempt by China to increase its influence in East or Southeast Asia. Yet, as the Chinese economy grows and China becomes a great power, it will naturally seek more control over its external environment. As Michael O'Hanlon and Bates Gill, both then at the Brookings Institution, perceptively noted, most of China's ambitions are not global and are no longer ideological; they are territorial and confined to exerting more regional influence over the islands and waterways to the south and southeast of its borders. The United States could accommodate such limited ambitions as long as they did not snowball—an unlikely scenario—into a conflict that drastically altered the power balance in East Asia. China has given no indication that it would like to make an attempt at imperial conquest of East Asia.

In the past, wars occurred when an established power refused to acknowledge the great power status of a rising nation—for example, Britain's refusal to acknowledge the kaiser's Germany in the late 1800s and early 1900s. The United States should not make the same mistake with a rising China. China should be allowed, as all great powers do, to develop a sphere of influence in its own region—that is, East Asia. Within limits, an expanded sphere of Chinese influence should not threaten U.S. vital interests, if defined less grandiosely than at present.

Unfortunately, the United States regards even the smallest change in the status quo in East Asia (unless the change expands the already overextended U.S. defense perimeter) with suspicion. The United States does have a vital interest in ensuring a diffusion of power in East Asia so that no hegemonic great power—like imperial Japan in the 1930s—arises. But, unlike the situation before World War II, when China was weak and the French and British colonial powers were spread too thin, centers of power in East Asia other than the United States exist to balance a rising China. Japan, alone or in combination with South Korea, Taiwan, Australia, and the Association of Southeast Asian Nations, could balance against China. The United States, instead of maintaining Cold War–era alliances and a forward military presence in the region, should gradually withdraw its forces from East Asia and allow those nations to be the first line of defense against China. Currently, those nations fail to spend enough on their security because the United States spends huge amounts on its military and is willing to subsidize their security for them. . . .

Only if the balance of power in East Asia broke down with the advent of an aggressive hegemonic power should the United States intervene militarily in the affairs of the region. That policy would be called a "balancer-of-last-resort" strategy. Such a strategy would minimize the danger of a confrontation with China.

U.S. Military Capabilities Compared With Those of China

The Bush administration's national security strategy attempts to ensure American primacy by outspending other nations on defense many times over, thus dissuading them from competing with the United States. The United States is already more powerful militarily relative to other nations of the world than the Roman, Napoleonic, or British Empire was at its height. According to the national security strategy, "Today, the United States enjoys a position of unparalleled military strength and great economic and political influence." And the Bush administration would like to keep such U.S. military dominance by profligate spending on military might that is deployed around the world. The history of international relations indicates that this strategy has little chance of succeeding. Historically, when threatened by a country that had become too powerful, nations banded together to balance against it. Of course, administration officials claim that the United States is a benevolent power and that other nations will feel no need to balance against it. Such countries as Russia, India, and especially China might disagree. For example, China accuses the United States of maintaining a policy of containment, and Russia has protested the expansion of the NATO alliance up to its borders. A good place for more sustainable and less threatening U.S. policies to start is in East Asia.

Forces and Defense Spending

Currently, the United States maintains about 100,000 military personnel in East Asia. That military presence is centered in Japan (41,000), South Korea (37,000), and afloat (19,000). At sea, the United States stations one carrier battle group and one Marine amphibious group forward in the region and will now ensure that a second carrier group will be there more of the time. The United States will also augment the number of nuclear submarines stationed in Guam. That military presence seems small compared to the military forces of China, which has active forces of 2.3 million.

Yet the U.S. military presence deployed forward in East Asia is only the tip of the iceberg. That presence is a symbol of U.S. interest in the region and of the world-dominant U.S. military juggernaut that could be brought to bear against the large, but largely antiquated, Chinese military during any war between the two nations.

The United States spends about $400 billion a year on national defense and alone accounts for about 40 percent of the world's defense spending. There is some dispute about how much China spends because not all of its defense spending (for example, funds for weapons research and procurement of foreign weapons) is reflected in the official Chinese defense budget. David Shambaugh, a prominent academic authority on the Chinese military, estimates total Chinese defense spending at about $38 billion per year. In the same ballpark, the International Institute of Strategic Studies' [IISS'] *Military Balance* estimates such spending at $47 billion per year. In contrast, the U.S. Department of Defense's estimate is predictably much higher—noting that annual Chinese military spending "*could* total $65 billion." Because Shambaugh and the IISS do not build weapon systems to combat threats and thus have no inherent conflict of interest, their independent estimates are probably less prone to threat inflation than is DoD's estimate.

China has had real (inflation-adjusted) increases in defense spending only since 1997. Chinese military expenditures are constrained by limits on the ability of China's central government to collect revenues and the concomitant budget deficit. Moreover, increases in military spending have been surpassed by rapid Chinese economic growth, leading to declines in defense spending as a proportion of gross domestic product.

The $38 billion to $47 billion range is roughly what other medium powers, such as Japan, France, and the United Kingdom, spend on defense. But the militaries of those other nations are much smaller and more modern than the obsolete Chinese military, which needs to be completely transformed from a guerrilla-style Maoist people's army into a modern force that emphasizes projection of power on the sea and in the air. . . . So the Chinese must spend much of their increases in official defense funding to prop up their sagging, oversized force and slowly convert it to a force that can project power. . . .

Consequently, China's spending to acquire weapons is equivalent only to that of countries with total defense budgets of $10 billion to $20 billion. Given that the United States, with a gargantuan budget for the research, development, and procurement of weapons—well over $100 billion per year—is leaving its rich NATO allies behind in technology (there is fear in NATO that U.S. capabilities are so far advanced that the U.S. armed forces would not be able to operate with allied militaries), it most surely is leaving China in the dust.

The Chinese Defense Industry

The Chinese defense industry remains state owned, is grossly inefficient, and has had an abysmal track record of developing and producing technologically sophisticated weaponry. . . . In fact, most of the significant technological progress in the Chinese military has resulted from weapons purchases from Russia. . . .

Even when the Chinese buy advanced weapon systems abroad, they have difficulty integrating them into their forces. For example, the Chinese have had problems integrating the Russian-designed Su-27 fighter into their air force. As in many other militaries of the Third World, deficiencies in Chinese training, doctrine, and maintenance for sophisticated arms do not allow the full exploitation of such systems.

Military Equipment

Although the best crude measure of a nation's military power is probably its defense spending (because it includes money spent for the all-important "intangibles," such as pay, training, ammunition, maintenance of equipment), a nation's military capital stock—the dollar value of its military hardware—is a measure of its force's modernity. The U.S. military's capital stock is almost $1 trillion. In contrast, despite the purchase of some sophisticated Russian weapons, the capital stock of the largely obsolete Chinese military is only one-tenth of that total—well under $100 billion. In fact, China has fewer top-of-the-line weapons than middle powers, such as Japan and the United Kingdom, and smaller powers, such as Italy, the Netherlands, and South Korea. A further measure of a military's true capability is based on how much is spent per soldier (for training, weapons, and the like). Even when calculated from the inflated DoD estimate of Chinese defense spending—$65 billion per year—China's spending is less than $33,000 per troop, whereas the United States spends $213,208 and Japan spends $192,649.

That disparity in value mirrors a wide gap in capabilities. In contrast to the thoroughly modern U.S. military, China's armed forces have been able to modernize only slowly and in pockets. According to DoD, the Chinese have a large air force—3,400 combat aircraft—but only about 100 are modern fourth-generation aircraft . . . Most Chinese aircraft incorporate technology from the 1950s or 1960s. In contrast, all of the more than 3,000 aircraft in the U.S. air services are fourth-generation aircraft . . . , and fifth-generation aircraft (F-22s and F-18E/Fs) are already beginning production. . . .

The Chinese army is still an oversized, outdated Maoist guerrilla army with insufficient airlift, logistics, engineering, and medical capabilities to project power very far. In fact, most of the Chinese army is good only for internal security purposes. The force's equipment is antiquated—for example, most tanks incorporate technology from the 1950s. . . . In contrast, the United States has the most potent and technologically sophisticated army in the world. . . .

According to DoD, the Chinese navy appears to have postponed indefinitely plans to buy an aircraft carrier. In addition, DoD notes that the Chinese navy's air defense against enemy aircraft, precision-guided munitions, and cruise missiles is limited by short-range weapons (only a few of China's ships have longer-range surface-to-air missiles) and a lack of modern air surveillance systems and advanced data links to communicate that "air picture" to other ships in the fleet. The purchase of a few SOVREMEN-NYY-class destroyers from Russia will not alter that state of affairs significantly. In modern war, ships are vulnerable to attack from the air, and those limitations make the Chinese navy a sitting duck in any conflict. In contrast, the U.S. fleet has global dominance with 12 large aircraft carriers . . . , the best submarines in the world, and the most sophisticated air defense capabilities afloat. . . .

The Chinese are slowly modernizing their small strategic nuclear arsenal to make it less vulnerable to a preemptive attack from the world's most potent nuclear force—the U.S. strategic arsenal of thousands of warheads. But even with such modernization, China's nuclear arsenal will pale in comparison with the robust U.S. nuclear force. The Chinese currently have only about 20 long-range missiles—housed in fixed silos—that can reach the United States. The missiles, their liquid fuel, and their warheads are stored separately, making them very vulnerable to a preemptive strike before they could be assembled and launched. . . .

The Chinese have only one ballistic missile submarine, which usually remains at the dock for repairs. Even at sea, to fire its missiles, the submarine must operate fairly close to the United States—where it would be more vulnerable to attack. In contrast, the United States has 14 ballistic missile submarines that are the most powerful weapon systems ever built and can launch their missiles at a target from across the ocean. The Chinese have a successor ballistic missile submarine in development, but they have never had much luck perfecting the technology. The only time China's small nuclear arsenal could become a problem for the United States would be in an emotional Chinese reaction to U.S. intervention in a crisis between China and Taiwan.

A Massive Military Buildup?

Even the Defense Intelligence Agency and high-ranking U.S. military officials seem to agree [that China is not engaged in a massive arms race]. According to the Defense Intelligence Agency, by 2010, even the best 10 percent of the Chinese military will have equipment that is more than 20 years behind the capabilities of the U.S. military (equivalent to U.S. equipment in the late 1980s). The other 90 percent of the Chinese military will have even more outdated equipment. . . . Even DoD has admitted that "the PLA [People's Liberation Army] is still decades from possessing a comprehensive capability to engage and defeat a modern adversary beyond China's boundaries." . . .

Although in the last few years the Chinese have been modernizing their military more rapidly than in the past, recent hikes in the U.S. budget for national defense have been extraordinary. The increase in the U.S. budget for national defense in 2003 alone is of approximately the same magnitude as the entire Chinese defense budget. . . . The United States spends more than $40 billion a year on research and development for weapons (again, roughly equal to total annual Chinese defense spending) and more than $60 billion yearly on weapons procurement. Thus, the speed of U.S. military modernization dwarfs the pace of improvements in parts of the antiquated Chinese forces. In fact, U.S. military modernization is outpacing even that of wealthy NATO allies—the next most capable militaries on the planet. . . .

In conclusion, even though the Chinese military is modernizing more rapidly than in the past, the speed of the modernization is less than that of the modernization of the already vastly superior U.S. force. In other words, despite all of the clamor in the press and in the U.S. government about Chinese military modernization, the U.S. military is way ahead and the gap is actually widening (the same situation holds when U.S. armed forces are compared with all of the other militaries in the world). When pressed, even anti-China hawks admit that Chinese military capabilities are far behind those of the United States. . . .

Chinese Military Modernization and Its Implications for Taiwan

China's economy is now four times the size of Taiwan's ($1.2 trillion versus about $300 billion) and is growing faster. . . . That economic disparity could, at least theoretically, be turned into a military disparity. But according to the Pentagon, Taiwan's strategy is to enhance key aspects of its military capabilities—counterblockade operations, air superiority over the Taiwan Strait, and defense against amphibious and aerial assault on the island. . . .

The Taiwanese certainly could do more than they currently do to ensure their security. Defense expenditures have actually been declining as a percentage of Taiwanese government spending. The Taiwanese fail to do more to enhance their own defenses because they believe that the United States will come to their aid if a crisis occurs with China. Although U.S. policy is ambiguous on that point, President [George W.] Bush made it less ambiguous by saying the United States would do "whatever it took" to defend Taiwan. . . .

President Bush's pledge to do "whatever it takes" to defend Taiwan is dubious. The security of Taiwan has never been vital to the United States, and dueling with a nuclear-armed power in any crisis over the small island would be ill-advised. . . .

Rather than provide an informal security guarantee for Taiwan, the United States should sell Taiwan the arms to defend itself. President Bush has authorized the sale of a greater number of weapon systems to Taiwan than President [Bill] Clinton approved. But Taiwan has been slow to come up with the money to buy many of them. Taiwan needs to do more for its own defense but will not if the United States continues to informally guarantee Taiwan's security.

Implications for the United States: China's Relations With Other Neighbors

According to Christopher A. McNally and Charles E. Morrison, authors of *Asia Pacific Security Outlook 2002,* China has steadily improved relations with countries sharing its land borders, but the Chinese still have issues with their maritime neighbors. In 2001 the leaders of China, Russia, Kazakhstan, Uzbekistan, Tajikistan, and the Kyrgyz Republic created the Shanghai Cooperation Organization that was designed to increase cooperation in regional security, economic relations, culture, science, education, and environmental protection. In addition, China signed the Treaty of Good-Neighborliness and Friendly Cooperation with Russia, which pledged, among other things, that the two nations would not use force in disputes and would not target missiles at each other. China's relations with former foes—India and Vietnam—have warmed with growing economic relations and high-level visits. In general, such arrangements contribute to the security of the regions involved. The military cooperation between China and Russia bears watching, but it is only exacerbated by the unstated U.S. policy of containing China with encircling alliances and a continuing forward military presence in East Asia.

According to James Holt, an analyst for the World Policy Institute, China's military is qualitatively inferior to that of Russia, India, Vietnam, and Taiwan and would lose any war against any of those nations. In particular, since the 1960s India has more than doubled the size of its military and modernized its armed forces to a greater extent than China. In addition, Holt argues that for the last 30 years China's military power has also been declining vis-à-vis that of the United States, Taiwan, Japan, and South Korea. Holt maintains that unless China at least doubles real military spending, its rate of weapons purchases in relation to the size of its armed forces is so low that its military will continue to decline relative to those of the United States, Taiwan, India, Japan, and South Korea. This need is caused by the low percentage of Chinese defense expenditures that is currently allocated to the acquisition of weapons.

In contrast to China's improving relations with its neighbors on land, its dispute with a maritime neighbor, the Philippines, over islands in the South China Sea continues. But the dispute is contained because both nations want to avoid spillover effects into their bilateral political and economic relationships, according to McNally and Morrison. China and the Association of Southeast Asian Nations recently signed an agreement to manage such territorial disputes in the South China Sea. In any event, according to DoD, the Chinese navy is inferior to other regional navies in most technologies. . . . Therefore, those navies—alone or in concert—should be able to contain any Chinese adventurism, should negotiations fail. In the end, however, who owns the small island chains in the South China Sea, or the resources under the waters surrounding them, is not important to the security of the United States. The United States should not interfere in efforts to negotiate a peaceful solution to the problem.

Conclusion

Although many alarmist articles in the press have trumpeted improvements in the Chinese military, those enhancements are pockets of modernization in a largely antiquated force. China's military modernization is more rapid than before but is not a massive Soviet-style military buildup. As the Chinese economy grows and China becomes a great power, the United States should accept that it, like other great powers, will want more influence over its region. If kept within bounds, that increased sphere of influence should not threaten vital U.S. interests.

But the United States, especially under the Bush administration's new expansive national security strategy of primacy and preemption, sees any change in the status quo in East Asia as a threat to its expansive list of vital interests. If the United States unnecessarily maintains, or even continues to expand, its defense perimeter to surround and contain China, the rising lower and the status quo power—both armed with nuclear weapons—may come into needless conflict. The United States must take a less grandiose view of its vital interests, redraw its defense perimeter, abrogate its Cold War–era alliances (including the informal alliance with nonstrategic Taiwan), and reverse its military buildup. Currently, the United States is unnecessarily modernizing its armed forces faster than is China, which is starting from an extremely low level of military modernity. China, whose highest priority is economic development, is now reacting to the expansion of the U.S. defense perimeter and the U.S. military buildup by increasing its own defense budget more rapidly. Thus, U.S. policy may be engendering the threat it most fears.

POSTSCRIPT

Do China's Armaments and Intentions Pose a Long-Term Threat?

There can be no doubt that China's power continues to develop. It has the world's largest military and an array of nuclear weapons. It is one of the world's 10 largest economies in terms of overall gross domestic product (GDP) and exports. It is the third largest producer of steel, ranks second in the production and consumption of energy, and is fourth in the production of oil.

Yet China remains a poor and, in many ways, weak country. Dividing its impressive GDP, exports, and production figures by its more than 1 billion people leaves China one of the poorest countries in the world by most per capita measures. China's nuclear and advanced conventional weapons systems are less numerous and less sophisticated technologically than those of the United States. Thus, an important question in this debate is whether to consider China a great power rival of the United States or a country struggling to develop. This topic is further complicated by the change in China's leadership in 2003, with a new generation of younger, arguably more pragmatic leaders headed by Hu Jintao (b. 1943), who replaced Jiang Zemin (b. 1926), as president and general secretary of the Communist Party. One study that discusses the future of China is Ross Terrill, *The New Chinese Empire: Beijing's Political Dilemma and What It Means for the United States* (Basic Books, 2003).

There will likely be outstanding issues between the United States and China. During 2001, for example, the two countries experienced high tensions after a U.S. intelligence plane and a Chinese interceptor collided just off China's coast. The 24-person U.S. crew was detained by China after the crippled plane was forced to land on Hainan Island. Soon thereafter, China raised major objections to the sale of modern U.S. warships and other military equipment to Taiwan, which China considers to be one of its provinces. China's determination to reunify the country by reincorporating Taiwan and President George W. Bush's public declaration that the United States would defend the island if Beijing tried to militarily accomplish its goal set the stage for a potential war between the world's superpower and what many believe is an emerging superpower. An analysis of the differing world views of Washington and Beijing can be found in Bates Gill, *Contrasting Visions: United States, China, and World Order* (Brookings Institution Press, 2003).

Several sources provide more information on China's military. Two that can serve as background to the Fisher and Eland readings are the report of the U.S.-China Economic and Security Review Commission, located at http://www.uscc.gov, and the U.S. Defense Department's annual review of China's forces for 2002, which can be found at http://www.defenselink.mil/news/Jul2002/d20020712china.pdf.

ISSUE 6

Would It Be an Error to Establish a Palestinian State?

YES: P. J. Berlyn, from "Twelve Bad Arguments for a State of Palestine," A Time to Speak, http://www.israel.net/timetospeak/bad.htm (December 2002)

NO: Rosemary E. Shinko, from "Why a Palestinian State," An Original Essay Written for This Volume (2004)

ISSUE SUMMARY

YES: P. J. Berlyn, an author of studies on Israel, primarily its ancient history and culture, refutes 12 arguments supporting the creation of an independent state of Palestine, maintaining that such a state would not be wise, just, or desirable.

NO: Rosemary E. Shinko, who teaches in the Department of Political Science at the University of Connecticut, contends that a lasting peace between Israelis and Palestinians must be founded on a secure and sovereign homeland for both nations.

The history of Israel/Palestine dates to biblical times when there were both Hebrew and Arab kingdoms in the area. In later centuries, the area was conquered by many others; from 640 to 1917 it was almost continually controlled by Muslim rulers. In 1917 the British captured the area, Palestine, from Turkey.

Concurrently, a Zionist movement for a Jewish homeland arose. In 1917 the Balfour Declaration promised increased Jewish immigration to Palestine. The Jewish population in the region began to increase slowly, then it expanded dramatically because of refugees from the Holocaust. Soon after World War II, the Jewish population in Palestine stood at 650,000; the Arab population was 1,350,000. Zionists increasingly agitated for an independent Jewish state. When the British withdrew in 1947, war immediately broke out between Jewish forces and the region's Arabs. The Jews won, establishing Israel in 1948 and doubling their territory. Most Palestinian Arabs fled (or were driven) from Israel to refugee camps in Gaza and the West Bank (of the Jordan River), two areas that had been part of Palestine but were captured in the war by Egypt and Jordan, respectively. As a result of the 1967 Six Day War between Israel and Egypt, Jordan, and Syria, the Israelis again expanded their territory by capturing several areas, including the Sinai Peninsula, Gaza, the Golan Heights, and the West Bank. Also in this period the Palestine Liberation Organization (PLO) became the major representative of Palestinian Arabs. True peace was not possible because the PLO and the Arab states would not recognize Israel's legitimacy and because Israel refused to give up some of the captured territory.

Since then, however, continuing violence, including another war in 1973, has persuaded many war-exhausted Arabs and Israelis that there has to be mutual compromise to achieve peace. Perhaps the most serious remaining sore point between the Arabs and Israelis is the fate of the Palestinians, who live primarily in the West Bank and Gaza.

In 1991 Israelis and Palestinians met in Spain and held public talks for the first time. Israeli elections brought Prime Minister Yitzhak Rabin's liberal coalition to power in 1992. This coalition was more willing to compromise with the Arabs than had been its more conservative predecessor. Secret peace talks occurred between the Israelis and Palestinians in Norway and led to the Oslo Agreement in 1993. Palestinians gained limited control over Gaza and parts of the West Bank and established a quasi-government, the Palestinian authority led by Yasser Arafat.

The peace process was halted, perhaps even reversed, when in 1995 Prime Minister Rabin was assassinated by a Jewish fanatic opposed to Rabin's policy of trying to compromise with the Palestinians. Soon thereafter, the conservative coalition headed by Prime Minister Benjamin Netanyahu came to power. He dismissed any possibility of an independent Palestine, made tougher demands on the PLO, and moved to expand Jewish settlements in the West Bank. With some 200,000 Jews already in the West Bank and East Jerusalem, these actions compounded the difficult issue of the fate of those people in a potentially Palestinian-controlled area.

Pressure from a number of quarters, including the United States, kept the Israelis and Palestinians talking. Meeting in 1997 at the Wye River Plantation in Maryland under the watchful eye of President Bill Clinton, Israel agreed to give the Palestinians control over additional areas of the West Bank, and the Palestinians agreed to work to protect Israel from Arab terrorist attacks and to remove language in the PLO charter that called for the destruction of the Jewish state. The immediate impact of the Wye River Agreement was negligible.

A liberal government under Prime Minister Ehud Barak failed to move the peace process forward and was replaced by a conservative government led by Ariel Sharon, who favors a very stern approach to the Palestinians. Under Sharon, the Israeli military has responded to terror attacks by conducting extensive military operations in Gaza and the West Bank. Whatever the justice or wisdom of that approach by Israel, it has not stemmed the suicide bombings and other attacks on Israelis by the Palestinians. In the following selections, P. J. Berlyn argues that creating an independent Palestine would be a grave error, while Rosemary E. Shinko maintains that there will be no end of the violence until Palestinians have their own independent homeland.

P. J. Berlyn

 **YES**

Twelve Bad Arguments for a State of Palestine

In 1991, the United States, during the administration of President George H. W. Bush, sponsored the Madrid Conference at which Israel is invited to meet with Jordan and other Arab States to negotiate peace. In a letter to the Government of Israel, the Government of the United States pledges:

> "In accordance with the United States, traditional policy, we do not support the creation of an independent Palestinian state. [. . . .] Moreover, it is not the United States' aim to bring the PLO [Palestine Liberation Organization] into the process or to make Israel enter a dialogue or negotiations with the PLO."

This pledge was indeed consonant with history, strategy, justice and common sense. It was not, however, to be honored. In 2001, President George W. Bush announced that it was a "vision of long standing" of U.S. policy to create a Palestinian State west of the Jordan River. Such a state would be under the rule of the PLO and must be recognized by Israel. The United States proceeded swiftly to have this newly-discovered long-standing vision ensconced in a resolution of the United Nations Security Council.

It is now conventional to suppose that the invention of a PLO State in the Land of Israel is wise, just and desirable, even inevitable. Among the platitudes strung into a mantra:

1] It will rectify an historic injustice to the Arabs.

On the contrary: Of all that Arabs have demanded for themselves since the end of World War I, they have been given 99.5 percent.

In 1921 the League of Nations defined Mandate Palestine, as The Jewish National Home, to be "open to close Jewish Settlement." In 1922, the British Mandatory Government subtracted the entire region east of the Jordan River,

From P. J. Berlyn, "Twelve Bad Arguments for a State of Palestine," *A Time to Speak,* http://www.israel.net/timetospeak/bad.htm (December 2002). Copyright © 2002 by P. J. Berlyn. Reprinted by permission.

more than 75 percent of the Jewish National Home, to create the Arab Kingdom of [Trans]-Jordan.

Great Britain then progressively restricted or banned Jewish immigration and settlement even west of the Jordan River, rigidly blockading the Land of their fathers to Jews trying to escape the gas chambers of Europe. At the same time, the British authorities permitted massive immigration of the Arabs into Mandatory Palestine, whose families now number among those who claim it as their ancient ancestral homeland.

In 1947, the United Nations attempted to whittle away the remnant of the Jewish National Home with a second partition to create a second Arab state in Palestine, this one west of the Jordan River where most of the Arab population were but recently arrived, with no roots and no history in the land. Had the Arabs accepted that offer, they would have had 83 percent of the Land of Israel-Jewish National Home. Instead, they went to war to get 100 percent of it.

The real injustice is depriving Israel of its historic homeland, in order to invent a 23rd Arab state where none ever existed.

2] It will end Israel's occupation of Palestinian territory.

On the contrary: There is no such thing as "Palestinian territory" and there is no "occupation" of what never belonged to any Arab nation. Furthermore, of the Arab residents of Judea, Samaria [biblical terms used for the West Bank] and Gaza, 98 percent now live under the rule of the PLO.

3] Israel must comply with United Nations resolutions.

On the contrary: The Arab attack on Israel in June 1967 left Israel in possession of Judea and Samaria, that Jordan had seized in the attack of 1947, Gaza, that Egypt had seized in the war of 1947, the Sinai and the Golan Heights. The UN Security Council, that had done nothing to prevent or even deplore the Arab attack, took it upon itself to pass a resolution to guide a future settlement. That was Resolution 242, that calls on Israel, in the context of a peace settlement, to withdraw from unspecified "territories" to "agreed and secure borders."

This was very specifically *not* a demand for a return to the borders of June 4, 1967 [prior to Israel's victory in the 1967 war]—which were themselves merely the ceasefire lines of the War of Independence, a war launched by the Arabs to destroy Israel in 1948.

The author of Resolution 242 was Lord Caradon, representative of the United Kingdom. He explained to the British Parliament: "It would be wrong to demand that Israel return to its positions of June 4, 1967, because those positions were undesirable and artificial."

The United States was a co-sponsor of Resolution 242, and its representative stated: "The notable omissions—which were not accidental—in regard to withdrawal are the words 'the' or 'all' and 'the June 5, 1967 lines'."

It is now widely and repeatedly alleged that this resolution demands that Israel withdraw from *all* the territories. *That is a lie.* It is alleged that this withdrawal is unconditional. *That is a lie.* It is even alleged that it calls for an independent Palestine-Arab state in the territories. *That is a lie.*

In accordance with the Israel-Egypt treaty of 1978, Israel withdrew from the entire Sinai peninsula, 91 percent of the "territories." That may well be considered more than sufficient to satisfy the terms of Resolution 242.

In more recent resolutions, the sponsors of No. 242 have reneged on their own positions. The United States has reversed all previous policy statements and promises to Israel and called for a PLO state. Great Britain has turned its own resolution upside down by demanding full withdrawal as well as a PLO state.

These fickle flip-flops show that the authors and sponsors of the resolution cannot be trusted to stand by their own creation. Israel thus betrayed should not be expected to bow to every new whim of the moment. Add to that betrayal the role of the United Nations as the world epicenter for hatred of Israel and Jews, and it is absurd to argue that Israel has any obligation to submit to its demands.

4] It will bring peace and stability to the Middle East.

On the contrary: It will establish a Middle Eastern national base for terror, that will spread incitement, bloodshed, and mortal danger not only to Israel but also to Arab regimes in the neighborhood.

The citizens supposed to build this peaceful and stable state will be the ones that the PLO regime has programmed to hatred and contempt, to yearn to earn martyrdom by murdering Jews. They will be the hysterics who run through the streets, some in costumes to rival the Ku Klux Klan, brandishing weapons and shrieking curses and threats.

5] It will satisfy the demands of the Palestinian Arabs, who will give up terrorism and war and settle down to building a society.

On the contrary: The PLO Charter of 1964 defines its sole purpose as the destruction of Israel. (That was three years *before* 1967, when there were no "occupied territories" to liberate.) Despite flimflam to the contrary, that Charter still stands unamended, and so does the goal.

The PLO openly and repeatedly proclaims that it will never settle for less than every inch between the Jordan River and the Mediterranean Sea—including all of what is now the State of Israel. That is how "Palestine" is shown on its maps and emblems. If it condescends to accept a smaller state it will be only as an interim measure, to facilitate the future destruction of Israel.

6] A State of Palestine will honor a pledge to respect Israel's right to exist.

On the contrary: The PLO has already made at least six formal agreements with Israel and has pledged itself to at least seven ceasefires. Not a single term or clause of any of them has been kept for a single day. To expect any other behavior in the future defies basic common sense.

Repeated statements by officials of the Fatah, Hamas, and other member bodies of the PLO declare over and over again that their goal is the end of Israel, the expulsion of the Jews, and an Arab Palestinian State from the Jordan River to the Mediterranean.

7] A State of Palestine will be demilitarized and thus no danger to Israel.

On the contrary: There were limitations in the Oslo Accords—a police force of no more than 8,000 and no heavy weapons. Today the PLO has a military force of at least 50,000 with heavy weapons. More are smuggled in constantly, from Egypt, Syria and elsewhere.

If Israel is deprived of the strategic defense line of the Jordan valley and the highlands to the west, then its width will be reduced to *nine miles.* The PLO state can become militarized in mere hours (as was the "demilitarized" Sinai in 1967), and forces from Iraq and Syria can sweep in without hindrance.

8] It will secure the human rights of the Palestinian Arabs.

On the contrary: The PLO regime in the areas it controls in Judea, Samaria and Gaza has nothing at all to its credit in human rights, and everything to its discredit. (This should have been expected from the example of its rule over southern Lebanon in the 1980s.) Overseers of human rights who used to keep captious watch on these areas when they were under Israel's administration have been on an extended vacation since September 1993.

9] It will solve the Arab refugee problem.

On the contrary: The PLO insists that it will not absorb these exploited people, but will demand their "return" to Israel—meaning the destruction of Israel. The residents of the UN refugee camps also insist they will settle for nothing less than "return", and the UN that runs the camps makes sure the residents do not budge from this determination.

10] It will encourage civic and economic development, raise the standard of living and bring contentment to the people.

On the contrary: In areas administered by the PLO, the standard of living drops and hardship increases. Economic development is strangled by graft and corruption, and revenues are squandered. The United States and the European Union have given large donations for development, but the bulk of the money melts away or ends up in private foreign bank accounts.

11] It will win the respect of world opinion for Israel.

On the contrary: "World opinion" is a jigsaw of many mismatched pieces. It includes a professional contempt for Israel, conspicuous especially in journalism and academia. It includes taste-makers of a Europe that have been trying to crush the Jews for two millennia and have not yet given up the habit. It has ancient roots in religious convictions, and in history, and new shoots of resentment and envy.

There are even a few examples here and there of good sense and fair play, but those are not found in the portion of "world opinion" that makes relentless demands on Israel—sometimes masquerading as "friendly advice in your own best interest." Israel is not obliged to satisfy those demands by making itself shrunken, demoralized, discredited, and vulnerable, nor would it be any better liked if it did.

That is not to say "world opinion" never approves of anything Israel does. It did welcome with delight the self-demeaning and self-destructive Oslo Accords [agreement between Israel and the PLO signed in 1993].

12] If a State of Palestine commits aggression against Israel, then Israel can fight its military forces and win back what it gave away.

On the contrary: The supporters of the Oslo Accords in 1993 also said "If they [the PLO] do not keep their commitment to peace, we will just take the land back." But those who said it took their words back—or ignored that they ever said them.

Now these areas are used as bases for terrorism against Israel. When the IDF [Israel Defense Force] goes in even briefly, to close down terror bases and weapons factories and dumps, the world—including even the United States—howls for Israel to "get out of Palestinian territory immediately." If those areas were to become territory of an Arab "State of Palestine" any defensive actions by Israel would be branded aggression against a sovereign state. It would be condemned and threatened even more harshly than when it moved against terrorism when these areas were held by Jordan and Egypt—without sovereignty.

If these areas of the Land of Israel became a PLO state, Israel would lose even minimal control. It could not restrict import of heavy weapons, destroy weapons factories and depots, intercept terrorist activities or arrest terrorists. It could not even prevent the entry of foreign troops from other Muslim countries.

Israel, drastically restricted geographically, will be exposed and vulnerable. When the PLO and its allies launch all-out war, the cost to Israel will be horrendous.

If Israel wins a battle for survival, it still will not be able to regain what it gave away. Even if a PLO State is defeated in battle, it will not cease to exist. In all of the Arab wars against Israel, outside powers have intervened to save them from total defeat. A PLO state can thus survive defeats and repeat its aggressions.

Giving up Judea-Samaria would also mean that Jews would be cut off from the cities and sites that are the heart of their historic homeland. Israel could not prevent Arabs from destroying ancient Jewish sites and relics. There would be no chance to make new discoveries that shed new light on the history of Israel and its people.

Some who are made aware of all these circumstances nevertheless say: "It is useless to oppose a State of Palestine—it is inevitable." Such passive submission, such moral indolence, is tacit consent to an act inimical to the Jewish people and the Land of Israel. It is a limp surrender of both the past and the future.

For 2000 years, the Jewish people did not despair of restoration to their Land. When the restoration has at last come, those who toss it away betray both their ancestors and their descendants.

Rosemary E. Shinko   ← **NO**

Why a Palestinian State

On July 8, 1937 the Palestine Royal Commission (Peel Commission) offered its recommendations to the British government regarding the disposition of the Palestinian question. The commission expressed serious reservations about the possibility of reconciliation between Arabs and Jews and thus they concluded, "only the 'surgical operation of partition' offers a chance of ultimate peace" (www.guardiancentury.co.uk). They proposed the establishment of two separate states—a sovereign Arab State and a sovereign Jewish State. United States President, Bill Clinton, reiterated these same sentiments in a speech he delivered on January 7, 2001. "I think there can be no genuine resolution to the conflict without a sovereign, viable, Palestinian state that accommodates Israel's security requirements and the demographic realities." Any settlement must ultimately be "based on sovereign homelands, security, peace and dignity for both Israelis and Palestinians."

Why is it then that P. J. Berlyn argues that the "invention" of a 23rd Arab state would be unwise, unjust and undesirable? Her arguments revolve around the following five main assertions: there is no such thing as "Palestinian territory" nor do the Arabs constitute a "Palestinian people," if such a state were "invented" the Arabs would be unable to fulfill the rights and duties associated with statehood, it would not be in the self-interest of the State of Israel to allow such a state to be created in its midst, and finally it would betray the sacrifices of the past and the promises of the future of the Jewish people.

What is a "state" and why does the possibility of the creation of a Palestinian State, in particular, provoke such a strong, emotional response from Ms. Berlyn? What does the term "state" signify? According to Hegel, "only those peoples that form states can come to our notice" because it is the state that provides the foundation for "national life, art, law, morality, religion, [and] science" (*Reason In History*). The political identity of most peoples is inextricably bound up with the notion of statehood (Rourke, 2001: 189). According to an international relations text, *International Politics on the World Stage,* written by John Rourke and Mark Boyer, "States are territorially defined political units that exercise ultimate internal authority and that recognize no legitimate external authority over them." The political implications of legitimacy that would flow from the establishment and recognition of a Palestinian state are extremely significant in this particular instance. As scholar Malcom Shaw notes in his 1999 Cambridge University Press book *International Law,* a state is recognized as having a "legal personality," which includes the capacity to possess and exercise certain rights and to perform specific duties. These rights and duties encompass the attributes of independence, legal equality, and peaceful coexistence. Thus a Palestinian State would claim the right to exercise jurisdiction over its population and territory, as well as the right to self-defense. Such a state would also have a concomitant duty not to intervene in the internal affairs of another state and a duty to respect the territorial integrity and sovereignty of other states.

The establishment and recognition of a Palestinian State would confirm the political legitimacy and legal equality of the Palestinian people. Historically they have been denied recognition as a people and the legitimacy of their claims to the territory of Palestine have been dismissed. Ms. Berlyn's arguments are designed to foster the sense of Palestinian illegitimacy with her bold assertion that the Arabs have "no roots and no history" in Palestine, and that there is "no such thing as a Palestinian territory." To round out her argument, she employs the term "invent" when referring to the establishment of a Palestinian State in order to further deny and delegitimate Palestinian claims to their homeland. All states are man made creations, all states are reflective of the political, legal, social, economic and historical conditions which led to their rise. All states, even the State of Israel, are a political creation of men and women. Ms. Berlyn's arguments, which attempt to dismiss a Palestinian presence and history, are an extension of the earlier Zionist attempts to portray Palestine as a "land without people for a people without land." A perception was fostered that the territory of Palestine was "empty" and that its only inhabitants were uncivilized, backward nomads.

Demographic realities, however, proved otherwise. "There were always real, live Palestinians there; there were census figures, land-holding records, newspaper and radio accounts, eyewitness reports and the sheer physical traces of Arab life in Palestine before and after 1948," according to Edward W. Said, and Christopher Hitchens in their 1988 study, *Blaming the Victims, Spurious Scholarship and the Palestinian Question.* In 1947 when the United Nations Committee on Palestine (UNSCOP) made its recommendation that Palestine be portioned into two separate states, there were 1.2 million Arabs as compared to 570,000 Jews living in the territory. Clearly the Arabs formed the majority of the population in Palestine. On what basis then can it be maintained that the Palestinians had no history, had no roots, and had no presence in Palestine? "The fact is that [when] the people of Israel . . . came home, the land was not all vacant," President Clinton commented in 2001. Statehood confirms presence, establishes legitimacy, confers recognition, and provides a focal point for a peoples' identity.

David Shipler in an October 15, 2000 article in the *New York Times* commented astutely that "Recognizing the authenticity of the other in that land comes hard in the midst of the conflict. Yet the conflict cannot end without that recognition." Ultimately legitimacy and recognition are the keys to the end of conflict and the establishment of peace in the Middle East. Peace cannot occur without the recognition of the Palestinian people's right to self-determination and without their consent to the government that exercises authority over them. "The six wars with the Arabs created a situation in which 3 million Jews came to control territories that contained nearly 2 million Arabs," scholar John G. Stoessinger observed in his 2001 book, *Why Nations Go to War.* The United Nations General Assembly also concluded that without "full respect for and the realization of these inalienable rights of the Palestinian people," namely the right to self-determination and the right to national independence and sovereignty, there would be no resolution of the question of Palestine (http://domino.un.org).

Statehood implies the capacity to maintain certain rights and the performance of specific duties. Ms. Berlyn maintains that even if the Palestinians were granted their own state they would not live up to the duties of a state because a Palestinian State would be committed to the destruction of Israel and merely serve as a base for further acts of terror. In her estimation a Palestinian State would not respect the territorial integrity and sovereignty of the State of Israel. Furthermore, she even questions the ability of such a state . . . to fulfill the requirements of statehood, including civic and economic development and the promotion of human rights. Fundamentally such a negative assessment rests on conjecture and an underlying sense of distrust born of conflicting claims of legitimacy to the same parcel of land. No one can claim to profess the future, and not even Ms. Berlyn can with any certainty predict the actions of a State of Palestine. One thing does however appear to be foreseeable, and that is the continued agitation of the Palestinians for recognition, self-determination and legitimacy. As Professor Stoessinger put it in *Why Nations Go to War,* "The shock inflicted on the Arab consciousness by the establishment of Israel and the resulting homelessness of a million native Palestinians grew more, rather than less, acute as Arab nationalism gathered momentum." The Arabs perceive Israel as the ever-expanding and ever-growing threat to their survival, thus a state is the only way to insure their continued existence as a people. Declaring that a Palestinian State would be unable to fulfill the requirements of statehood is merely a thinly veiled ethnocentric critique, which smacks of patronization and cultural superiority. What precisely do the Palestinians lack that would deem them ill-suited to exercise self-rule and incapable of founding a government that rests on consent which would secure their rights to life, liberty and property? The Lockean assertion that the only legitimate form of government is that which rests on consent is as true for the Jews as the Arabs of Palestine.

Israel is exposed and vulnerable as a result of the Palestinians' unrelenting quest for legitimacy and recognition. Israel's self-interest may ultimately rest with the establishment of a separate Palestinian State in order to diffuse the longstanding animosities and hatreds that have arisen between the two peoples. We have seen where the denial of legitimacy has taken us, and it has not nurtured the seeds of peace. Peace can only be established in the wake of the recognition of legitimacy and equality between the two peoples. In order to secure the national character and the cultural identity of the Israelis, the national character and the cultural identity of the Palestinian Arabs must likewise be secured. This can only occur with the establishment of a separate, sovereign Palestinian State.

POSTSCRIPT

Would It Be an Error to Establish a Palestinian State?

The Middle East's torment is one of the most intractable problems facing the world. In addition to the ancient territorial claims of Jews and Palestinian Arabs, complexities include long-standing rivalries among various religious and ethnic groups and countries in the region. To learn more about the history of the current conflict, consult Ian J. Bickerton and Carla L. Klausner, *A Concise History of the Arab-Israeli Conflict,* 4th ed. (Prentice Hall, 2002) and Walter Laqueur and Barry Rubin, eds., *The Israel-Arab Reader: A Documentary History of the Middle East Conflict* (Viking Penguin, 2001).

Complicating matters for Israel is the fact that the country is divided between relatively secular Jews, who tend to be moderate in their attitudes toward the Palestinians, and Orthodox Jews, who regard the areas in dispute as land given by God to the Jewish nation and who regard giving up the West Bank and, especially, any part of Jerusalem as sacrilege. Furthermore, there are some 200,000 Israelis living in the West Bank, and removing them would be traumatic for Israel. The issue is also a matter of grave security concern. The Jews have suffered mightily throughout history; repeated Arab terrorism represents the latest of their travails. It is arguable that the Jews can be secure only in their own country and that the West Bank (which cuts Israel almost in two) is crucial to Israeli security. If an independent Palestine centered in the West Bank is created, Israel will face a defense nightmare, especially if new hostilities with the Palestinians occur.

Thus, for the Israelis the "land for peace" choice is a difficult one. Some Israelis are unwilling to cede any of what they consider the land of ancient Israel. Other Israelis would be willing to swap land for peace, but they doubt that the Palestinians would be assuaged. Still other Israelis think that the risk is worth the potential prize: peace.

Although Israeli elections turn on more than the "security question," it is a major factor and has played an important role in determining who governs Israel and what the policy on the Palestinian question is. On this matter, Israelis have swung back and forth. For more on Israel's political system, read Alan Dowty, *The Jewish State: A Century Later* (University of California Press, 2001).

For another view that is sympathetic to a Palestinian state, see Edward W. Said and Christopher Hitchens, eds., *Blaming the Victims: Spurious Scholarship and the Palestinian Question* (W. W. Norton, 2001). The opposite perspective can be found in the online article "The Stillborn Palestinian State," by Boris Shusteff, http://www.freeman.org/m_online/jun02/shusteff1.htm (June 2002).

ISSUE 7

Was the War With Iraq Unjustified?

YES: John Mueller, from "Should We Invade Iraq? A Reason On-line Debate," *Reason* (January 2003)

NO: Brink Lindsey, from "Should We Invade Iraq? A Reason Online Debate," *Reason* (January 2003)

ISSUE SUMMARY

YES: Professor of political science John Mueller characterizes U.S. policy toward Iraq as an overreaction and an unhealthy reflection of the U.S. sense that the United States is an indispensable country.

NO: Brink Lindsey, a senior fellow at the Cato Institute, asserts that Iraq's potential nuclear capability combined with Saddam Hussein's malevolence provided sufficient justification for a U.S.-led invasion.

On August 2, 1990, Iraq invaded Kuwait and quickly conquered the tiny emirate. There was an immediate and strong reaction by many countries, particularly the United States, to the aggression by Iraqi leader Saddam Hussein. Within days, President George Bush announced that the United States would send approximately 250,000 troops to Saudi Arabia to protect that country from a possible further move by Iraq. Working with its allies, especially the British, and through the United Nations, the United States also undertook a diplomatic effort to build a coalition to apply economic, then military, pressure on Iraq to withdraw from Kuwait. The UN passed a series of resolutions denouncing Iraq's actions and taking economic and other measures against Iraq. These moves culminated in November 1990 with the passage of Resolution 678 by the UN Security Council. The resolution demanded that Iraq withdraw from Kuwait by January 15, 1991, and authorized all UN members to use "all necessary means" after that date to compel Iraqi adherence to the UN's demands.

Iraq refused to comply, and on January 17, 1991, a coalition of more than a dozen countries, with a vast preponderance of American forces, moved to expel Iraq from Kuwait. After a six-week air campaign, over a half million coalition troops moved forward and overwhelmed the already devastated and dispirited Iraqi forces in a matter of days.

The terms of peace for Iraq were embodied in Resolution 687, which was passed by the UN Security Council in April 1991. The most important stipulations of the resolution were that Iraq was required to give up all of its remaining weapons of mass destruction (WMDs), both chemical and biological, and was barred from producing any more such weapons, including nuclear. UN arms inspectors were to be granted unhindered access anywhere in the country to ensure that Iraq was disarming, and a trade embargo was to remain in place until Iraq was found to be in full compliance with the weapons of mass destruction and inspection clauses of Resolution 687.

A cat and mouse game ensued. Iraq claimed to be complying with Resolution 687 but refused to allow UN inspectors unimpeded access to suspected weapons sites and plants. Economic sanctions continued, and there were periodic attacks by U.S. warplanes and cruise missiles.

Events began to move toward a final crisis in August 1998 when Iraq expelled the UN arms inspectors. That was followed in December by a four-day U.S. air war campaign against Iraq based on what President Bill Clinton called a "stark, sobering, and profoundly disturbing report" by the chief UN arms inspector about Iraq's refusal to cooperate.

Although the United States continued to argue for maintaining pressure on Iraq to resume arms inspections, the U.S. attitude hardened considerably in the aftermath of the September 11, 2001, terrorist attacks. In November President George W. Bush demanded that Iraq readmit the weapons inspectors. Iraq countered that it would not do so until the UN abolished all economic sanctions on Iraq and ended the "no-fly zones" in northern and southern Iraq.

In September 2002 Bush addressed the UN, denounced Iraq as constituting a "grave and gathering danger," and declared that if Iraq did not cooperate with Security Council resolutions related to the arms inspections, "action will be unavoidable." Just days later, Iraq indicated it would accept new inspections "without conditions," and Tariq Aziz argued that "all the reasons for an attack have been eliminated." A White House spokesperson described the move as "a tactical step . . . that will fail."

Reflecting that skepticism, in October the U.S. Congress passed a resolution authorizing the use of U.S. force against Iraq if necessary. In November, under U.S. sponsorship, the Security Council adopted Resolution 1441, which stated that Iraq must allow UN inspectors the unconditional right to search anywhere in Iraq for WMDs and that it must make an "accurate, full and complete" declaration of its WMD capabilities and program.

According to the reports filed by Blix, the progress of the inspectors in Iraq was mixed. In essence, his reports said that Iraq was cooperating in part but not fully and that questions remained. This led the United States to argue in December 2002 that Iraq was in "material breach" of Resolution 1441 and others.

It is at that point that the debate between John Mueller and Brink Lindsey that is excerpted in the following selections took place. Mueller argues that the war is unneeded and may well be costly. Lindsey contends that the war is justified and worth the likely cost.

John Mueller  **YES**

Should We Invade Iraq?

What's the Rush?

The Devil du Jour Is a Feeble Tyrant

In preparing for a war against Iraq, military planners seem to anticipate a walkover. The Iraqi military performed badly in the Gulf War of 1991: Saddam Hussein promised the mother of all battles, but his troops delivered instead the mother of all bugouts. And the planners note that Iraq is even weaker now.

Moreover, the regime appears to enjoy very little support. Saddam Hussein lives in such fear of his own military forces that he keeps them out of Baghdad. It is generally anticipated that most of the military will not fight for him—indeed, that there may be substantial defections to the invaders even among the comparatively coddled Republican Guard.

In addition, the regime really controls only a shard of the country. The Kurds have established a semi-independent entity in the north, and the hostility toward Saddam's rule is so great in the Shiite south that government officials often consider the region hostile territory.

Advocates of a war with Iraq insist such a venture is necessary because Iraq's feeble, wretched tyranny is somehow a dire and gathering threat to the entire area and even to the United States. Saddam's inept, ill-led, exhausted, and thoroughly demoralized military force, it is repeatedly argued, will inevitably be used by its leader for blackmail and regional dominance, particularly if it acquires an atomic bomb or two.

Exactly how this might come about is not spelled out. The notion that Israel, with a substantial nuclear arsenal and a superb and highly effective military force, could be intimidated out of existence by the actions or fulminations of this pathetic dictator can hardly be taken seriously. And the process by which Saddam could come to dominate the oil-producing states in the Middle East is equally mysterious and fanciful. Apparently, he would rattle a rocket or two, and everyone would dutifully jack up the oil price to $90 a barrel.

Saddam's capacity for making daffy decisions is, it is true, considerable. But he seems mostly concerned with self-preservation—indeed, that is about the only thing he is good at. And he is likely to realize that any aggressive mili-

From John Mueller, "Should We Invade Iraq? A Reason Online Debate," *Reason*, vol. 34, no. 8 (January 2003). Copyright © 2002 by The Reason Foundation, 3415 S. Sepulveda Blvd., Suite 400, Los Angeles, CA 90034. Reprinted by permission.

tary act in the region is almost certain to provoke a concerted, truly multilateral counterstrike that would topple his regime and remove him from existence. Even if he ordered some sort of patently suicidal adventure, his military might very well disobey, or simply neglect to carry out, the command. His initial orders in the Gulf War, after all, were to stand and fight the Americans to the last man. When push came to shove, his forces treated that absurd order with the contempt it so richly deserved.

During the last half-century American policy makers have become hysterical over a number of Third World dictators, among them Egypt's Nasser, Indonesia's Sukarno, Cuba's Castro, Libya's Qaddafi, and Iran's Khomeini. In all cases, the threat these devils du jour posed to American interests proved to be highly exaggerated. Nasser and Sukarno are footnotes, Castro a joke, and Qaddafi a mellowed irrelevance, while Khomeini's Iran has become just about the only place in the Middle East where Americans are treated with popular admiration and respect.

Significantly, Iran is also just about the only place in the area where the United States has been unable to meddle during the last 20 years. And it is possible there is a lesson here.

With characteristic self-infatuation, American leaders like to declare their country to be "the world's only remaining superpower" or "the indispensable nation." But this self-proclaimed status doesn't mean that it is obligatory or possible or wise for the United States to seek to run the world.

Or even the Middle East. American interests there are limited. There is a romantic and sentimental attachment to Israel, of course, but that country seems fully capable of taking care of itself. In time, perhaps, and probably after a change of leadership on both sides, mediation efforts between Israel and the Palestinians can become productive again. But for now at least the conflict is so deep that there is little any outsider (even an "indispensable" one) can do about it.

Quite a bit of oil comes from the Middle East, of course, but discussions of the American interest on that score tend to ignore simple economics. The area already is dominated by an entity, OPEC [Organization of Petroleum Exporting Countries], which would dearly love to hike the price for the commodity. It is constrained from doing so not by warm and cuddly feelings toward its customers but by the grim economic realization that such a policy would reduce demand, intensify the search for new petroleum sources, and bring about a worldwide inflation that would raise the prices of imported commodities even more than any gains obtained by an increase in the oil price. Whatever happens in the region, this fundamental market reality is likely to mellow and correct incidental distortions.

In the meantime, monarchs in a number of countries may gradually be coming to the realization that they are out of date, rather in the way Latin American militarists more or less voluntarily decided during the last quarter century to relinquish control to democratic forces. If this does happen, however, the process will be impelled, as in Latin America, primarily by domestic forces, not outside ones.

A humanitarian argument could be made for a war against Iraq—to liberate its people from a vicious tyranny and from the debilitating and destructive effects of the sanctions which the United States apparently is incapable of relaxing while Saddam Hussein remains in power. Such a war would have to be kept inexpensive in casualties, and the United States would have to be willing to hang on for quite some time to help rebuild the nation, something experience suggests is unlikely.

But calls for war do not stress this argument. Instead, they raise alarms about vague, imagined international threats that, however improbable, could conceivably emanate from a miserable and pathetic regime. In due course, nature (there have been persistent rumors about cancer) or some other force will remove our devil du jour. The situation calls for patient watchfulness, not hysteria.

Suicide Watch

Betting on Saddam's Recklessness

It may be useful to parse the argument for a preventive war against Iraq as developed by Brink Lindsey into two considerations: the military threat Iraq presents or is likely to present, and the regime's connection to international terrorism.

The notion that Iraq presents an international military threat seems to be based on three propositions:

1. Iraq will have a small supply of atomic weapons in a few years.
2. Once it gets these arms, Saddam Hussein won't be able to stop himself from engaging in extremely provocative acts such as ordering the military invasion of a neighbor or lobbing missiles at nuclear-armed Israel—acts that are likely to trigger a concerted multilateral military attack upon him and his regime.
3. If Saddam issues such a patently suicidal order, his military—which he himself distrusts—will dutifully carry it out, presumably with more efficiency, effectiveness, and élan than it demonstrated in the Persian Gulf War.

I will leave it to those more expert in the field to assess the first proposition. At worst we have a window of a few years before the regime is able to acquire atomic arms. Some experts seem to think it could be much longer, while others question whether Saddam's regime will ever be able to gather or make the required fissile material. Effective weapons inspections, of course, would reduce this concern.

The second proposition rests on an enormous respect for what I have called Saddam's "daffiness" in decision making. I share at least part of this respect. Saddam does sometimes act on caprice, and he often appears to be out of touch—messengers bringing him bad news rarely, it seems, get the opportunity to do so twice. At the same time, however, he has shown himself capable of pragmatism. When his invasion of Iran went awry, he called for retreat to the prewar status quo; it was the Iranian regime that kept the war going. After he invaded Kuwait in 1990, he quickly moved to settle residual issues left over from the Iran-Iraq War so that he had only one enemy to deal with.

Above all, Saddam seems to be entirely nonsuicidal and is primarily devoted to preserving his regime and his own personal existence. His brutal killing (and gassing) of Kurds was carried out because they were in open rebellion against him and in effective or actual complicity with invading Iranians during the Iran-Iraq War. Much of his obstruction of arms inspectors seems to arise from his fear that agents among them will be used fatally to triangulate his whereabouts—a suspicion that press reports suggest was not exaggerated. If Saddam does acquire nuclear arms, accordingly, it seems most likely that he will use them as all other leaders possessing such weapons have since 1945: to deter an invasion.

The third proposition is rarely considered in discussions of the war, but it is important. One can't simultaneously maintain that Iraq's military forces will readily defect and can easily be walked over—a common assumption among our war makers—and also that this same pathetic military presents a serious international threat.

The argument connecting Iraq to terrorism is mostly based on arm waving. As Lindsey notes, international terrorists are based all over the world—in fact, just about everywhere except Iraq. Their efforts are hardly likely to be deflated if Iraq's regime is defeated. Indeed, it seems likely that an attack will supply them with new recruits, inspire them to more effort, and provide them with inviting new targets in the foreign military and civilian forces that occupy a defeated, chaotic Iraq. Lindsey suggests that a war is required to make it "clear that the United States means business in dealing with terrorism." I would have thought this was already extremely clear.

Terrorism, like crime, has always existed and always will. It cannot be "crushed," but its incidence and impact can be reduced, and some of its perpetrators can be put out of business. But this is likely to come about through patient, diligent, and persistent international police work rather than costly wars based on tenuous reasoning.

Deterring the Egomaniac Dictator

War Is Not Necessary to Keep a Street Thug in Check

Brink Lindsey wants to argue that Saddam Hussein is reckless, but even he concedes that "no country, not even one as rash as Iraq, would dare to use weapons of mass destruction against the United States because of the threat of overwhelming retaliation." That is, it is entirely possible to deter Iraq. This deterrent would surely hold for an attack on Israel, which has an enormous retaliatory capacity and an even greater incentive to respond than the U.S. I would suggest that it holds as well for just about any substantial military provocation that Saddam might consider.

It is true that much of the world managed to contain its outrage when Iraq invaded Iran in 1980. But that was because the attack was directed against Khomeini's seemingly expansionary theocracy, which was seen to be a bigger

threat at the time. It is simply not true that "the world" was "all too happy to acquiesce in Kuwait's disappearance" when Iraq invaded it in 1990. There was almost universal condemnation of the attack, even from Iraq's erstwhile friend
and ally, the Soviet Union, and the debate was over tactics: whether to use war
immediately to push back the aggression or to wait to see if sanctions could do
the trick.

Reaction to a third Saddam adventure would surely follow the Kuwait pattern, except that the troops would now go all the way to Baghdad. Moreover, as
I've suggested, Saddam's army, which even he finds unreliable, would be unlikely to carry out patently suicidal orders even if they were issued—as it showed
in the Gulf War of 1991.

Lindsey's appreciation for Saddam's egomania is fully justified. It's just
that egomania is standard equipment for your average Third World tyrant. Indonesia's Sukarno haughtily withdrew from the United Nations and set up his
own competing operation in Djakarta (only China joined); Egypt's Nasser (Saddam's sometime inspiration), who planned to unite and dominate the Arab
world, died quietly in bed after being humiliated by Israeli arms; Khomeini's
global revolution has essentially been voted out even in its Iranian homeland;
and Cuba's Castro probably *still* hopes to become the new Simón Bolívar
[highly esteemed general] of Latin America. Self-important street thugs like
Saddam Hussein love to flail and fume in the company of sycophants, but that
doesn't make them any less pathetic.

We are left with the warning that Saddam will give weapons of mass destruction to shadowy terrorists to deliver for him. Lindsey is unusual in suggesting that Saddam might do this with nuclear weapons (which, of course, he
doesn't have and perhaps never will have). Most observers assume he would
selfishly keep them himself to help deter an attack on Iraq.

The case is more plausible for chemical or biological weapons—which,
however, have proven to be so difficult to deploy effectively that it is questionable whether they should be considered weapons of "mass destruction" at all, as
Gregg Easterbrook pointed out in the October 7 [2002] issue of *The New Republic*. But terrorists may be after these weapons anyway, and the question is
whether it is worth a war to eliminate one of many potential sources. Moreover,
as Daniel Benjamin noted in the October 31 [2002] *Washington Post*, the best
CIA assessment is that Saddam and Al Qaeda are most likely to bed together if
his regime is imminently threatened by the preventive war (it would be in no
reasonable sense an act of pre-emption) that Lindsey so ardently advocates.

NO ↵

Brink Lindsey

Should We Invade Iraq?

No More 9/11s

The Case for Invading Iraq

John Mueller tries to make light of Iraq. *Feeble, inept, pathetic,* and *daffy* are some of the adjectives he uses to describe the blood-soaked, predatory regime now in power there. The implication is that only the paranoid could find in Saddam Hussein's buffoonery any cause for serious concern.

Well, I beg to differ. Iraq is no joke: The crimes that the Ba'athist regime there has committed and may intend to commit in the future are deadly serious business. Under the reign of Saddam Hussein, Iraq has invaded two of its neighbors, lobbed missiles at two other countries in the region, systematically defied U.N. resolutions that demand its disarmament, fired on U.S. and coalition aircraft thousands of times over the past decade, and committed atrocious human rights abuses against its own citizens, including the waging of genocidal chemical warfare against Iraqi Kurds. In short, this is a regime that is responsible for hundreds of thousands, perhaps millions, of deaths.

Meanwhile, Iraq has a long record of active support for international terrorist groups. Indeed, it apparently has staged terrorist attacks of its own directly against the United States. I am speaking of Iraq's likely involvement in the attempted assassination of former President Bush in Kuwait in 1993.

Most ominously, Iraq has been engaged for many years in the monomaniacal pursuit of weapons of mass destruction (WMD). It reportedly has significant stockpiles of biological weapons, and its aggressive, large-scale nuclear program is thought to be at most a few years away from success. The fact that Iraq has been willing to endure ongoing sanctions, and thus the loss of hundreds of billions of dollars in oil revenue, rather than dismantle its WMD programs shows the ferocity of its commitment to maximizing its destructive capabilities.

In light of the above, I would support military action against Iraq even if 9/11 had never happened and there were no such thing as Al Qaeda. After all, I supported the Gulf War back in 1991 in the hope of toppling Saddam Hussein's regime before it fulfilled its nuclear ambitions. Unfortunately, quagmire was

From Brink Lindsey, "Should We Invade Iraq? A Reason Online Debate," *Reason,* vol. 34, no. 8 (January 2003). Copyright © 2002 by The Reason Foundation, 3415 S. Sepulveda Blvd., Suite 400, Los Angeles, CA 90034. Reprinted by permission.

plucked from the jaws of victory in that conflict, and so today we are faced with concluding its unfinished business. In my view, standing by with "patient watchfulness" while predatory, anti-Western terror states become nuclear powers is irresponsible and dangerous folly.

As for the headline question, "What's the rush?," [in Mueller's essay] my reply is: North Korea. In 1994 President [Bill] Clinton, with the help of former President [Jimmy] Carter, swept the Korean threat under the rug and trusted that "nature," or something, would deal with that "devil du jour." Now North Korea's psychopathic regime informs us that it has nuclear weapons, a fact that vastly complicates any efforts to prevent the situation from getting even worse. We can look forward to similar complications with Iraq unless we act soon.

The case for action against Iraq is further strengthened by the unfortunate facts that 9/11 did happen and Al Qaeda does exist. Here is the grim reality: Radical Islamism is in arms against the West, and its fanatical followers have pledged their lives to killing as many of the infidel as they possibly can. American office workers in New York and Washington, French seamen in Yemen, Australian tourists in Bali, Russian theatergoers in Moscow—nobody is safe. However exactly this conflict arose, it is now in full flame. And let there be no mistake: This is a fight to the death. Either we crush radical Islamism's global jihad, or thousands, even millions, more Americans will die.

Iraq occupies a strategic position in the war against Islamist terror along several dimensions. First, Iraq's WMD programs threaten to stock the armory of Al Qaeda & Company. Saddam Hussein's regime has a long and inglorious history of reckless aggression and grievous miscalculation. The decision to use terrorist intermediaries to unleash, say, Iraqi bioweapons against the United States strikes me as an entirely plausible scenario, assuming that Iraq's leadership can convince itself that the attack could be carried out with "plausible deniability." Given that more than a year has gone by since [2002's] anthrax letter scare and we still have no idea who was responsible, the threat posed by Iraq's WMD programs is far from idle. It is, in fact, intolerable.

Second, the resolution—one way or another—of our longstanding conflict with Iraq will have vitally important repercussions in the larger war against terror. If we proceeded to remove the Ba'athist regime from power, we would make it clear that the United States means business in dealing with terrorism and its sponsors. All those countries that continue, more than a year after 9/11, to demonstrate their incapacity or unwillingness to root out the terrorists in their midst (e.g., Iran, Pakistan, Saudi Arabia, Syria, Lebanon, Yemen) would have newly strengthened incentives to do the right thing. If, on the other hand, all the tough talk against Iraq turned out to have been hot air, U.S. credibility would sustain a major blow. Al Qaeda would be emboldened by perceived American weakness, and countries that have to balance fear of the United States against fear of Islamists at home would be inclined to take U.S. displeasure less seriously.

Finally, regime change in Iraq offers the opportunity to attack radical Islamism at its roots: the dismal prevalence of political repression and economic stagnation throughout the Muslim world. The establishment of a reasonably liberal and democratic Iraq could serve as a model for positive change throughout the region. Of course, the successful rebuilding of Iraq will not be easy, but we cannot shrink from necessary tasks simply because they are hard. And we cannot simply assume that "nature" will bring freedom to a region that has never known it on a time scale consistent with safeguarding American lives.

Mueller's "What, me worry?" attitude captures perfectly the prevailing opinion about Afghanistan circa September 10, 2001. The Taliban were more a punch line than a serious foreign policy issue; only the most fevered imagination could see any threat to us in that miserable, dilapidated country. The next day, 3,000 Americans were dead.

We can't let that happen again.

Nasty Realities

Evading Them Won't Make Us Safe

John Mueller sees correctly that the Iraq problem has two aspects: 1) regional security and 2) global terrorism. Unfortunately, he fails to grasp the nasty realities of either.

Mueller's assessment of the regional threat posed by a nuclear Iraq is nothing short of fantastic. He pooh-poohs the possibility that Iraq might invade one or more of its neighbors and argues that Saddam Hussein "is primarily devoted to preserving his regime and his own personal existence." Huh? Try telling that to Iran and Kuwait.

Mueller needs to read Mark Bowden's superb, chilling profile of Saddam in the May 2002 issue of *The Atlantic*. Bowden makes clear that Saddam sees himself as a world-historical figure, a man destined to lead pan-Arabia back to greatness. Perversely, every brush with disaster and death "has strengthened his conviction that his path is divinely inspired and that greatness is his destiny." Why on earth should we suppose that a nuclear arsenal—built in reckless defiance of the United States and the world—would temper rather than inflame Saddam's raging megalomania?

Mueller blithely assumes that any future Iraqi aggression would be "likely to trigger a concerted multilateral military attack upon him and his regime" and thus "patently suicidal." Excuse me, but there was no multilateral response to Iraq's attack on Iran, and the world would have been all too happy to acquiesce in Kuwait's disappearance had the first President Bush not stepped in and forced the issue. What makes Mueller think the world would rush in to confront a nuclear-armed Iraq? That task, inevitably, would fall to the United States. Mueller's counsel boils down to this: The United States should avoid war with a relatively weak Iraq today so that it can tangle with a nuclear adversary tomorrow.

What about the nexus between Iraq and terrorism, which Mueller dismisses as so much "arm waving"? Allow me to quote Bowden's article once more, this time from a scene in which Saddam is addressing Iraqi military leaders who run terrorist training camps: "He told [them] that they were the best men in the nation, the most trusted and able. That was why they had been selected to meet with him, and to work at the terrorist camps where warriors were being trained to strike back at America. The United States, he said, because of its reckless treatment of Arab nations and the Arab people, was a necessary target for revenge and destruction. American aggression must be stopped in order for Iraq to rebuild and to resume leadership of the Arab world."

This meeting occurred back in 1996—before the recent heating up of the conflict. So much for Saddam's live-and-let-live foreign policy.

Bellicose rhetoric is one thing; the ability to back it up is something altogether more serious. Here is the ultimate threat, the one that Mueller can't even bring himself to discuss: Iraqi biological or nuclear weapons might someday be put in the hands of terrorist groups. If that were to happen, America could experience horrors that would dwarf those unleashed on September 11.

Opponents of action against Iraq argue that we can rely on deterrence to protect us from such atrocities: No country, not even one as rash as Iraq, would dare to use weapons of mass destruction against the United States because of the threat of overwhelming retaliation. That argument has considerable force with respect to a direct attack by Iraq, but it fails completely to confront the possibility that Iraq could use terrorist intermediaries to do its dirty work while masking its own involvement. How is deterrence supposed to work when WMD lack a return address?

Recall, again, [2002's] anthrax attacks. We still don't know who was responsible, or whether there was any foreign state involvement. Just this week, a *Washington Post* article cast considerable doubt on the FBI's favored theory that the murders were the work of a disgruntled American scientist—and suggested that an Iraqi role remains a live possibility.

Go back a few more years, to the 1993 plot to assassinate former President Bush in Kuwait. It appears that the attack was an Iraqi operation, but as Seymour Hersh showed in a 1993 *New Yorker* article in which he reviewed the less-than-airtight case in depth, the fact is we're not really sure.

Welcome to the shadowy world in which we now live. A world in which deterrence no longer suffices. A world in which the judicious use of American power to pre-empt looming threats may be all that stands between us and catastrophe.

Here is what we know about the current Iraqi regime: It has weapons of mass destruction and is actively seeking to add to its arsenal. It is rabidly hostile to the United States. It has an established track record of predatory conduct and a demonstrated willingness to take extreme risks in pursuing its predatory ambitions. There is not another country on earth that matches Iraq's combination of destructive capacity, anti-American animus, and recklessness in projecting power. In a shadowy world, this much is clear: We are not safe while the present regime rules Iraq.

Weighing the Risks

There's No Invisible Hand to Protect Us

I argue that Iraq is a serious threat to the surrounding region and to us. John Mueller disagrees. I contend that toppling the current Iraqi regime will aid in the broader campaign against Islamist terrorism. Mueller worries that an invasion of Iraq will backfire.

Risks of action, risks of inaction: Which are greater? Solid facts are few and far between; we're forced to make our way based on hypotheticals and maybes and historical analogies. How can we have any confidence that we are weighing the risks intelligently?

One point in my favor is that I am actually weighing the risks. That's why I support military action against Iraq: I believe the risks of inaction outweigh the risks of action.

I am not a reflexive hawk. I opposed our recent military adventures in Panama, Haiti, Somalia, and the Balkans. I would not support military action against, say, Burma, merely because its government is despicable. Odious as it is, the Burmese regime poses no significant threat to its neighbors or to us. I would not have supported making war on China in the 1960s, even though its rulers were wildly anti-American and seeking to develop a nuclear arsenal. Despite the threat China posed to us, the risks of acting were far too great (especially the possibility of an escalation with the Soviets) and the price of victory against such a formidable and fanatical adversary would have been far too high. In that situation, deterrence and diplomacy (in particular, playing the Chinese and Soviets against each other) were the better options.

So on the general question of preventive war—whether to make war now in order to avoid a worse war later—my position is: It depends on the circumstances. The decision whether to go to war should turn on a pragmatic assessment of relative risks. Sometimes the balance will tilt in favor of action, sometimes not. In the particular case of Iraq in 2002, I believe the balance tilts strongly toward action.

Many who oppose invading Iraq . . . reject the kind of pragmatic assessment that I think is called for. They believe that preventive war is just a bad idea, period—that it's wrong, or at least reckless, to fire the first shot unless you're absolutely sure the other guy is about to squeeze the trigger.

When I'm debating the Iraq question with someone like that, we're talking past each other. I'm explaining the reasons that led me to my conclusion. He's marshaling evidence in support of a predetermined conclusion.

Not that there's anything wrong, in general, with predetermined conclusions—they're called principles. But all principles aren't created equal. Some are sound, some are iffy, and some are downright worthless.

What about the principle of no preventive wars? Specifically, what is the basis for assuming that preventive wars always make matters worse? In economic policy, there are solid grounds for the principle of no government meddling with markets. Market competition has enormous advantages over government action in making use of and coordinating dispersed information, in encouraging innovation, in supplying appropriate incentive structures, and so on. Accordingly, anyone arguing that government intervention in the marketplace can improve economic performance has an extremely difficult case to make.

Many libertarians slide easily from noninterventionism in domestic affairs to noninterventionism abroad, believing they're on equally firm footing with both positions. But they're not, because the fact is that there's no invisible hand in foreign affairs. There are no equilibrating mechanisms or feedback loops in the Hobbesian jungle of predatory dictatorships and fanatical terrorist groups that give us any assurance that, if the United States were only to stand aside, things would go as well for us in the world as they possibly could.

Accordingly, it seems to me that a no-exceptions policy against preventive war rests ultimately on an untenable assumption: that unrousable passivity on the part of the greatest and most powerful country that ever existed will somehow yield the most favorable achievable conditions in the world—that, in an intricately interconnected world, leaving everything outside our physical borders to the wolves will ensure that everything turns out for the best.

I don't buy it. Hostile regimes bent on relentless expansion and pursuing weapons of mass destruction are a threat to global security. Hostile regimes that could put weapons of mass destruction into the hands of terrorists are a direct threat to the lives of Americans. If regimes fitting either of these descriptions don't change their ways, military action against them should be an option.

Iraq's current regime fits both descriptions. It is not going to change its ways. The risks of war are real but manageable. Let's act before it's too late.

POSTSCRIPT

Was the War With Iraq Unjustified?

The question of whether the war with Iraq was justifiable or unjustifiable contains a host of important issues. One is how to proceed with countries that defy UN resolutions. There can be no argument that Iraq willingly or fully cooperated at any point with the inspections or met the "accurate, full and complete" provisions of Resolution 1441. Although many people argued against the war because they did not believe Iraq posed a major threat, that is not the point. Rather, the issues are whether a country that does not comply with UN resolutions should be allowed to remain in defiance and, if not, when it should be confronted with coercive force. For some background on the UN inspection process in Iraq, read Jean E. Krasno and James S. Sutterlin, *The United Nations and Iraq* (Praeger, 2002).

A second issue relates to the use of force without UN consent. The UN Charter, which the United Nations and all other member nations have agreed to uphold, allows countries to take unilateral military action to defend themselves and when authorized by the UN or another international organization. Part of the controversy surrounding the war is whether or not the United States and others violated the UN Charter. It is something of a legal debate, in which a reasonable case can be made for both the argument that Iraq's long-standing violation of numerous UN resolutions provided sufficient authority for the U.S.-led coalition to act and the argument that the attack was illegal without a final, specific authority by the Security Council.

Another part of the issue relates to when self-defense begins. The George W. Bush administration alleged that Iraq had ties to terrorists, that this constituted a threat to the United States, and that over time Iraqi WMDs and missile technology could also pose a threat. Therefore, under the Bush doctrine of preemption, the action against Iraq was self-defense. Where is the line? An analysis of these matters of the charter and the use of force is Thomas M. Franck's *Recourse to Force: State Action Against Threats and Armed Attacks* (Cambridge University Press, 2003).

A final point relates to the choice of readings from before the war began. That was done in an effort to remove hindsight from deciding whether or not the war was justified. Certainly the costs in blood and money are part of the argument, but neither proponents nor opponents of the war with Iraq or any other conflict could be sure what those costs would be. Therefore, arguing about the wisdom of going to war needs to be free of such information. For a discussion of the costs of war versus crisis prevention in hot spots such as Iraq, see Michael E. Brown and Richard N. Rosecrance, *The Costs of Conflict* (Rowman & Littlefield, 1999).

ISSUE 8

Should North Korea's Arms Program Evoke a Hard-Line Response?

YES: William Norman Grigg, from "Aiding and Abetting the 'Axis,'" *The New American* (February 24, 2003)

NO: Fred Kaplan, from "Appeasement, Please: The Case for Paying North Korea's Nuclear Blackmail," *Slate,* http://slate.msn.com/id/2076213 (December 31, 2002)

ISSUE SUMMARY

YES: William Norman Grigg, senior editor of *The New American*, argues that North Korea is a dangerous country with an untrustworthy regime and that it is an error for the United States to react to North Korea's nuclear arms program and other provocations by offering it diplomatic and economic incentives to be less confrontational.

NO: Fred Kaplan, author of the "War Stories" column in *Slate*, maintains that the horrendous costs of a conflict with North Korea mean that the United States and others would be better off trying to assuage North Korea than confronting it.

The global effort to control the spread of nuclear weapons centers on the Nuclear Non-Proliferation Treaty (NPT). It was first signed in 1968, then renewed and made permanent in 1995. The 85 percent of the world's countries that have agreed to the NPT pledge not to transfer nuclear weapons or to assist in any way a nonnuclear state to manufacture or otherwise acquire nuclear weapons. Under the NPT, nonnuclear countries also agree not to build or accept nuclear weapons and to allow the UN's International Atomic Energy Agency (IAEA) to monitor their nuclear facilities to ensure that they are used exclusively for peaceful purposes.

For all its contributions, the NPT is not an unreserved success. India and Pakistan both tested nuclear weapons in 1998; Israel's possession of nuclear weapons is an open secret; and other countries, such as Iran, have unacknowledged nuclear weapons programs. Then there is North Korea.

The immediate background of this issue dates to early 1993, when North Korea announced its withdrawal from the NPT. Fear that North Korea would try to develop nuclear weapons was heightened by media reports that the CIA believed that North Korea probably already had one or two nuclear weapons.

For a year and a half after North Korea's statement about withdrawing from the NPT, there was diplomatic maneuvering designed to persuade North Korea to continue to abide by the NPT. North Korea eventually agreed to remain a party to the NPT, to suspend work on the nuclear reactors it had under construction, to dismantle its current nuclear energy program over 10 years, and to allow the IAEA inspections to resume. The United States and its allies—principally Japan and South Korea—pledged that they would spend approximately $4 billion to build in North Korea two nuclear reactors that were not capable of producing plutonium for bomb building. The allies also agreed to help meet North Korea's energy needs by supplying it with about 138 million gallons of petroleum annually until the new reactors are on-line.

The issue flared anew to near crisis proportions beginning in late 2002 when North Korea expelled IAEA inspectors, dismantled their monitoring equipment, once again renounced its adherence to the NPT, and moved to restart its Yongbyon nuclear power plant, a facility capable of producing weapons-grade plutonium and uranium.

The alarm over North Korea's actions was even greater than it had been in 1993. The director of the CIA estimated in early 2002 that Pyongyang could reprocess the 8,000 containers of nuclear fuel containers stored at Yongbyon to build a half dozen nuclear weapons within several months and then create another two or more nuclear weapons a year to add to its suspected existing small arsenal. He also warned that the United States could face a "near term" intercontinental ballistic missile (ICBM) threat from North Korea's extensive missile program.

Washington at first responded to North Korea's actions with threats. However, faced with some harsh realities, the Bush administration was forced to retreat from its initial tough stance. One reality was that U.S. options were limited at the time because it was on the edge of war with Iraq. Second, even if it did not use nuclear weapons, North Korea is still capable of inflicting massive damage on South Korea, which has population centers (including its capital, Seoul) within artillery range of the border. Third, almost all other concerned countries, including South Korea and Japan, favored a placating rather than hostile reaction. Fourth, it is possible, given North Korea's alleged nuclear weapons inventory and its missile capabilities, that an overly aggressive response could lead to a nuclear war, including an attack on Japan. As a result, Washington moved to downplay the confrontation, to express assurances that it would not attack North Korea, and to suggest that aid might be resumed or even increased if North Korea abided by the NPT.

There the matter stood when the following selections were written. In the first, William Norman Grigg castigates the Bush administration for its willingness to appease North Korea and for its focus on what Grigg considers a much less dangerous Iraq. In the second, Fred Kaplan maintains that North Korea should be offered carrots, not threatened with sticks.

 **YES**

Aiding and Abetting the "Axis"

Saddam Hussein's Iraqi regime may be close to building a nuclear weapon. Kim Jong-Il's North Korean hell state, according to intelligence estimates, currently possesses two nukes, and will shortly develop the capacity to produce an entire arsenal. Under threat of war, Saddam has allowed UN weapons inspectors to canvass Iraq for evidence of weapons of mass destruction. Last December [2002], North Korea summarily evicted UN weapons inspectors from its Yongbyon nuclear plant, and disabled surveillance equipment used to monitor the suspected weapons production facility.

Crippled by the 1991 UN-led Gulf War, intermittent bombings by U.S. and British aircraft, and 12 years of devastating sanctions, Saddam's military poses little threat to Iraq's neighbors, let alone the United States. North Korea, on the other hand, boasts the world's fourth-largest military; it has 37,000 U.S. troops within easy striking range of its artillery. Seoul, the South Korean capital, is 34 miles away from the demilitarized zone and well within striking distance of North Korean artillery tubes. And Kim's regime has successfully tested the Taepo Dong, a missile capable of hitting Japan; the missile's next generation may be able to strike Alaska.

Moreover, North Korea brazenly and unrepentantly sponsors and participates in international terrorism. Adept in using infiltrators and sleeper agents, Pyongyang poses a real threat of nuclear terrorism against the region—and conceivably even the United States.

Of these two members of the "axis of evil," North Korea is—by any rational calculation—a far greater threat than Iraq. Yet in dealing with Pyongyang, the president displays none of the stiff-spined, bellicose rectitude that characterizes his treatment of Baghdad. Crusading for war against a prostrate Iraq, Mr. [George W.] Bush strikes poses of jut-jawed, Churchillian resolution; confronting an insurgent, nuclear-equipped North Korea, he essays a credible Neville Chamberlain impersonation.

Why is this so? How could the same president who identified North Korea as a member of an "axis of evil" now stand ready to lavish that terrorist regime with aid, trade, and technology? Mr. Bush, recall, has condemned not only ter-

From William Norman Grigg, "Aiding and Abetting the 'Axis,'" vol. 19, no. 4 (February 24, 2003). Copyright © 2003 by *The New American*. Reprinted by permission.

rorism but those countries supporting terrorism. "[W]e will pursue nations that provide aid or safe haven to terrorism," he said in a nationally televised address to a joint session of Congress shortly after the 9-11 terrorist attacks. "Every nation, in every region, now has a decision to make. Either you are with us, or you are with the terrorists. From this day forward, any nation that continues to harbor or support terrorism will be regarded by the United States as a hostile regime."

How does Mr. Bush reconcile this tough stance with aiding North Korea, the most militant of the three "axis of evil" regimes he named? And how does he reconcile that stance with counting as allies in the war against terrorism Russia and Communist China, who are the puppet-masters behind the three "axis" nations? Based on Mr. Bush's own definition, would not his policies qualify his own administration as "a hostile regime"?

Power Behind the Axis

Last December 12th, while the attention was focused on Baghdad and Pyongyang, Russian President (and KGB veteran) Vladimir Putin made what the *New York Times* described as "a quick but high-profile visit to Beijing" for a summit with Communist Chinese ruler Jiang Zemin. "China and Russia will be good neighbors, friends and partners forever," proclaimed Jiang during the quickie summit, held to reiterate the Sino-Russian "Good Neighborly Treaty of Friendship and Cooperation" signed in 2001.

One tangible item of business in the December 2002 Beijing meeting was a joint declaration urging the U.S. to normalize relations with North Korea "on the basis of continued observation of earlier reached agreements, including the framework agreement of 1994." Under that agreement, the U.S. and key allies—particularly South Korea and Japan—would pay at least $4 billion to supply North Korea with light-water nuclear reactors (which would be used to produce weapons-grade plutonium) and unspecified amounts to provide Pyongyang with heavy fuel oil and upgrades to its decrepit power grid. In exchange, North Korea supposedly agreed to "freeze" its nuclear program, and submit to international inspections beginning in 1999. In predictable fashion, Kim Jong-Il and his cohorts eagerly accepted these incredible concessions while covertly continuing their "frozen" nuke research.

Incredible as it may seem, the Bush administration allowed oil shipments to North Korea to continue after Pyongyang announced in October 2002 that the 1994 agreement was "nullified." "Can you imagine the uproar if Bill Clinton had let the deliveries to go forward [sic] if he had been told the agreement was nullified?" commented a Democratic congressional aide to the October 23rd [2002] *Washington Post*.

According to the CIA, North Korea attempted to buy equipment for a uranium weapons program from Communist China in 2001. During the same year, Beijing provided crucial missile-related technology to Pyongyang, and Russia concluded a defense agreement setting the stage for arms sales and weapons technology transfers to North Korea. This is curious behavior for powers hailed by President Bush as valued allies in the "war on terror"—and North Korea was hardly the only beneficiary of this treacherous Sino-Russian support. The CIA report, as summarized by *Washington Times* defense affairs analyst Bill Gertz, "identified Russia, China, and North Korea as major suppliers of chemical, biological and nuclear-arms goods and missile systems to rogue states or unstable regions."

A Terror Regime

North Korea is a museum-quality exhibit of Communism in the full flower of its malignancy. In congressional testimony . . . , Norbert Vollertsen, a German physician who lived in North Korea for a year and a half as a humanitarian volunteer, described how that nation's wretched hospitals are filled with people "worn out by compulsory drills, the innumerable parades, the assemblies from 6:00 in the morning and the droning propaganda. They are tired and at the end of their tether. Clinical depression is rampant. Alcoholism is common because of mind numbing rigidities and hopelessness of life."

Mass starvation is a hallmark of Communism, and North Korea has preserved this tragic tradition as well. Since 1992, at least one million North Korean subjects—and perhaps as many as four million, or one-quarter of the population—have died from starvation. And as has been the case in Soviet Russia, Red China, Ethiopia, and Zimbabwe, famine has been used as a weapon of social control. "North Korea is a terror regime," testified Dr. Vollertsen. "They are committing genocide there. . . . They are using food as a weapon against their own people. . . . North Korea [represents] the real killing fields of the 21st Century."

The Bush administration, citing humanitarian concerns, has repeatedly promised to continue providing food shipments to North Korea via the UN's World Food Program. But such aid actually compounds the humanitarian crisis by helping to prop up Kim's regime, which rations the food through the country's Public Distribution System (PDS). A North Korean subject's access to food and other necessities is strictly defined by his loyalty to the regime. According to Sophie DeLaunay of the humanitarian group Doctors Without Borders, "the three class labels—'core,' 'wavering,' and 'hostile'—continue to be used to prioritize access to jobs, region of residence, and entitlement to items distributed through the Public Distribution System. . . ."

"There are two worlds in North Korea," observed Dr. Vollertsen. "The world for the senior military, the members of the [ruling] party and the country's elite. . . . In the world for these ordinary people in a hospital one can see young children, all of them too small for their age, with hollow eyes and skin stretched tight across their faces, wearing blue-and-white striped pajamas like the children in Auschwitz and Dachau in Hitler's Nazi Germany." In September 1995, Kim Jong-Il issued orders to arrest wandering homeless children found outside their home counties and imprison them in the North Korean gulag.

While the common people starve, North Korea's Communist oligarchy lives in royal splendor. Seeking to co-opt Dr. Vollertsen, the Communist government awarded him a "friendship medal" and offered him unprecedented access to the "festivities . . . [of] all those who are in charge of power in the foreign ministry."

In that company, the German physician saw the country's elite "enjoying a nice lifestyle with fancy restaurants, diplomatic shops with European food, nightclubs and even a casino. . . ." The North Korean *nomenklatura* [leadership class] does little to disguise its privileged status. The October 5, 1999 *South China Morning Post* reported that Kim's regime purchased a $20 million fleet of 200 Mercedes-Benz S500 class cars for its leadership.

Gangster State

An unavoidable consequence of Communist central planning, the North Korean famine has been exacerbated by the regime's investment in narco-terrorism. The February 15, 1999 issue of *U.S. News & World Report* observed that up to 17,000 acres of farmland have been locked up by state-mandated opium farming, which began in the mid-1980s under dictator Kim Il-Sung, the present dictator's late father.

Kim Jong-Il has "ordered a major expansion of the drugs-for-export program," noted the magazine, which also reported that "U.S. food aid to the regime—over $77 million worth this year—may be needed in part because farm acreage is used to grow poppies for opium." To that figure can be added millions of additional dollars stolen by the regime from charitable aid sent to North Korea by private and religious relief organizations.

"Interviews with law enforcement officials, intelligence analysts, and North Korean defectors suggest that the regime is now dramatically expanding its narcotics production and that much of the criminal activity is controlled at the highest levels of government," reported *U.S. News*. "[I]t is clear that the worldwide network of North Korean embassies, coupled with the use of diplomatic pouches and immunity, offers the ideal cover for a criminal enterprise. . . ."

"Authorities in at least nine countries have nabbed North Korean diplomats with a virtual pharmacy of illegal drugs: opium, heroin, cocaine, hashish," continued the report. In July 1998, two North Korean diplomats were arrested in Cairo with six suitcases containing 506,000 tablets of Rohypnol, the so-called "date rape drug." During the same month, Japanese authorities intercepted a North Korean methamphetamine shipment worth $170 million.

Pyongyang is also deeply involved in counterfeiting. According to South Korea's National Intelligence Service, North Korea has printed vast quantities of counterfeit bills, including $15 million in "super notes"—bogus bills that are very difficult to detect—using new counterfeiting technology. The South Korean report charges that the counterfeiting operation was authorized at the highest levels of the North Korean government and cites as evidence that, in 1999, an aide to Kim Jong-Il was caught trying to exchange $30,000 in counterfeit notes in Vladivostok.

In typical Communist fashion, the North Korean gangster regime often sends politically suspect subjects to the gulag on spurious criminal charges.

This was the case with Sun-Ok Lee, a survivor of Pyongyang's gulag. Lee was convicted of spurious embezzling charges—and eventually escaped from North Korea to bear witness of the regime's unfathomable crimes against its most innocent subjects.

"In the 'reform institute' in Kaechon where I was held, there were 200 women housewives as prisoners," recalled Lee in congressional testimony. "In the case of these women, if any is pregnant, the baby would be killed. If the baby's mom was a political criminal, inside her the baby is the same political criminal. So the seed of a political criminal should not be allowed to be born."

Lee personally witnessed instances in which gulag officers would murder newborn infants by "stepping on the baby's neck with his boots once he or she was born. If the mom would cry for help to save her child, it was an expression of dissatisfaction against the party. So such a woman would be dragged out of the building and put to public execution by firing squad."

True Face of Evil

Such is the nature of the regime directly supported by our "allies" Russia and Communist China—and which the Bush administration is courting with humanitarian aid and promises of economic and technical assistance.

The Bush administration's treatment of North Korea exemplifies the utter phoniness of the "war on terrorism." Of the three "axis of evil" states, North Korea is undoubtedly the most oppressive and aggressive, and it poses the most immediate threat to U.S. citizens. Yet the administration has chosen to temporize in its dealing with Pyongyang in order to focus on the unnecessary, UN-authorized confrontation with Iraq.

And indeed, the North Korean hell state is a direct product of our nation's tragic entanglement with the UN. . . . [I]n the early stages of the Korean War the U.S.-led coalition liberated the entire peninsula from Communist hands—only to see the UN reverse this victory. That betrayal, and its tragic consequences, serves as a compelling illustration of the utter foolishness of fighting a "war on terrorism" through the UN.

NO

Fred Kaplan

Appeasement, Please: The Case for Paying North Korea's Nuclear Blackmail

On Oct. 18, 1962, the third day of the Cuban Missile Crisis, President John F. Kennedy, sitting in the Cabinet room with his advisers, wondered aloud why [Soviet leader] Nikita Khrushchev had launched this adventure. He figured that it must be part of some bargaining scheme and that, to make him get rid of the missiles, we had to come up with some way of letting the Soviet leader save face, of "giving him some out." It would be good to know if anyone inside President [George W.] Bush's White House is thinking along similar lines in the current crisis—or, as Secretary of State Colin Powell prefers to call it, "serious situation"— with North Korea. True, this is not the Cuban Missile Crisis; Kim Jong-Il is not Nikita Khrushchev; North Korea is not the U.S.S.R. Still, few would dispute that Kim's latest outrageous move—which will have him churning out A-bombs by the dozen in six months' time, unless something is done stop him—amounts to a desperate bargaining ploy, a time-tested way of frightening everyone around him (*nukes!*) and extorting them into giving him what he needs.

As any review of North Korea's diplomatic record would indicate, this is par for the course. From its very beginnings, North Korea has thrived—in many ways, has survived—on a diplomacy of permanent crisis: shrill invective, out-landish (but not quite incredible) threats, gross intimidation, and seemingly fearless brinkmanship. Korea, as one proverb has it, is a "shrimp among whales," and North Korea's rulers (there have been only two—Kim Jong-Il and his father, Kim Il Sung) have been masters at the art of turning their own weak-ness into strength and their foes' strength into weakness. In the game of high-way chicken, North Korea is the shrewd lunatic who very visibly throws his steering wheel out the window, forcing the other, more responsible driver to veer off the road. North Korea's long-chosen path of severe secrecy and isola-tion—Saddam Hussein's Iraq is practically a Western democracy by compari-son—helps assure its success at this game. Neither its friends nor foes really know what the hell is going on inside the inner sanctum. Richard Nixon tried to intimidate North Vietnam by pretending to be a "madman." Kim Jong-Il, at least by the standards of normal international relations, is a madman.

So, what is a country like the United States to do? On the one hand, as many Bush officials have noted, it's a bad idea in principle to pay off blackmailers. On the other hand, what choice do we have? Kim can sustain this crisis far longer than we can. First, his regime thrives on it. Second, he doesn't need to worry about domestic political groups or foreign allies because he doesn't have any. Third, if all else fails and the United States doesn't go along with his demands, he ends up with nukes, which he can use for further diplomatic games or sell and barter for much-needed currency, fuel, and food.

By contrast, look at our situation on each point. First, with the war on terrorism still brewing and a war with Iraq . . . , the last thing Bush needs is a nuclear stand-off in northern Asia. Second, South Korea has just elected a new president on a platform of friendlier relations with the North; Japan, China, and Russia aren't keen for confrontation, either; yet, in any successful counter-brinkmanship strategy, we would need the seamless support of all these players. Third, we really don't want North Korea to possess, or be able to pass around, a handful, much less a cargo-full, of nuclear weapons. Nor, alas, is the Osirak gambit much of an option. Unlike Iraq, when Israel bombed its nuclear reactor [at Osirak] in 1981, North Korea is already believed to have a couple of nukes, and it definitely has 11,000 artillery tubes (and who knows how many reloadable shells) less than a minute's flight-time from downtown Seoul. The risk of retaliation—and endangering tens of thousands of South Korean citizens and American soldiers—is commonly regarded as too high.

In short, we have almost no means of leverage in this game, and we might as well face that fact while those spent fuel rods, though unlocked and unmonitored, are still in place.

What does North Korea say it wants from this adventure? A non-aggression pact with the United States (thus ending the 1950–53 war) and a resumption of our obligations under the 1994 Agreed Framework negotiated with President Clinton. The events leading up to that accord were similar to today's. North Korea removed the fuel rods from its experimental nuclear reactor and threatened to pull out of the Non-Proliferation Treaty. Partly as the result of an unauthorized visit to Kim Il Sung by ex-President Jimmy Carter, a settlement was reached whereby the United States would provide food, fuel, and a light water nuclear reactor (which cannot in any way be used to make a bomb) in exchange for North Korea's continued compliance with the treaty.

This arrangement, administered by a jerry-rigged but highly competent entity called the Korean Peninsula Energy Development Organization, or KEDO (headed by a U.S. diplomat and staffed by Americans, Japanese, and South Koreans), worked well for a while and even helped relax regional tensions generally. It hit a big obstacle when a North Korean submarine wound up in South Korean waters. It started to unwind when President Bush, upon entering office, made clear he had no interest in continuing this entente. It fell apart altogether when Bush, in his post-9/11 address on terrorism, accused North Korea, Iran, and Iraq of forming an "axis of evil." It crumbled to bits last October [2002] when, after much probing and interrogation, American diplomats got North Korean officials to admit that they had restarted their nuclear program. In response, the

United States stopped shipping fuel and food—to which North Korea replied by unsealing the fuel rods, disconnecting the IAEA's [International Atomic Energy Agency's] cameras and ordering the inspectors home.

Who's ultimately responsible for this breakdown is, in some ways, an academic question. Neither side can claim to be purely an innocent bystander or victim. But would it be so terrible—would it really be "appeasement," as many conservative commentators have thundered—to offer a resumption of KEDO, *simultaneous with* a resumption of North Korea's responsibilities under the Non-Proliferation Treaty? If Bush wants to take control of the negotiations, as opposed to letting Kim define the terms and then manipulate them, he could go further and offer a whole package of economic investments, tied not just to denuclearization but to a gradual opening of North Korean society—which, in the long run, would be in our interests.

In the longer run still, the United States—if not Bush, then whoever follows—must devise a nuclear proliferation policy, because North Korea, though unique in many ways, does point a scraggly finger toward the future. In the 1960s and '70s, many arms-control scholars warned that 20 or 30 countries would acquire nuclear weapons in the next decade. It didn't happen, not because those countries were unable to do so, but rather because the Cold War was an international security system. The United States and the U.S.S.R. each extended the deterrent of its nuclear arsenal to its circle of allies. With the U.S.S.R. vanquished, this "nuclear umbrella" has folded up as well, and it will become harder and harder to keep particularly insecure powers from building their own nukes—especially since, as North Korea is now demonstrating, you only need a few nukes to be suddenly taken seriously.

POSTSCRIPT

Should North Korea's Arms Program Evoke a Hard-Line Response?

The confrontation with North Korea over its nuclear weapons program raises a host of important, long-term issues that extend far beyond the immediate concern. One is that fashioning an appropriate response to another country's action is often difficult because one cannot be sure what is motivating the other country. North Korea is perhaps the most secretive country on Earth. During the 1993–1995 crisis, one U.S. official commented, "The fact of the matter is that we don't really understand what they are doing." Knowledge of what caused North Korea in late 2002 to ignite the crisis once again could provide better direction for policymakers. For example, a hard-line response would arguably be appropriate if, as some people think, North Korea was merely taking advantage of the fact that the United States was snarled up in the Iraq crisis, using that opening to renew its efforts to acquire nuclear weapons in order to increase its power, and using that augmented power to cow South Korea and Japan and perhaps to force a U.S. military withdrawal from the region.

By contrast, a soft-line approach might be better if North Korea was trying to play a "nuclear chip" to garner more aid to ease the horrendous economic conditions (including widespread starvation) in the country. Arguably, a very soft approach would be best if North Korea was restarting its nuclear weapons program out of a defensive fear that it would be attacked by the United States. Americans are hesitant to see their country as the aggressor, but that reluctance is not shared worldwide. After all, President George W. Bush had labeled North Korea a member of an "axis of evil," and Washington was moving toward war with Iraq, one of the other "evil axis" countries. It would not be unreasonable, then, for North Korea to think it was next on the American "hit list." To gain further insight into the politics of the Korean peninsula, see Marcus Noland and C. Fred Bergsten, *Avoiding the Apocalypse: The Future of the Two Koreas* (Institute for International Economics, 2000). Newer and on CD is the U.S. government's *Twenty-First Century Complete Guide to North Korea and the Regime of Kim Jong-il: DPRK Nuclear and Missile Programs, With Material From the DOD, Military, Congress, the White House, CIA Factbook, and Library of Congress Country Studies—State Department and U.S. Policy on North Korea*, Core Federal Information Series (Progressive Management, 2003).

Another long-term issue is how to contain the spread of nuclear weapons. At least some countries ask why it is acceptable for some countries to have nuclear weapons and unacceptable for others. A particularly sore point in some regions of the world is the U.S. silence on Israel's all-but-official nuclear arsenal. When the NPT was being renewed in the mid-1990s, many nonnuclear countries were reluctant to do so. They wanted countries that had nuclear weapons

132

to set a timetable for dismantling their arsenals because, as Malaysia's delegate to the conference noted, without such a pledge renewing the treaty would be "justifying nuclear states for eternity" to maintain their monopoly. These objections were partly overcome when the states with nuclear weapons pledged to conclude a treaty to ban all nuclear testing. The United States reneged on that pledge when the Senate refused to ratify the resulting Comprehensive Test Ban Treaty in 1999.

A third extended issue is what effect the North Korean nuclear program will have on nuclear proliferation in general and, more particularly, in Northeast Asia. China and Russia already have such weapons, as does another major player in the area—the United States. If North Korea also has an arsenal, can defense planners in South Korea and Japan reasonably be asked not to build their own capabilities? The U.S. position is that those countries should rely on the U.S. "nuclear umbrella" to deter an attack, but if you were Japanese or South Korean, would you leave your defense to others? If Japan and South Korea do acquire nuclear weapons, what argument can be made to restrain other countries? As George Tenet, the director of the CIA, has commented, "The desire for nuclear weapons is on the upsurge. Additional countries may decide to seek nuclear weapons as it becomes clear their neighbors and regional rivals are already doing so. The 'domino theory' of the 21st century may well be nuclear." Two studies on nuclear nonproliferation are Carl Ungerer and Marianne Hanson, eds., *The Politics of Nuclear Non-Proliferation* (Allen & Unwin, 2002) and Scott D. Sagan and Kenneth N. Waltz, *The Spread of Nuclear Weapons: A Debate Renewed,* 2d ed. (W. W. Norton, 2003).

International Development Exchange (IDEX)

This is the Web site of the International Development Exchange (IDEX), an organization that works to build partnerships to overcome economic and social injustice. The IDEX helps people gain greater control over their resources, political structures, and the economic processes that affect their lives.

http://www.idex.org

United Nations Development Programme (UNDP)

This United Nations Development Programme (UNDP) site offers publications and current information on world poverty, the UNDP's mission statement, information on the UN Development Fund for Women, and more.

http://www.undp.org

Office of the U.S. Trade Representative

The Office of the U.S. Trade Representative (USTR) is responsible for developing and coordinating U.S. international trade, commodity, and direct investment policy and leading or directing negotiations with other countries on such matters. The U.S. trade representative is a cabinet member who acts as the principal trade adviser, negotiator, and spokesperson for the president on trade and related investment matters.

http://www.ustr.gov

The U.S. Agency for International Development (USAID)

This is the home page of the U.S. Agency for International Development (USAID), which is the independent government agency that provides economic development and humanitarian assistance to advance U.S. economic and political interests overseas.

http://www.info.usaid.gov

World Trade Organization (WTO)

The World Trade Organization (WTO) is the only international organization dealing with the global rules of trade between nations. Its main function is to ensure that trade flows as smoothly, predictably, and freely as possible. This site provides extensive information about the organization and international trade today.

http://www.wto.org

Third World Network

The Third World Network (TWN) is an independent, nonprofit international network of organizations and individuals involved in economic, social, and environmental issues relating to development, the developing countries of the world, and the North-South divide. At the network's Web site you will find recent news, TWN position papers, action alerts, and other resources on a variety of topics, including economics, trade, and health.

http://www.twnside.org.sg

PART 3

Economic Issues

*I*nternational economic and trade issues have an immediate and personal effect on individuals in ways that few other international issues do. They influence the jobs we hold and the prices of the products we buy—in short, our lifestyles. In the worldwide competition for resources and markets, tensions arise between allies and adversaries alike. This section examines some of the prevailing economic tensions.

- Is Free Economic Interchange Beneficial?

- Should the Rich Countries Forgive All the Debt Owed by the Poor Countries?

- Are Patents on HIV/AIDS Drugs Unfair to Poor Countries?

ISSUE 9

Is Free Economic Interchange Beneficial?

YES: Anne O. Krueger, from Remarks at the 2002 Eisenhower National Security Conference on "National Security for the Twenty-First Century: Anticipating Challenges, Seizing Opportunities, Building Capabilities" (September 26, 2002)

NO: José Bové, from "Globalisation's Misguided Assumptions," *OECD Observer* (September 2001)

ISSUE SUMMARY

YES: Anne O. Krueger, first deputy managing director of the International Monetary Fund, asserts that the growth of economic globalization is unstoppable and that supporting it is one of the best ways to improve global conditions.

NO: José Bové, a French farmer and antiglobalization activist, contends that multinational corporations, government leaders, and others are engaged in a propaganda campaign to sell the world on the false promise of economic globalization.

The impact of international economics on domestic societies has expanded rapidly as world industrial and financial structures have become increasingly intertwined. Foreign trade wins and loses jobs, for example, and people in most countries depend on petroleum and other imported resources to fuel cars, homes, and industries. Inexpensive imports into industrialized countries from less economically developed countries also help to keep inflation down and the standard of living up. In fact, global exports grew from $53 billion in 1948 to $7.5 trillion in 2001. It is likely that the clothes you are wearing and the television you watch were made in another country.

In addition to trade, the trend toward globalization includes such factors as the growth of multinational corporations (MNCs), the flow of international investment capital, and the increased importance of international exchange rates. There are now at least 40,000 MNCs that conduct business (beyond just sales) in more than one country. Of these, just the 500 largest global corporations (the so-called Global 500) produced $9.2 trillion in goods and services, about 30 percent of the world's total economic production, in 2000. Foreign investment is also immense and growing. In 1990 the total new flow of foreign direct investment (FDI, significant enough to buy ownership of a company or real property) was $199 billion; that figure was up to $885 billion in 2000.

The issue here is whether this economic globalization and integration is a positive or negative trend. For about 60 years, the United States has been at the center of the drive to open international commerce. The push to reduce trade barriers that occurred during and after World War II was designed to prevent a recurrence of the global economic collapse of the 1930s and the war of the 1940s. Policymakers believed that protectionism had caused the Great Depression, that the ensuing human desperation provided fertile ground for the rise of dictators who blamed scapegoats for what had occurred and who promised national salvation, and that these fascist dictators had set off World War II. In sum, policymakers thought that protectionism caused economic depression, which caused dictators, which caused war. They believed that free trade, by contrast, would promote prosperity, democracy, and peace.

Based on these political and economic theories, American policymakers took the lead in establishing a new international economic system. As the world's dominant superpower, the United States played the leading role at the end of World War II in establishing the International Monetary Fund (IMF), an international organization that was set up in the 1940s to promote the international monetary stability necessary for extensive global economic interchange; the World Bank; and the General Agreement on Tariffs and Trade (GATT). The latest GATT revision talks were completed and signed by 124 countries (including the United States) in April 1994. Among the outcomes was the establishment of a new coordinating body, the World Trade Organization (WTO).

During the entire latter half of the twentieth century, the movement toward economic globalization has been strong, and there have been few influential voices opposing it. Most national leaders, business leaders, and other elites continue to support economic interdependence. The people in various countries have largely followed the path set by their leaders. In the following selection, Anne O. Krueger, a ranking official of the IMF, takes the view that the movement toward globalization remains beneficial.

More recently, the idea that globalization is either inevitable or necessarily beneficial has come under increasing scrutiny and has met increasing resistance. The strongest critique of globalization as it is occurring comes from analysts who are often referred to as "economic structuralists." They believe that the way global politics work is a function of how the world is organized economically. Structuralists contend that countries are divided into "haves" and "have-nots" and that the world is similarly divided into have and have-not countries. Moreover, structuralists believe that, both domestically and internationally, the wealthy haves are using globalization to keep the have-nots weak and poor in order to exploit them. Representing this view, José Bové argues in the second selection that economic globalization benefits the few at the expense of the many.

Anne O. Krueger

 **YES**

Supporting Globalization

Introduction

. . . Supporting globalization is one of the best investments we can make to improve today's security environment. Globalization is the process of integration
of nations through the spread of ideas and the sharing of technological advances, through international trade, through the movement of labor and capital across national boundaries. It is a process that has been going almost
throughout recorded history and that has conferred huge benefits. Globalization involves change, so it is often feared, even by those who end up gaining
from it. And some do lose in the short run when things change. But globalization is like breathing: It is not a process one can or should try to stop; of course,
if [there] are obvious ways of breathing easier and better one should certainly
do so.

Like any process, globalization has been subject to ebbs and flows. It
gained impetus during the period of great discoveries in the 15th century, and
in later centuries from dramatic falls in the costs of communication and transportation. For instance, the fortunes of the House of Rothschild were helped by
their being the first to use carrier pigeons to carry business news between London and Brussels. The invention of the telegraph and the laying of the transatlantic cable cut settlement times between New York and London from ten days
to three days. Stop and think about what it must have been like when the first
telegraph wire went through: it must have been as big a breakthrough as the
ones in more recent times that we rave about.

After World War II, globalization received a boost from the dramatic lowering of trade barriers among the major industrialized nations. Over the last
fifty years the process of integration has accelerated among the industrialized
nations and also started to embrace many nations in the developing world.

Growth and Globalization in the 20th Century

Economic growth in the last fifty years has been faster than it was in earlier centuries. In the nineteenth century, for instance, the leading nations such as

From Anne O. Krueger, Remarks at the 2002 Eisenhower National Security Conference on "National
Security for the Twenty-First Century: Anticipating Challenges, Seizing Opportunities, Building
Capabilities" (September 26, 2002). Copyright © 2002 by The International Monetary Fund.
Reprinted by permission.

Britain grew at an average annual rate of 1.5 percent per capita. But in this present era of globalization, many countries have achieved per capita annual growth rates of 5 to 8 percent. Thus, in the heyday of their growth, the Asian Tigers achieved every decade what rapidly growing countries had previously achieved in a century.

Why has growth been faster? A big reason is growth and globalization have gone hand-in-hand: Access to a buoyant international market has greatly facilitated faster growth. It has permitted a degree of reliance on comparative advantage and division of labor that was not possible in the nineteenth century. Not only has there been rapid growth in world trade, but it has taken place in an environment where the support facilities are readily available from other trading nations. These facilities—communications, wholesalers, finance and insurance—would have been expensive for poor countries to provide themselves and would have put them at a disadvantage competitively. Of course, technology transfer helps as well to boost growth rates. Latecomers to development have the advantage of ready access to all the blueprints developed over the past several hundred years in the more advanced nations. And latecomers also derive benefits from the very large declines in the costs of transport and communications. Korea, for example, shifted from being a 70 percent rural economy to a 70 percent urban economy in the course of three decades. Such a shift would not have been possible without the support of an international economy. More recently, over the last decade, joining the international economy has helped some regions in India make the transition to an information-based economy.

The impact of the faster growth on living standards has been phenomenal. We have observed the increased well-being of a larger percentage of the world's population by a greater increment than ever before in history. Growing incomes give people the ability to spend on things other than basic food and shelter, in particular on things such as education and health. This ability, combined with the sharing among nations of medical and scientific advances, has transformed life in many parts of the developing world. Infant mortality has declined from 180 per 1000 births in 1950 to 60 per 1000 births. Literacy rates have risen from an average of 40 percent in the 1950s to over 70 percent today. World poverty has declined, despite still-high population growth in the developing world. Since 1980, the number of poor people, defined as those living on less than a dollar a day, has fallen by about 200 million, much of it due to the rapid growth of China and India.

If there is one measure that can summarize the impact of these enormous gains, it is life expectancy. Only fifty years ago, life in much of the developing world was pretty much what it used in be in the rich nations a couple of centuries ago: "nasty, brutish and short." But today, life expectancy in the developing world averages 65 years, up from under 40 years in 1950. Life expectancy was increasing even in sub-Saharan Africa until the effects of years of regional conflicts and the AIDS epidemic brought about a reversal. The gap between life expectancy between the developed and developing world has narrowed, from a gap of 30 years in 1950 to only about 10 years today.

There have been gains in other spheres of life as well. Economic growth has raised the demand for democracy and representation. As a result, many more people around the world live free. A large part of the world's population now lives under governments they have elected. According to Freedom House, the proportion of countries with some form of democratic government rose from 28% in 1974 to 62% in 2000. Electoral democracies now represent 120 of the 192 or so existing countries and constitute nearly 60% of the world's population. People have also been given much more opportunity to vote with their feet. They go where there are more opportunities and chances to build a better life.

Why the Protests?

There is a clear contradiction between these manifest benefits of growth and globalization and the outcry against them. The protests are particularly bewildering because the gains have come about without many, or indeed any, of the feared side effects coming to pass.

Take the perennial concern that rapid growth depletes our fuel resources and once that happens growth will come to a complete dead stop. World oil reserves today are higher than in 1950. Then the world's known reserves of oil were expected to be enough for only 20 more years of consumption. That is, we were expected to run out by 1970. It did not happen. Today, our known reserves are enough to keep us going for another 40 years at our present rate of consumption. There is no doubt that by the time 2040 rolls around research and development will have delivered new breakthroughs in energy production and use.

Nor have we done irreparable harm to the environment. The evidence shows quite convincingly that economic growth brings an initial phase of deterioration in some aspects, but followed by a subsequent phase of improvement. The turning point at which people begin choosing to invest in cleaning up and preventing pollution occurs at a per capita GDP [gross domestic product] of 5000 dollars.

What about the impact of growth and globalization on labor and social conditions in the developing world? Conditions in so-called "sweatshop" factories in developing countries should be compared to the other choices available to people in those countries. For instance, the growth of the footwear industry in Vietnam has translated into a five-fold increase in wages in a short period of time; while still a pittance by our standards, the higher wages have completely transformed for the better the lives of those workers and their families. Insisting that such workers be given a "decent wage" by our standards would completely erode any competitive advantage of businesses using unskilled labor on the international market. Likewise, child labor is sometimes prevalent in developing countries because the alternatives are so much worse: starvation or malnutrition, forced early marriages (for girls) or prostitution, or life on the streets as a beggar. There is ample evidence that parents in developing countries, like parents everywhere, choose schooling for their young when they can afford to do so, and the quickest path to that outcome is through more rapid economic growth.

Concerns that globalization is associated with a loss of control are also misplaced. In fact, it is poverty that is a state of almost total loss of control.

Growth has done more to give people a control over their lives than any alleged loss of control to multinationals or fickle flows of foreign capital.

Some anti-globalizers allege that the gains of globalization are not universally shared. Inequality of outcomes is said to be the Achilles Heel of globalization. This characterization is misleading in several respects. At the very outset, one has to wonder about the preoccupation with inequality. As Chinese Premier Deng Tsiao Ping famously remarked: "I have a choice. I can distribute wealth or I can distribute poverty." Poor people are desperate to improve their material conditions in absolute terms rather than to march up the income distribution. Hence it seems far better to focus on impoverishment than on inequality. And there is no doubt that growth reduces the incidence of impoverishment. Empirical studies show clearly that the incomes of those at the bottom of the income distribution rise one-for-one with growth. To take one specific example, identifying who lost in absolute terms in Korea during its period of high growth in the 1960s or 1970s is a difficult task. The losers were largely older peasants, and in their case too their offspring often sent remittances from their urban jobs so that rural living standards were rising rapidly.

Even if one is going to worry about inequality, there is no evidence that globalization has any systematic impact on a country's income inequality. Many countries have experienced fast growth by opening up to the world economy, but without changes in inequality. In any event, countries that are concerned about inequality of incomes within their countries can, and do, address it through government policies.

If one looks not at within-country inequality but at world inequality, the news is actually very encouraging. The evidence, though difficult to piece together, suggests that world inequality is declining. This is happening in large part because of the phenomenal growth of China and India. Because the majority of the poor reside in these two countries, their growth helps to reduce inequality of world incomes, even though many smaller countries have had stagnant incomes.

What the protestors are against is not growth or globalization, but change and fear of the impacts of change. There once used to be warnings that train travel in excess of 15 miles per hour would be hazardous to health. Many of the alleged hazards of growth and of joining the international economy are similar to such warnings. To be sure, there are losers whenever there is change. But as we get richer, we can afford greater protection and insurance against any negative results of free economic interchange.

Conclusion

In summary, 20th century growth and globalization have brought great benefits to the majority of the world's population. In the nineteenth century, while governments helped build railroads and so on, there was no presumption that they were responsible for growth and therefore had to do something when it faltered. In the 20th century, in contrast, development has been a conscious goal of policy and is deemed to be the responsibility of policymakers. The underlying beliefs on which development policies were based in the first decades after 1945 were a deep-seated distrust of markets and a strong commitment to placing the government in charge of the commanding heights of the economy.

Gaining access to growth in the 20th century has required putting in place institutions and implementing policies that ran counter to these beliefs. In that sense, the barriers to rapid growth in the 20th century were not economic but political. The Asian tigers were the first to shake off these beliefs and enjoy rapid growth and, much later, China and India have followed in their footsteps.

This perspective leads me to conclude on an optimistic note on the prospects for continued economic growth and globalization in the 21st century. Many more countries are choosing economic policies today without the ideological baggage that slowed their growth in the 20th century. These countries can continue to bank on the kind of supportive international economy that was instrumental in helping many of them in the 20th century. Though global economic conditions have been weak [recently], a recovery is underway and the medium-term prospects for global economic growth remain good. And the gains from international division of labor and specialization are far from exhausted. If heavily-populated countries such as India, China and Vietnam can all join the global economy at around the same time and can all experience rapid growth simultaneously, it should surely be possible to make room for the others.

Countries in which citizens have a say in their government, and where citizens can concentrate on—and succeed in—improving their material fortune and their families' lives, are not countries likely to present security threats.

NO ↰

José Bové

Globalisation's Misguided Assumptions

Humanity is grappling with a formidable creed, which, like so many others, is totalitarian and planetary in scope, namely free trade. The gurus and zealous servants of this doctrine ("responsible" people) are saying that the Market is the only god, and that those who want to combat it are heretics ("irresponsible" people). So we find ourselves faced with a modern-day obscurantism—a new opium on which the high priests and traffickers are sure they can make populations dependent. Recent articles in the international press supporting new trade rounds and the like are quite clear on the dogma that some people would like to impose on the men and women of this planet.

More and more people are coming out against the free market credo advocated by the WTO [World Trade Organization], the damage inflicted by it being so plain to see, and the falsehoods on which it is based so blatant.

The first falsehood is the market's self-regulating virtues, which form the basis of the dogma, but this ideological mystification is belied by the facts. In the field of agriculture, for example, since 1992 the major industrialised countries have embraced global markets with open arms—the United States enacted the FAIR (Federal Agriculture Improvement and Reform) Act, a policy instrument that did away with direct production subsidies, instead "decoupling" aid and allowing farmers to produce with no restrictions whatsoever—but this has done nothing to calm the wild swings in the markets.

It has, in fact, done quite the opposite, since markets have experienced unprecedented instability since the trade agreements [modifying and extending the General Agreement on Tariffs and Trade] signed in Marrakech in 1995. The most spectacular effect of this American "decoupling" has been the explosion of emergency direct subsidies to offset declining prices. These subsidies reached a record high of more than $23 billion in 2000 (four times more than the amount budgeted in the 1996 Farm Act).

So, contrary to free-marketers' assertions, markets are inherently unstable and chaotic. Government intervention is needed to regulate markets and adjust price trends, to guarantee producers' incomes and thus ensure that farming activity is sustained.

The second blatant untruth is that competition generates wealth for everyone. Competition is meaningful only if competitors are able to survive. This is especially true for agriculture, where labour productivity varies by a factor of a thousand to one between a grain farmer on the plains of the Middle West and a spade-wielding peasant in the heart of the Sahel.

To claim that the terms of competition will be healthy and fair, and thus tend towards equilibrium if farm policy does not interfere with the workings of a free market, is hypocritical. How can there be a level playing field in the same market between a majority of 1.3 billion farm workers who harvest the land with their hands or with harnessed animals, and a tiny minority of 28 million mechanised farmers formidably equipped for export? How can there be "fair" competition when the most productive farmers of rich countries receive emergency subsidies and multiple guarantees against falling prices on top of their direct and indirect export bonuses?

The third falsehood is that world market prices are a relevant criterion for guiding output. But these prices apply to only a very small fraction of global production and consumption. The world wheat market accounts for only 12% of overall output and international trade takes place at prices that are determined not by aggregate trade, but by the prices of the most competitive exporting country.

The world price of milk and dairy products is determined by production costs in New Zealand, while New Zealand's share of global milk production averaged only 1.63% between 1985 and 1998. The world price of wheat itself is pegged to the price in the United States, which accounted for only 5.84% of aggregate world output from 1985 to 1998.

What is more, these prices are nearly always tantamount to dumping (i.e., to selling below production costs in the producing and importing countries) and are only economically viable for the exporters thanks to the substantial aid they receive in return.

The fourth falsehood is that free trade is the engine of economic development. For free marketers, customs protection schemes are the root of all evil: they claim that such systems stifle trade and economic prosperity, and even hinder cultural exchanges and vital dialogue between peoples. Yet who would dare to claim that decades of massive northbound coffee, cocoa, rice and banana exports have enriched or improved the living standards of farmers in the south? Who would dare to make such a claim, looking these poverty-stricken farmers straight in the eyes? And who would dare to tell African breeders, bankrupted by competition from subsidised meat from Europe, that it is for their own good that customs barriers are falling?

To achieve their ends, the proponents of free trade exploit science in the name of so-called "modernism," asserting that the development of any scientific discovery constitutes progress—as long as it is economically profitable. They cannot bear the idea that life can reproduce on its own, free of charge, whence the race for patents, licences, profits and forcible expropriation.

Obviously, when talking about agriculture, it is impossible not to evoke the farce of GMOs [genetically modified organisms]. Nobody is asking for them, yet they must be the answer to everyone's dreams! There is pressure on us to concede that genetically modified rice (cynically dubbed "golden rice") is going to nourish people who are dying of hunger and protect them from all sorts of diseases, thanks to its new Vitamin A-enriched formula. But this will not solve the problems of vitamin deficiencies, because a person would have to eat three kilograms of dry rice every day, whereas the normal ration is no more than 100 grams.

The way to fight malnutrition, which affects nearly a third of humanity, is to diversify people's diets. This entails rethinking the appalling state of society, underpinned by free market economics which strives to keep wages in southern countries as low as possible in order to maximise profits. It is therefore a good idea to throw some vitamins into the rice that is sold to poor people, so that they don't die too quickly and can continue working for low wages, rather than helping them build a freer and fairer society. Jacques Diouf, Director-General of FAO [Food and Agriculture Organization of the United Nations], recently pointed out that "to feed the 8 003 million people who are hungry, there is no need for GMOs" (*Le Monde*, 10 May). No wonder the Indian farmers of Via Campesina, an international small farmers' movement, destroy fields of genetically modified rice.

The FAO is not the only international institution to question some of the certainties and radical WTO positions regarding the benefits of free markets. The highly free-market OECD [Organization for Economic Co-operation and Development] acknowledges in a recent report entitled *The Well-being of Nations* that the preservation and improvement of government services (healthcare, education) are a key factor underlying the economic success of nations.

We therefore have every reason to oppose the dangerous myth of free trade. Judging by the substantial social and environmental damage free trade has inflicted, before anything else, it is necessary for all of us—farmers and non-farmers alike—to make it subject to three fundamental principles: food sovereignty—the right of peoples and of countries to produce their food freely, and to protect their agriculture from the ravages of global "competition;" food safety—the right to protect oneself from any threat to one's health; and the preservation of bio-diversity.

Along with adherence to these principles must come a goal of solidarity-based development, via the institution of economic partnership areas among neighbouring countries, including import protection for such groups of countries having uniform structures and levels of development.

The WTO wants to take its free-market logic even further. Next November [2002], in the seclusion of a monarchy that outlaws political parties and demonstrations—Qatar—it will attempt to attain its goals. But if major international institutions are becoming increasingly critical and are casting doubt on these certainties, then mobilised citizens can bring their own laws to bear on trade.

Between the absolute sovereign attitude of nationalists and the proponents of free trade, there are other roads. To echo the theme of the World Social Forum that took place in Porto Alegre last January [2001], "another world is possible!"—a world that respects different cultures and the particularities of each, in a spirit of openness and understanding. We are happy and proud to be part of its emergence.

POSTSCRIPT

Is Free Economic Interchange Beneficial?

There can be no doubt that the global economy and the level of interdependence have grown rapidly since World War II. To learn more about this, read Robert Gilpin and Jean M. Gilpin, *Global Political Economy: Understanding the International Economic Order* (Princeton University Press, 2001).

For most of the period since the 1940s, the drive to create ever freer international economic interchange sparked little notice, much less opposition. That has changed, and one of the most remarkable shifts in political momentum in recent years has been a marked increase in resistance to globalization. Not long ago, meetings of international financial organizations such as the IMF and the WTO used to pass unnoticed by nearly everyone other than financiers, scholars, and government officials. Now their meetings often occasion mass protests, such as the riots that broke out in Seattle, Washington, in 1999 at a meeting of the WTO. Similarly, when the heads of the leading industrialized countries, known as the Group of Seven (G-7, Canada, France, Germany, Great Britain, Italy, Japan, and the United States) held their annual meeting in Genoa, Italy, in 2001, they were confronted by an estimated 100,000 protesters.

One of the oddities about globalization, economic or otherwise, is that it often creates a common cause between those of marked conservative and marked liberal views. More than anything, conservatives worry that their respective countries are losing control of their economies and, thus, a degree of their independence. Echoing this view, archconservative 2000 presidential candidate Patrick Buchanan warned that unchecked globalism threatens to turn the United States into a "North American province of what some call The New World Order."

Conservatives also worry that increased economic interdependence can endanger national security. If, for example, a country becomes dependent on foreign sources for vehicles, then it may well have no ability to produce its own military vehicles in times of peril if cut off from its foreign supplier or, worse, if that supplier were to become an international antagonist.

Some liberals share the conservatives' negative views of globalization but for different reasons. This perspective is less concerned with sovereignty and security; it is more concerned with workers and countries being exploited and the environment being damaged by MNCs that shift their operations to other countries to find cheap labor and to escape environmental regulations. Referring to the anti-WTO protests in 1999, U.S. labor leader John J. Sweeney told reporters, "Seattle was just the beginning. If globalization brings more inequality, then it will generate a violent reaction that will make Seattle look tame." For a view that globalization needs to be restrained, consult William Greider, *The Manic Logic of Global Capitalism* (Simon & Schuster, 1997).

For all these objections, the continued thrust among governments—with strong support in business, among economists, and from other influential groups—is to continue to promote expanded globalism. "Turning away from trade would keep part of our global community forever on the bottom. That is not the right response," President Bill Clinton warned just before leaving office.

For now, the upsurge of feelings against globalism has pressed policymakers and analysts to consider what reforms are necessary to continue globalization while instituting reforms that will help quiet the opposition. There are also those who argue that economic globalization is not unstoppable and can even be reversed. For an example, see the International Forum on Globalization's *Alternatives to Economic Globalization [A Better World Is Possible]* (Barrett-Koehler, 2003).

ISSUE 10

Should the Rich Countries Forgive All the Debt Owed by the Poor Countries?

YES: Romilly Greenhill, from "The Unbreakable Link—Debt Relief and the Millennium Development Goals," a Report From Jubilee Research at the New Economics Foundation (February 2002)

NO: William Easterly, from "Debt Relief," *Foreign Policy* (November/December 2001)

ISSUE SUMMARY

YES: Romilly Greenhill, an economist with Jubilee Research at the New Economics Foundation, contends that if the world community is going to achieve its goal of eliminating world poverty by 2015, then there is an urgent need to forgive the massive debt owed by the heavily indebted poor countries.

NO: William Easterly, a senior adviser in the Development Research Group at the World Bank, maintains that while debt relief is a popular cause and seems good at first glance, the reality is that debt relief is a bad deal for the world's poor.

Like all countries, the world's poor states, or less developed countries (LDCs), are determined to improve their economic circumstances. However, most LDCs find it difficult to raise capital internally because their poor economic bases leave little for them to tax and leave almost nothing for savings and investment by the countries' citizens. Moreover, a great deal of the goods and services that these countries need to develop has to be imported, and most businesses and countries will not accept payment in the generally unstable currencies of the LDCs.

Therefore, to move forward the LDCs not only need significant amounts of development capital, but those funds have to be American dollars, Japanese yen, European euros, or one of the other "hard currencies" acceptable in world markets. A key concern for LDCs is where to obtain hard currency to use for development. There are four main sources: loans, investment, trade, and aid. This issue takes up some of the controversy raised by problems with the first of these four sources: loans.

Loans flow to LDCs from international organizations (such as the World Bank); from other, wealthier countries; and from private investors (such as banks). There is nothing unusual about such transactions. Indeed, throughout the 1800s American businesses borrowed huge sums to finance the building of the nation's railroads and steel mills and for other means of economic development. And when it is running a deficit, the U.S. government still borrows tens of billions of dollars annually from foreign investors to finance the national debt.

The background to the current debt problem can be traced to the 1970s, when a number of factors converged to persuade the LDCs to borrow heavily to finance their development needs. The reasons are complex, but suffice it to say that LDCs were in dire need of funds and the lenders in the economically developed countries (EDCs) had surplus capital, which they urged the LDCs to borrow. By 1982 LDC international debt had sharply risen to $849 billion, and by 2000 it stood at $2.1 trillion.

For a variety of reasons, some of which were the fault of the LDCs (poor planning, corruption, political instability) and some of which were not the fault of the LDCs (a lagging world economy and other barriers to their exports), many of the LDCs have found themselves severely burdened by their debt service (principal and interest payments). There are two levels to this problem.

One level has to do with the drain that the debt service puts on struggling LDCs. Export earnings are one source of funds to pay off debt, but in 2000 the LDCs owed more than 137 percent of their export revenue. In dollar figures, this mean that the LDCs paid out $348 billion (22.5 percent of their annual export earnings) that year to meet their principal and interest obligations. Supporters of debt reduction or total forgiveness argue that those funds should be reinvested in the LDCs rather than sent abroad.

The second level occurs when countries cannot meet their debt service. In essence, they are bankrupt. Such collapses harm the borrowers (whose credit is ruined and whose economies sometimes collapse), the lenders (who lose their money), and, arguably, the global community, which must continue to deal with all the ills that are rooted in the grinding poverty of a good percentage of the world's people and countries.

Some analysts, including Romilly Greenhill, the author of the first of the following selections, argue that in both the short and long runs it would be best to wipe out the debt of the heavily indebted poor countries. There are various ways to do this. One limited debt-relief program that was instituted in the early 1990s was the Brady Plan (after Secretary of the Treasury Nicholas Brady). Under it, banks forgave over $100 billion of what the LDCs owed (and took deductions on their corporate taxes), lowered interest rates on continuing loans, and made new loans at low rates. In return, international organizations (the World Bank, the International Monetary Fund) and the governments of the capital exporting countries guaranteed the loans and increased their own lending to the LDCs.

Not everyone agrees with such across-the-board debt-relief plans. William Easterly is one such opponent, and in the second selection, he maintains that forgiving debts will only encourage the poor practices that led to the debt crisis in the first place.

Romilly Greenhill

YES

The Unbreakable Link—Debt Relief and the Millennium Development Goals

At the start of the new millennium, the world's leaders met in the United Nations General Assembly to set out a new global vision for humanity. In their Millennium Declaration, the statesmen and women recognised their 'collective responsibility to uphold the principles of human dignity, equality and equity at the global level.' They pledged to 'spare no effort to free our fellow men, women and children from the abject and dehumanising conditions of extreme poverty.'

From these fine words, a set of goals was born: to eliminate world poverty by the year 2015; to achieve universal primary education; to promote gender equality and empower women; to reduce child mortality; improve maternal health; to combat HIV/AIDs and other diseases; and to ensure environmental sustainability. . . .

Since then, the Millennium Development Goals [MDGs]—as they were subsequently named—have been adopted by all major donor agencies as guiding principles for their strategies for poverty eradication. . . .

Moreover, since the adoption of the MDGs in the year 2000, events have conspired to reinforce the urgent need for poverty reduction in the world. According to Gordon Brown, the aftermath of September 11th has shown that 'the international community must take strong action to tackle injustice and poverty. . . . [and to] achieve our 2015 Millennium Development Goals.'

But meeting the 2015 targets requires resources. Ernest Zedillo, in his report of the High Level Panel for Financing for Development, has assessed that total additional resources of $50bn per year will be needed to meet these targets worldwide, over and above the current level of spending in key areas. This estimate is based on detailed costings in some of the key goal areas by UN bodies such as UNICEF, the World Health Organisation, and others such as the World Bank.

The UN Millennium Declaration was not the only remarkable event of the year 2000. Equally notable—though perhaps more poignant—was the winding down of the Jubilee 2000 campaign, described by Kofi Annan as 'the voice of

From Romilly Greenhill, "The Unbreakable Link—Debt Relief and the Millennium Development Goals," A Report From Jubilee Research at the New Economics Foundation (February 2002). Copyright © 2002 by The New Economics Foundation. Reprinted by permission. Notes omitted.

the world's conscience and indefatigable fighters for justice.' The Jubilee 2000 coalition had campaigned for the cancellation of the un-payable debts of the poorest countries by the end of 2000, under a fair and transparent process. Their petition—the largest ever—had been signed by 24 million people world-wide.

The central message of the Jubilee 2000 campaign was that human rights should not be subordinated to money rights. Poor countries prepared to commit resources to meeting the basic needs and economic rights of their populations should not be prevented from doing so because of the need to pay back debts to rich creditor countries and institutions.

The Jubilee 2000 campaign had won a commitment to a $110bn write off of un-payable debts. This was to be achieved partly through an extension of the World Bank's Heavily Indebted Poor Countries (HIPC) initiative, and partly through additional bilateral commitments from creditors such as the UK.

But it is now clear that the HIPC initiative is not delivering enough either to produce the promised 'robust exit' from unsustainable debts or to meet internationally agreed poverty reduction goals. . . . [B]y the end of 2001—a full year after the millennium deadline called by the Jubilee 2000 coalition—only four countries had passed through all the hoops of the HIPC initiative. Out of the 42 countries included in the process, almost half of these had not even reached 'decision point', after which they receive some interim relief on their debt service payments. Moreover, even when relief is provided, research by Jubilee Plus has shown that debt burdens remain unsustainable. . . .

We show in this report that if poor country governments are to have sufficient resources to meet the MDGs, as well as to meet other essential expenditure needs and pro-poor investments, *the 42 HIPC countries as a whole cannot afford to make any debt service payments.* In fact, we find that *even if all the debts of these 42 countries are cancelled, the HIPCs will need an additional $30bn in aid each year if there is any hope of meeting goal 1 while for the other goals, a total of $16.5bn will be needed.*

These figures are based on actual debt service payments for 1999—before most of the HIPCs had received any substantial debt service relief from the HIPC initiative. But as the latest figures from the World Bank have shown, even when all 42 countries have fully passed through the HIPC initiative, the savings will only amount to a paltry $3.5bn per year. It is clear that much deeper, and faster, debt relief must be provided. . . .

Debt Service Payments Take Resources From the MDGs

Calculating the resources needed to meet the MDGs in each country is no easy task. Data on the number of poor people in each country, the current level of indicators such as HIV and malarial prevalence, or even the number of children in school, is often not available, or not reliable. Moreover, working out the exact amount that will need to be spent across different countries to meet common objectives requires making heroic assumptions about costs in each country.

Some of the goals—such as 'reversing the loss of environmental resources' are inherently very difficult to evaluate. . . .

Goal 1: Eradicate extreme poverty and hunger

- Halve, between 1990 and 2015, the proportion of people whose income is less than one dollar a day
- Halve, between 1990 and 2015, the proportion of people who suffer from hunger.

Eradicating mass poverty is often seen as the most fundamental of the MDGs. In the simplistic world of the donor community, extreme poverty is defined as living on less than one dollar a day. . . .

Of all the MDGs, this goal is also the most difficult to relate to debt service payments. It is clear that debt repayments are taking resources that could be spent to reduce poverty, but quantifying the exact linkages is much more difficult. . . .

Our overall point is simply that by using standard and widely accepted economic models, we can show that in a world of finite development resources, debt repayments will be traded off with limited poverty reduction expenditures.

Goal 2: Achieving universal primary education

- Ensure that, by 2015, children everywhere, boys and girls alike, will be able to complete a full course of primary schooling

Access to primary education is a basic human right. Education benefits individuals, their families, and also society as a whole, by enabling greater participation in democratic processes. Education serves to empower individuals, helps them to take advantage of economic opportunities, and improves their health and that of their family.

Yet, in 2000, *one in three* children across the developing world did not complete the 5 years of basic education which UNICEF believes is the minimum required to achieve basic literacy. We are clearly a long way from achieving the Millennium Development Goal of achieving Universal Primary Education by 2015.

UNICEF has calculated the amount that countries will need to spend in order to meet the MDGs. They found that almost all of the HIPCs will need to increase spending on education. . . .

[T]he HIPC countries will only need to spend $6.5bn each year in order to ensure that every child gets an education sufficient to ensure basic literacy. While large relative to the incomes of HIPCs, on a global scale this figure is miniscule—representing, for example, less than half of one percent of the projected US defence budget of $1,600bn over the next five years. And only $1.2bn of this is additional to what governments are currently spending—although

[some] countries . . . will need much larger increases in spending than some of the other HIPCs.

Goal 4: Reducing child mortality

- Reduce, by two-thirds, between 1990 and 2015, the under-five mortality rate

Goal 5: Improving maternal health

- Reduce by three quarters, between 1990 and 2015, the maternal mortality ratio

Goal 6: Combating HIV/AIDS, malaria and other diseases

- Have halted by 2015, and begun to reverse, the spread of HIV/AIDS
- Have halted by 2015, and begun to reverse, the incidence of malaria and other related diseases.

A tragedy is unfolding in Africa. Within the last 24 hours, 5,500 Africans were killed by HIV/AIDS. One in five of all adults in Africa are infected by the virus, while 17 million Africans have died from AIDS since the start of the epidemic. AIDS has so far left 13 million children orphaned, a figure which will grow to 40 million by 2010 if no action is taken.

Moreover, AIDS is not the only killer. Other diseases such as malaria, TB, childhood infectious diseases, maternal and prenatal conditions and micronutrient deficiencies abound. Average life expectancy in Africa has *fallen* since 1980, from 48 to 47—and in individual countries, the fall is much more extreme. Life expectancy in Zambia is now only 38 years, down from 50 years in 1980, while Sierra Leone has a life expectancy of only 37 years. And even these figures mask the catastrophic impact on children. In Africa, 161 children out of every 1,000 children will die before their fifth birthday; in Niger, this figure is as high as one in four.

Yet, the Global Commission on Macroeconomics and Health has estimated that *eight million* lives could be spared each year if a simple set of health interventions needed to meet the MDGs were put in place.

. . . [T]he 39 HIPCs will between them need to spend $20bn each year on health if the MDGS are to be met—almost three times their 1999 levels of debt service. This figure may sound large, but it is only slightly more than the $17bn spent each year in Europe and the US on pet food. As with education, larger countries will need bigger increases in health spending. . . .

The need for more debt relief is evident. The . . . vast improvements in the lives of millions of people in poor countries are achievable, with an increase in expenditure totalling only 0.1% of GDP of the rich donor and creditor countries. Yet, despite this overwhelming imperative, the poorest countries are still paying debt service of $8bn per year.

Target 10: Halve, by 2015, the proportion of people without sustainable access to safe drinking water

Like education and health care, access to safe water is a basic right. Safe water is vital for proper health and hygiene, including the prevention of water borne diseases. Distances travelled to fetch water result in a huge loss of time for poor people, particularly women and children. Yet, *one billion* people currently lack safe drinking water and almost *three billion—half the world's population—*lack adequate sanitation. *Two million* children die each year from water-related diseases. As the Vision 21 Framework for Action states, this situation is 'humili-ating, morally wrong and oppressive.'. . .

In order to meet the MDG of halving the proportion of people without sus-tainable access to safe drinking water, . . . the HIPCs would have to spend only $2.4bn per year on water and sanitation—less than Europe spends on alcohol over ten days.

Target 11: By 2020, to have achieved a significant improvement in the lives of at least 100 million slum dwellers

. . . Hundreds of millions of the urban poor in developing countries cur-rently live in unsafe and unhygienic environments where they face multiple threats to their health and security. . . . [T]he HIPCs will need to spend 1% of GDP annually on improving slum conditions. In total, this comes to *$1.7bn* for all the 39 HIPCs considered.

Other Goals and Targets

Goal 3: Promoting gender equality and empowering women

- Eliminate gender disparity in primary and secondary education prefer-ably by 2005 and to all levels of education no later than 2015

Target 9: Integrate the principles of sustainable development into country poli-cies and programmes and reverse the loss of environmental resources

Providing basic health, education and water to the populations of poor countries is clearly vital and should be given preference over debt service pay-ments. But at the same time, other dimensions of development—such as pro-moting gender equality and protecting environmental resources, are also needed if development is to be sustainable in the long run.

Unfortunately, however, these goals are inherently difficult to cost, and are therefore difficult to compare with debt service payments.

Total Required to Meet MDGs

. . . [T]he total funds required each year to meet MDGs 2 to 7 are not exorbitant. . . . [A] mere $30.6bn per year is required. This figure may be small in global terms. But . . . it represents 18% of GDP for the 42 HIPCs as a whole, and a staggering 355% of their debt service. . . .

Linking Debt Servicing to the Millennium Development Goals

Even without servicing their external debts, it is clear that the 39 HIPCs face a formidable challenge if they are to raise the level of resources required to meet the MDGs.

While it is true that governments can raise their own revenues by taxing their domestic populations, in most of the HIPC countries the extreme poverty experienced means that governments find it very difficult to raise the kind of resources needed.

Debt servicing worsens this position by diverting preciously needed resources, which could be used for saving lives, and educating children, towards rich country creditors.

Moreover, governments cannot be spending all their revenues on social expenditures. Crucial expenditures such as maintaining law and order, public administration, essential infrastructures such as roads, policing and defence are also needed. . . .

Given the current ability of the HIPC governments to raise revenues, and given the essential expenditures needed to meet the MDGs and for other essential expenditures, we now ask: how much can the HIPCs afford to pay to their rich country creditors in debt service payments? . . .

Our analysis shows that, as a whole, *the HIPCs have no spare resources available that could be used for debt servicing.* In fact, even with 100% debt cancellation, *the HIPCs will require an additional $16.5bn if goals 2 to 7 are to be met,* and this is without the additional $30bn needed for goal 1. . . .

Conclusion: The Need for a Sabbath Economics

Our conclusion is clear. If the Millennium Development Goals are to be met, all of the HIPCs will need full cancellation of all of their debts. This is not an act of charity, but a moral imperative. While eight million die each year for want of the funds spent by the rich countries on their pets; when millions of children stay out of school for want of half a percent of the US defence budget; and when the amount spent on alcohol in a week and a half in Europe would be adequate to provide sanitation to half the world's population, something is very wrong.

Maybe it is time, once again, to call on biblical principles. The 'Jubilee' principle—which provides ways of reversing the relentless flow of resources from the poor to the rich, and narrowing the gap between—formed the foundation of one of the most successful global campaigns ever. But there are others.

The central tenets of 'Sabbath Economics' are that the world is abundant and provides enough for everyone—but that human communities must restrain their appetites and live within limits. For Sabbath Economics, disparities in wealth and power are not natural, but come about through sin, and must be mitigated within the community through redistribution.

We do not have to believe in God—or indeed any religion—to accept these principles. It is enough for us to recognise that more than a billion people do not need to live in poverty while their debts continue to be repaid. The current HIPC initiative does not and cannot do enough to bring down the unsustainable debt burden of the world's poorest countries. If the Millennium Development Goals are to be met, there is no alternative but to provide a new framework for debt relief—one which respects the human rights of the poor.

NO ↵

William Easterly

Debt Relief

"Jubilee 2000 Sparked the Debt Relief Movement"

No Sorry, Bono, but debt relief is not new. As long ago as 1967, the U.N. Conference on Trade and Development argued that debt service payments in many poor nations had reached "critical situations." A decade later, official bilateral creditors wrote off $6 billion in debt to 45 poor countries. In 1984, a World Bank report on Africa suggested that financial support packages for countries in the region should include "multi-year debt relief and longer grace periods." Since 1987, successive G-7 [Group of Seven—Canada, France, Germany, Great Britain, Italy, Japan, and the United States] summits have offered increasingly lenient terms, such as postponement of repayment deadlines, on debts owed by poor countries. . . . In the late 1980s and 1990s, the World Bank and International Monetary Fund (IMF) began offering special loan programs to African nations, essentially allowing governments to pay back high-interest loans with low-interest loans—just as real a form of debt relief as partial forgiveness of the loans. The World Bank and IMF's more recent and well-publicized Highly Indebted Poor Countries (HIPC) debt relief program therefore represents but a deepening of earlier efforts to reduce the debt burdens of the world's poorest nations. Remarkably, the HIPC nations kept borrowing enough new funds in the 1980s and 1990s to more than offset the past debt relief: From 1989 to 1997, debt forgiveness for the 41 nations now designated as HIPCs reached $33 billion, while new borrowing for the same countries totaled $41 billion.

So by the time the Jubilee 2000 movement began spreading its debt relief gospel in the late 1990s, a wide constituency for alleviating poor nations' debt already existed. However, Jubilee 2000 and other pro–debt relief groups succeeded in raising the visibility and popularity of the issue to unprecedented heights. High-profile endorsements range from Irish rock star Bono to Pope John Paul II and the Dalai Lama to Harvard economist Jeffrey Sachs; even retiring U.S. Sen. Jesse Helms has climbed onto the debt relief bandwagon. In that respect, Jubilee 2000 (rechristened "Drop the Debt" before the organization's campaign officially ended on July 31, 2001) should be commended for putting the world's poor on the agenda—at a time when most people in rich nations simply don't care—even if the organization's proselytizing efforts inevitably oversimplify the problems of foreign debt.

"Third World Debts Are Illegitimate"

Unhelpful idea   Supporters of debt relief programs have often argued that new democratic governments in poor nations should not be forced to honor the debts that were incurred and mismanaged long ago by their corrupt and dictatorial predecessors. Certainly, some justice would be served if a legitimate and reformist new government refused to repay creditors foolish enough to have lent to a rotten old autocracy. But, in reality, there are few clear-cut political breaks with a corrupt past. The political factors that make governments corrupt tend to persist over time. How "clean" must the new government be to represent a complete departure from the misdeeds of an earlier regime? Consider President Yoweri Museveni of Uganda, about the strongest possible example of a change from the past—in his case, the notorious past of Ugandan strongman Idi Amin. Yet even Museveni's government continues to spend money on questionable military adventures in the Democratic Republic of the Congo. Would Museveni qualify for debt relief under the "good new government" principle? And suppose a long-time corrupt politician remains in power, such as Kenyan President Daniel Arap Moi. True justice would instead call for such leaders to pay back some of their loot to development agencies, who could then lend the money to a government with cleaner hands—a highly unlikely scenario.

Making debt forgiveness contingent on the supposed "illegitimacy" of the original borrower simply creates perverse incentives by directing scarce aid resources to countries that have best proved their capacity to mismanage such funds. For example, Ivory Coast built not just one but two new national capitals in the hometowns of the country's previous rulers as it was piling up debt. Then it had a military coup and a tainted election. Is that the environment in which aid will be well used? Meanwhile, poor nations that did not mismanage their aid loans so badly—such as India and Bangladesh—now do not qualify for debt relief, even though their governments would likely put fresh aid resources to much better use.

Finally, the legitimacy rationale raises serious reputation concerns in the world's financial markets. Few private lenders will wish to provide fresh financing to a country if they know that a successor government has the right to repudiate the earlier debt as illegitimate. For the legitimacy argument to be at all convincing, the countries in question must show a huge and permanent change from the corruption of past regimes. Indeed, strict application of such a standard introduces the dread specter of "conditionality," i.e., the imposition of burdensome policy requirements on developing nations in exchange for assistance from international financial institutions. Only rather than focusing solely on economic policy conditions, the international lending agencies granting debt relief would now be compelled to make increasingly subjective judgments regarding a country's politics, governance structures, and adherence to the rule of law.

"Crushing Debts Worsen Third World Poverty"

Wrong in more ways than one Yes, the total long-term debt of the 41 HIPC nations grew from $47 billion in 1980 to $159 billion in 1990 to $169 billion in 1999, but in reality the foreign debt of poor countries has always been partly fictional. Whenever debt service became too onerous, the poor nations simply received new loans to repay old ones. Recent studies have found that new World Bank adjustment loans to poor countries in the 1980s and 1990s increased in lock step with mounting debt service. Likewise, another study found that official lenders tend to match increases in the payment obligations of highly indebted African countries with an increase in new loans. Indeed, over the past two decades, new lending to African countries more than covered debt service payments on old loans.

Second, debt relief advocates should remember that poor people don't owe foreign debt—their governments do. Poor nations suffer poverty not because of high debt burdens but because spendthrift governments constantly seek to redistribute the existing economic pie to privileged political elites rather than try to make the pie grow larger through sound economic policies. The debt-burdened government of Kenya managed to find enough money to reward President Moi's home region with the Eldoret International Airport in 1996, a facility that almost nobody uses.

Left to themselves, bad governments are likely to engage in new borrowing to replace the forgiven loans, so the debt burden wouldn't fall in the end anyway. And even if irresponsible governments do not run up new debts, they could always finance their redistributive ways by running down government assets (like oil and minerals), leaving future generations condemned to the same overall debt burden. Ultimately, debt relief will only help reduce debt burdens if government policies make a true shift away from redistributive politics and toward a focus on economic development.

"Debt Relief Allows Poor Nations to Spend More on Health and Education"

No In 1999, Jubilee 2000 enthused that with debt relief "the year 2000 could signal the beginning of dramatic improvements in healthcare, education, employment and development for countries crippled by debt." Unfortunately, such statements fail to recognize some harsh realities about government spending.

First, the iron law of public finance states that money is fungible: Debt relief goes into the same government account that rains money on good and bad uses alike. Debt relief enables governments to spend more on weapons, for example. Debt relief clients such as Angola, Ethiopia, and Rwanda all have heavy military spending (although some are promising to make cuts). To assess whether debt relief increases health and education spending, one must ask what such spending would have been in the absence of debt relief—a difficult question. However, if governments didn't spend the original loans on helping the poor, it's a stretch to expect them to devote new fiscal resources toward helping the poor.

Second, such claims assume that the central government knows where its money is going. A recent IMF and World Bank study found that only two out of 25 debt relief recipients will have satisfactory capacity to track where government spending goes within a year. At the national level, an additional study found that only 13 percent of central government grants for non-salary education spending in Uganda (another recipient of debt relief) actually made it to the local schools that were the intended beneficiaries.

Finally the very idea that the proceeds of debt relief should be spent on health and education contains a logical flaw. If debt relief proceeds are spent on social programs rather than used to pay down the debt, then the debt burden will remain just as crushing as it was before. A government can't use the same money twice—first to pay down foreign debt and second to expand health and education services for the poor. This magic could only work if health and education spending boosted economic growth and thus generated future tax revenues to service the debt. Unfortunately, there is little evidence that higher health and education spending is associated with faster economic growth.

"Debt Relief Will Empower Poor Countries to Make Their Own Choices"

Not really   Pro–debt relief advocacy groups face a paradox: On one hand, they want debt relief to reach the poor; on the other, they don't want rich nations telling poor countries what to do. "For debt relief to work, let the conditions be set by civil society in our countries, not by big world institutions using it as a political tool," argued Kennedy Tumutegyereize of the Uganda Debt Network. Unfortunately, debt relief advocates can't have it both ways. Civil society remains weak in most highly indebted poor countries, so it would be hard to ensure that debt relief will truly benefit the poor unless there are conditions on the debt relief package.

Attempting to square this circle, the World Bank and IMF have made a lot of noise about consulting civil society while at the same time dictating incredibly detailed conditions on debt relief. The result is unlikely to please anyone. Debt relief under the World Bank and IMF's current HIPC initiative, for example, requires that countries prepare Poverty Reduction Strategy Papers. The World Bank's online handbook advising countries on how to prepare such documents runs well over 1,000 pages and covers such varied topics as macroeconomics, gender, the environment, water management, mining, and information technology. It would be hard for even the most skilled policymakers in the advanced economies to follow such complex (no matter how salutary) advice, much less a government in a poor country suffering from scarcity of qualified managers. In reality, this morass of requirements emerged as the multilateral financial institutions sought to hit on all the politically correct themes while at the same time trying hard to make the money reach the poor. If the conditions don't work—and of course they won't—the World Bank and IMF can simply fault the countries for not following their advice.

"Debt Relief Hurts Big Banks"

Wrong   During the 1970s and early 1980s, large commercial banks and official creditors based in rich nations provided substantial loans at market interest rates to countries such as Ivory Coast and Kenya. However, they pulled out of these markets in the second half of the 1980s and throughout the 1990s. In fact, from 1988 to 1997, such lenders received more in payments on old loans than they disbursed in new lending to high-debt poor countries. The multilateral development banks and bilateral lenders took their place, offering low-interest credit to poor nations. It's easy to understand why the commercial and official creditors pulled out. Not only did domestic economic mismanagement make high-debt poor countries less attractive candidates for potential loans, but with debt relief proposals in the air as early as 1979, few creditors wished to risk new lending under the threat that multilateral agencies would later decree loan forgiveness.

The IMF and World Bank announced the HIPC initiative of partial and conditional forgiveness of multilateral loans for 41 poor countries in September 1996. By the time the debt relief actually reached the HIPCs in the late 1990s, the commercial banks and high-interest official creditors were long gone and what was being forgiven were mainly "concessional" loans—i.e., loans with subsidized interest rates and long repayment periods. So really, debt relief takes money away from the international lending community that makes concessional loans to the poorest nations, potentially hurting other equally poor but not highly indebted nations if foreign aid resources are finite (as, of course, they are). Indeed, a large share of the world's poor live in India and China. Neither nation, however, is eligible for debt relief.

"Debt Relief Boosts Foreign Investment in Poor Nations"

A leap of faith   It is true that forgiving old debt makes the borrowers more able to service new debt, which in theory could make them attractive to lenders. Nevertheless, the commercial and official lenders who offer financing at market interest rates will not want to come back to most HIPCs any time soon. These lenders understand all too well the principle of moral hazard: Debt relief encourages borrowers to take on an excessive amount of new loans expecting that they too will be forgiven. Commercial banks obviously don't want to get caught with forgiven loans. And even the most charitable official lenders don't want to sign their own death warrants by getting stuck with forgiven debt. Both commercial and official lenders may want to redirect their resources to safer countries where debt relief is not on the table. Indeed, in 1991, the 47 least developed countries took in 5 percent of the total foreign direct investment (FDI) that flowed to the developing world; by 2000 their portion had dropped to only

2.5 percent. (Over the same period, the portion of global FDI captured by all developing nations dropped as well, from 22.3 to 15.9 percent.) Even capital flows to now lightly indebted "safe" countries might suffer from the perception that their debts also may be forgiven at some point. Ultimately, only the arms of multilateral development banks that provide soft loans—with little or no interest and very long repayment periods—are going to keep lending to HIPCs, and only then under very stringent conditions.

"Debt Relief Will Promote Economic Reform"

Don't hold your breath   During the last two decades, the multilateral financial institutions granted "structural adjustment" loans to developing nations, with the understanding that governments in poor countries would cut their fiscal deficits and enact reforms—including privatization of state-owned enterprises and trade liberalization—that would promote economic growth. The World Bank and IMF made 1,055 separate adjustment loans to 119 poor countries from 1980 to 1999. Had such lending succeeded, poor countries would have experienced more rapid growth, which in turn would have permitted them to service their foreign debts more easily. Thirty-six poor countries received 10 or more adjustment loans in the 1980s and 1990s, and their average percentage growth of per capita income during those two decades was a grand total of zero. Moreover, such loans failed to produce meaningful reforms, and developing countries now cite this failure as justification for debt relief. Yet why should anyone expect that conditions on debt forgiveness would be any more effective in changing government policies and behavior than conditions on the original loans?

Partial and conditional debt forgiveness is a *fait accompli.* Expanding it to full and unconditional debt forgiveness—as some groups now advocate—would simply transfer more resources from poor countries that have used aid effectively to those that have wasted it in the past. The challenge for civil society, the World Bank, IMF, and other agencies is to ensure that conditional debt forgiveness really does lead to government reforms that enhance the prospects of poor countries.

How can we promote economic reform in the poorest nations without repeating past failures? The lesson of structural adjustment programs is that reforms imposed from the outside don't change behavior. Indeed, they only succeed in creating an easy scapegoat: Insincere governments can simply blame their woes on the World Bank and IMF's "harsh" adjustment programs while not doing anything to fundamentally change economic incentives and ignite economic growth. It would be better for the international financial institutions to simply offer advice to governments that ask for it and wait for individual countries to come forward with homegrown reform programs, financing only the most promising ones and disengaging from the rest. This approach has worked in promoting economic reform in countries such as China, India, and Uganda. Rushing through debt forgiveness and imposing complex reforms from the outside is as doomed to failure as earlier rounds of debt relief and adjustment loans.

POSTSCRIPT

Should the Rich Countries Forgive All the Debt Owed by the Poor Countries?

It is easy to say, "Well, don't borrow if you cannot pay it back" or "If you borrowed it, you have to pay it back even if you suffer." Those who would disagree with that approach might say it contains something of Victor Hugo's *Les Miserables* (1862). The novel's main character, Jean Valjean, is convicted of theft and spends 19 years doing hard labor in the quarries. The injustice is that all he stole was some bread, and he only did so to feed the children of his impoverished sister. Advocates of debt relief would argue that, like Jean Valjean, the LDCs cannot be faulted too harshly for borrowing to try to meet rudimentary development needs.

The debt relief advocates would also argue that if the LDCs were allowed to invest what they pay in debt service in their economies, then they might be able to develop more quickly and become better contributing members of the global community. Here again, the "charity" of forgiving the debt echoes *Les Miserables*. Returning to that story, the charity of a priest helps Jean Valjean, and he builds on that to become the effective and revered mayor of a small French town. The point is that whether Greenhill or Easterly is correct, the story of heavily indebted poor countries is not one of deadbeat nations simply looking to get out of their debts. A worthwhile study on the debt issue is Nancy Birdsall, John Williamson, and Brian Deese, *Delivering on Debt Relief: From IMF Gold to a New Aid Architecture* (Institute for International Economics, 2002).

What to do about the debt also has to be seen in the context of the poverty that a great deal of the world experiences and the immense gap between the 15 percent of so of the world's population who live in the wealthy EDCs and those who live in the LDCs. For example, the two dozen wealthiest countries in 2000 had an average per capita gross national product (GNP) of $27,680; the LDCs' per capita GNP was $1,230. That figure somewhat obscures the fact that there are over 50 countries where the annual per capita GNP is $750 or less. Another stark figure is that over 1 billion people, or more than 20 percent of the combined populations of the LDCs, live on the functional equivalent of a dollar a day.

What to do about the gulf between the rich and the poor and how to relieve the suffering of the billions who live below or near the poverty line are complex issues, and there is an extensive and important literature on development. Two good studies are Michael Pettis, *The Volatility Machine: Emerging Economies and the Threat of Their Financial Collapse* (Oxford University Press, 2001) and William Easterly, *The Elusive Quest for Growth: Economists' Adventures and Misadventures in the Tropics* (MIT Press, 2001).

ISSUE 11

Are Patents on HIV/AIDS Drugs Unfair to Poor Countries?

YES: Doctors Without Borders/Médecins Sans Frontières, from "Equitable Access: Scaling up HIV/AIDS Treatment in Developing Countries" (December 2002)

NO: Alan F. Holmer, from "The Case for Innovation: The Role of Intellectual Property Protection," Statement Before the Economist's Second Annual Pharmaceuticals Roundtable (November 20, 2002)

ISSUE SUMMARY

YES: The international nongovernmental organization Médecins Sans Frontières (Doctors Without Borders) argues that the prevalence of HIV/AIDS is a crisis and that to meet that crisis, the high cost of medicines protected by patents needs to be reduced.

NO: Alan F. Holmer, president and chief executive officer of the industry trade group Pharmaceutical Research and Manufacturers of America, contends that the cost of research to develop new medicines is very high and that selling medicines at artificially low prices will harm the development of new pharmaceutical products.

AIDS (acquired immunodeficiency syndrome) has become the greatest transnational health disaster since the bubonic plague spread throughout Europe beginning in 1347. Thought to have originated in Africa, AIDS now threatens people everywhere. According to the *AIDS Epidemic Update—December 2002* report issued jointly by the UN and the World Health Organization (WHO), more than 65 million people have been infected by HIV (human immunodeficiency virus, the AIDS virus), and nearly one-third of those have died of AIDS-related diseases since the epidemic began. The morbid statistics are getting worse. In 2002, 5 million new cases were reported, and 3 million people died. This increased the number of people who are HIV-positive to 42 million.

Although HIV infections and the development of AIDS represent a pandemic that has assaulted people everywhere, its incidence is not spread evenly across the globe. Sub-Saharan Africa has been the hardest hit region. Seventy percent of those who are afflicted with HIV/AIDS are in sub-Saharan Africa, and these 29.5 million people constitute nearly 5 percent of the region's population. The disease is killing some 2.4 million people a year there, and there are some 11 million AIDS orphans, some of whom are themselves infected and will meet an early death.

Although it is hard to find bright spots in the grim saga of HIV/AIDS, there are some. One is that the global community has increased efforts to prevent new infections through the promotion of safe sex, to improve the medical care of those who have contracted the disease, and to seek medicines that will cure or prevent the disease. The Joint United Nations Programme on HIV/AIDS (UNAIDS), an umbrella organization for the WHO and other international agencies, has established the Global Fund to Fight AIDS, Tuberculosis and Malaria, and it is spending over $500 million annually on HIV/AIDS programs.

Yet there are major trouble spots. Because of the poverty of many of the most infected countries, the delivery of HIV/AIDS prevention and treatment programs in sub-Saharan Africa and elsewhere is still limited. This problem is worsened by the high cost of antiretroviral drugs. The result, according to one UN report, is that less than 4 percent of the people in poor countries who need antiretroviral drug treatment were receiving it at the beginning of 2002.

This brings us to the focus of this debate. National patents (intellectual property) protection was extended to international trade by the World Trade Organization's (WTO) Agreement on Trade-Related Aspects of Intellectual Property Rights (the TRIPS Agreement). Among other provisions, the TRIPS Agreement requires WTO members (which includes almost all countries) to protect patents for 20 years and to recognize the exclusive right of pharmaceutical companies to sell the drugs they have developed.

There are provisions in the TRIPS Agreement that allow countries to ignore these provisions in certain cases of urgent need and to compel their own companies to sell drugs at a set price. Also, at a meeting of the WTO members at Doha, Qatar, in late 2001, the organization adopted regulations that permitted greater flexibility in the TRIPS Agreement. Furthermore, pharmaceutical companies have sent large shipments of drugs at less than normal market prices.

Critics charge, however, that these efforts are not sufficient. Among other things, most of the neediest countries do not have their own drug industries, and WTO rules still place major obstacles in the way of drug companies in countries such as India, preventing them from ignoring patents and producing cheap generic versions of drugs for shipment to sub-Saharan Africa and elsewhere. In the following selection, the organization Doctors Without Borders/ Médecins Sans Frontières argues that patent rights should not prevent affordable drugs from reaching those who need them. In the second selection, Alan F. Holmer argues that abandoning the protections afforded by patents and the TRIPS Agreement would be detrimental to the future of new drug development and, therefore, the long-term health of the world's population.

Doctors Without Borders/
Médecins Sans Frontières

Equitable Access: Scaling Up HIV/AIDS Treatment in Developing Countries

HIV/AIDS Treatment Is Feasible in Developing Countries—Where Is the Political Will to Scale Up?

95% of the 42 million people who have HIV/AIDS worldwide live in developing countries. Yet an estimated 6 million people in developing countries are in immediate need of life-sustaining antiretroviral therapy (ART), and only around 300,000 are receiving them.

For many years, Doctors Without Borders/Médecins Sans Frontières (MSF) has been caring for people living with HIV/AIDS in developing countries. Medical projects include prevention efforts, voluntary counselling and testing, psychosocial support, and prophylaxis and treatment of opportunistic infections. Some of these programmes focus on reducing the transmission of the virus from pregnant mothers to their children. Starting in 2001, MSF has also been providing antiretroviral treatment to patients with HIV/AIDS in Cambodia, Cameroon, Guatemala, Honduras, Kenya, Malawi, South Africa, Thailand, Uganda and Ukraine. We now treat almost 2,300 patients in these countries, including 100 children. In the next year, MSF will double its intake of patients in existing projects, and will open new projects in Burkina Faso, Burma, Ethiopia, Indonesia, Laos, Mozambique, Peru, Rwanda, Zambia, and Zimbabwe.

Our experience has demonstrated that providing effective treatment is not only feasible in resource-poor settings, but has concrete clinical benefits and dramatic effects on the lives of individuals and their communities. Consolidated data from seven MSF projects in South Africa, Malawi, Kenya, Cameroon, Cambodia, Thailand and Guatemala showed that, among the 743 patients followed, the probability of survival was 93% at six months. At six months, patients who were weighed had gained an average of 3kg; patients who had CD4 cell counts taken had an increase of 104 cells/mm^3 on average. And in the three projects that systematically tested viral load at six months of treatment, 82% of patients showed undetectable levels of virus in their blood (<100 copies/ml).

Compliance rates were also impressive, with 95% of patients reporting taking their medicines properly at six months. In addition to the clinical benefits of providing ART, the provision of treatment has also strengthened prevention efforts by, for example, providing an incentive for people to come forward for voluntary counselling and testing, promoting openness about HIV and reducing stigma. . . .

While we know that treatment is possible, scaling up to national level has only been successfully implemented in a handful of countries.

Obstacles to Access

The greatest obstacle to access is lack of money. The Global Fund to Fight AIDS, Tuberculosis and Malaria clearly illustrates the extent of the funding deficit: as of October 2002, only US$2.1 billion had been pledged—a fraction of the estimated amount required each year to tackle AIDS alone. A recent analysis of country pledges examined how closely country contributions matched their share of the total need. The three wealthiest countries—the US, Japan and Germany—all contributed 7% or less of their share. The UK contributed 13%, France 11%, Switzerland 4%, and Austria, Denmark and Finland, 1%. If the countries had all funded at the level required to make up US$9 billion, it would have represented a mere 0.035% of each country's Gross Domestic Product (GDP).

Donors have skirted their responsibilities and repeatedly broken promises made over the last two years. The US has pledged US$200 million to the Global Fund to Fight AIDS, TB and Malaria in 2003, for example, representing just one-sixth of what the Global Fund itself estimates is needed for that year. No US funds have yet been appropriated for 2003. Meanwhile, the initial 2003 European Commission proposal for the EU [European Union] Programme for Action called for a 50% cut from the 2002 allocation to the Global Fund.

Where political will is matched with funding, the fortification of health care systems and the wide availability of affordable medicines, countries can achieve dramatic results with their HIV/AIDS treatment programmes. In Brazil, for instance, universal access to free AIDS treatment led to a 54% reduction in AIDS deaths between 1995 and 1999, and government savings totalled US$472 million between 1997 and 1999, thanks to prevented hospitalisations and a reduction in the burden of opportunistic infections. Other countries, including Cameroon, Uganda, Senegal, and Thailand are beginning to scale up. With a concerted effort and sufficient international support, many countries confronted with low or medium prevalence levels, such as those in Eastern Europe or Central America, would be able to move toward universal coverage. In some high prevalence countries, a more incremental approach will be needed.

Moving Towards Equitable Access

Drug budgets constitute a significant part of overall programme costs. For this reason, MSF supports the implementation of "Equitable Access" to keep drug prices down. Equitable pricing is based on the principle that the poor should have access to, and pay less for, essential medicines. Drug prices must be fair, equitable and affordable, both to individuals and the health systems that serve them. To ensure this, drugs must be bought from lowest cost reliable suppliers through competition and local production. WHO [World Health Organization] and UNICEF [United Nations Children's Fund] should offer technical support, including pre-qualification of medicines, bulk purchasing and assistance in overcoming patent barriers to access more affordable medicines.

Stimulating Generic Competition

An important element of equitable pricing is generic competition, a strategy that has proven to be the most effective means of lowering drug prices. During the last two years, originator companies have only consistently cut the prices of their drugs when faced with generic competition and international public pressure. . . . While in May 2000, originator prices were over US$10,000 per person per year, by July 2002, they had dropped to just over US$700 per person per year. In the meantime, generic prices sloped down to US$209 per person per year. To reduce prices further from current levels, it is essential to increase competition further up the line, i.e. between raw material manufacturers, which are still primarily in monopoly positions.

For developing countries that do not have strong regulatory authorities, creating a reliable process to assess the quality of generics of HIV/AIDS related medicines may be difficult. For this reason, WHO's pre-qualification system is an essential service. When countries are forced to buy drugs at a premium from originator companies because they do not have systems to assure quality of generics, drug prices are kept artificially high. Pre-qualification facilitates the ability of poor countries to pursue the best offers on the world market.

No single strategy will be sufficient to achieve and sustain equitable drug prices. Rather, what is needed is a comprehensive system of mutually supportive strategies that include generic competition, a systematic, transparent approach to differential pricing, bulk procurement, and local production.

Differential Pricing

Differential pricing policies refer to drug makers' price reductions for those countries classified as Least Developed (UNCTAD [United Nations Conference on Trade and Development] classification) and medium Human Development Index (UNDP [United Nations Development Programme] classification). In the past, differential pricing has been successfully implemented for vaccines and oral contraceptives, with drugs costing as much as 200 times less for poor countries.

So far, the most publicised effort to reduce prices has been the industry-led Accelerating Access Initiative (AAI). In the AAI, UNAIDS [Joint United Nations Programme on HIV/AIDS] and WHO help countries develop HIV/AIDS plans and play a facilitator role between countries and companies to negotiate discounted prices. But in this system, companies are free to set the rules, and they offer very different levels of discounts to selected countries and types of purchasers. Country eligibility restrictions for these prices also vary widely between companies. Unfortunately for people with HIV/AIDS in Central America, for instance, originator drug companies have decided, with the exception of Merck, to handle discounts on a case-by-case basis. The governments of Guatemala and Honduras, who have chosen to buy exclusively from originator companies, are paying up to twice as much more than MSF, which uses a mix of generics and originator products in its pilot projects.

UNAIDS and WHO, which have been helping countries negotiate within the context of the AAI, have now started encouraging the involvement of generic companies. In addition to encouraging competition where possible, WHO needs to replace ad hoc price reductions from originator companies with an international approach that specifies levels of reductions and ensures that all developing countries benefit from reduced prices.

Making Use of the Doha Declaration on the TRIPS [Trade-Related Aspects of Intellectual Property Rights] Agreement and Public Health

At the 4th Ministerial Conference of the World Trade Organisation (WTO) in Doha in November 2001, 142 countries adopted an historic Declaration on the TRIPS Agreement and Public Health, which firmly placed public health needs above commercial interests in international trade negotiations. Although the Declaration is a strong political and legal document, countries will only benefit if they enact pro-public health intellectual property rights legislation and start routinely issuing compulsory licences to ensure the availability of more afford-able medicines.

One year after the Doha Declaration, this advance is at risk of being com-promised in favour of the economic interests of developed countries. In partic-ular, the key issue of "production for export", which remained unresolved in Doha, could potentially hamper poor countries' access to medicines. In 2005, once the TRIPS Agreement is fully implemented, countries such as India and Thailand, which produce generic medicines and export them to other develop-ing countries will no longer be able to do so for new drugs. The Doha declara-tion granted least developed countries the option of a ten-year extension to the TRIPS implementation deadline, which means that they would not be obliged to provide patent protection for pharmaceutical products until 2016, and can therefore use generic versions of antiretrovirals [ARVs] and other new drugs. But how will these countries find suppliers of generics after 2005, when the pro-ducing countries must become TRIPS-compliant? Unless an export exception is allowed, non-producing countries will have only a theoretical right to use af-fordable generic medicines—in practice, they will not have access to them.

Along with WHO, the European and Belgian parliaments, the French gov-ernment and other NGOs [nongovernmental organizations], MSF is supporting a patent exception rule to allow countries to produce for export in order to sup-ply countries that cannot manufacture their own pharmaceutical products. However, the US and the EU are opposing such a solution. In addition, the US is attempting to undo the gains in Doha through negotiations on bilateral and re-gional trade agreements such as the Free Trade Area of the Americas (FTAA). The US is pushing to impose "TRIPS-plus" requirements, which directly contradict the spirit and letter of the Doha Declaration.

Regional and Local Production Through Licensing and Technology Transfer

The capacity to produce quality ARVs exists today in many developing countries. Brazil and Thailand have dramatically increased affordability of ARVs by producing them locally within government manufacturing organisations. Local production is a long-term, sustainable strategy that has the added benefit of stimulating the economic development and autonomy of developing countries. Existing production capacity should be enhanced and used to produce the drugs that are needed. This can be achieved through voluntary licensing agreements with originator companies—or compulsory licensing if originator companies choose not to cooperate—and technology transfer. In practice, a patent-holder is more likely to grant a voluntary licence if a country has a strong compulsory licensing system. Under TRIPS, developed countries are obligated to transfer technology to least developed countries.

Adapting Treatment to Clinical Reality

There is an urgent need to simplify treatment and monitoring protocols for resource-poor settings. This entails the quick introduction of affordable monitoring tools and fixed-dose, once-a-day formulations, as well as operational research to explore adapting treatment to clinical reality.

Fixed-Dose Combinations (FDCs)

FDCs can improve compliance and avoid development of resistance, and are recommended by the WHO's 12th Expert Committee on the Use of Essential Medicines. Ongoing pharmaceutical industry development focused on the needs of people with HIV/AIDS in the US and Europe will lead to new combinations that could potentially be used in developing countries. Of particular interest are the ongoing studies of once-a-day formulations and single once-a-day formulations being approved. Generic producers of ARVs are particularly well placed to develop fixed-dose combinations of recommended regimens. In fact, the only currently affordable fixed-dose formulations in developing countries are being produced by generic manufacturers. MSF supports WHO's call to make once-a-day fixed-dose combinations of ARVs available at US$70 per patient per year.

Moving Toward a Simplified Approach

On a parallel track to new formulation development, institutions and physicians working in resource-poor environments should be supported to develop simplified treatment protocols. Research focused on ways to better use clinical markers to reduce dependence on biological monitoring should be supported, for example. At the same time, tests that require less investment and operator expertise . . . should be further developed.

Among the additional research priorities, there is an urgent need for paediatric formulations adapted for use in developing countries. For example, syrups are often very difficult to store and are packaged in a manner that makes dosing complicated. Solid dosage formulations are often not available in strengths designed for children. One example is the antifungal drug flucona-zole, which does not exist as a syrup and is difficult to dose with available capsule strengths.

It is not an option to rely exclusively on the private sector, which has no incentive to adapt treatments to developing countries. Researchers in developing countries, along with groups of people living with HIV/AIDS, must have the support of their governments and the international community to further define and implement a research agenda.

Additional Challenges to Scaling Up

Clearly, there are numerous barriers to scaling up access to treatment in poor countries that are beyond the scope of this document, including those related to weak health systems and poor access to health care in general. But two additional factors contribute to the current bottleneck: insufficient training of health care workers and the lack of support for local NGOs and associations of people living with HIV/AIDS.

In addition to increased political commitment from donor countries and developing countries to address the needs laid out above, several other strategies are needed to scale up treatment in resource-limited settings. These include large-scale training initiatives for nurses, counsellors, medical officers, physicians and other care providers, and supporting community mobilisation. Led by people living with HIV/AIDS, local initiatives should provide treatment, carry out HIV/AIDS educational programmes including "treatment literacy" components, and develop advocacy initiatives to strengthen the role of people living with HIV/AIDS in formulating and implementing national and international AIDS programmes and policies.

An example of local mobilisation, the Pan-African HIV/AIDS Treatment Access Movement (PATAM), was launched in August 2002 to mobilise communities, political leaders and all sectors of society to ensure access to ARV treatment as a fundamental part of comprehensive care for all people with HIV/AIDS in Africa. Local and regional networks such as PATAM and its member organisations should be supported.

AIDS is the world's most disastrous pandemic and it requires bold unhesitating action. There is no reason to remain silent or polite about the more than 8,000 people we lose to AIDS every day. We must strive collectively to make effective HIV/AIDS treatment a reality for the millions of people who need it. To reach the millions in need, governments and the international community must immediately scale up treatment. This can be done by taking measures to significantly increase funding, continue bringing down the price of AIDS drugs, introduce easier-to-use combinations of drugs, support and carry out operational research to adapt care models to resource-poor settings, train health-care workers, and strengthen local and regional NGOs and networks of people living with HIV/AIDS.

TIME TO TREAT: TRANSFORMING AIDS TREATMENT FROM RIGHT TO REALITY

My name is Fred Minandi. I am 42 years old and I am a farmer in Malawi. I have got a wife, who is also a farmer, and two children. The eldest one is married while the other one is still at school. I have the chance to be one of the patients who is benefiting from ARVs [antiretroviral drugs] in one of the MSF projects described by Jean-Michel previously. My village is in a rural district and I have a grass-thatched house with one bedroom. I am getting money because of my garden where I grow crops for food.

Malawi is probably one of the poorest countries with the highest incidence of HIV in Africa. Fifteen percent of the adult population is infected with HIV. Where I am living, 24% of the pregnant women who are tested are HIV positive. Ten percent of the youth from the secondary school are also infected. These figures have been given to me, but what I see in my village is young people too sick to be able to cultivate their gardens and to feed their families. Families are spending a lot of money trying to find drugs, which could help the sick ones, without result. I have seen parents dying, leaving orphans to be cared for by the elderly who have no means and no hope of giving them a future.

Personally, I started to be sick in 1997 and I have been on and off during four years up to a point that it was difficult for me to do anything or work. I was tested in 2001 and I have been lucky to live where MSF had just started ARV treatment. I am one of the first patients to get ARVs for free in Malawi, and if I am here today able to speak with you it is because of this treatment.

There are some people who say that in Africa people will not be able to take drugs because they cannot tell time. I can assure you that I have no watch, but since I started taking my triple therapy in August last year, I haven't missed one dose.

Why? Because Margot, the MSF nurse, took a bit of time to explain to me how these drugs were working and that if I was not serious enough to take them very regularly, the drugs could not work any more. She explained the side effects and that it will be better after a while and it was tough. I got nauseous at first but after two months it disappeared. In fact, for me these drugs are much easier to take. Today, my life is attached to these drugs and I don't think I can forget them.

All my friends who are in the People Living With AIDS [PLWA] support group are feeling the same. We are 60 in this group and we meet regularly every month to be able to speak together. We explain to the ones who get treatment what to do if they have side effects and explain to them that it will stop. We help each other if one is too sick to come to collect their drugs. For some of us who have nobody to tell about our HIV and our treatment, we become guardians. It means that we are going with him to the compliance session and helping him to take the drugs at home. We try to support the family if one of us dies but mainly we speak about how we have improved and how the treatment gave us our life back and the improvement we notice almost each and every day. We try now to have some small income-generating activities to be able to support some of us who are poor. We inform our community about all the medical services, which are available for everybody in the district: the HIV clinic, the program for the pregnant women infected, the voluntary counseling center, where to get condoms, where to get information.

I had 107 CD4 [medical indicator for the body's natural resistance capacity to infections] when I started the treatment and today I have got 356 CD4 and I am very proud. When I was sick then, I knew I had HIV, but I would never admit it or speak about it. Speaking about it would have not changed anything for me except making me depressed. My neighbours were seeing me becoming weaker and weaker every day. Of course, they all knew what I had, but nobody asked me. They just gradually started to not come see me. Most of the people are like that in Malawi: they don't speak because they don't want to know. It is why my country is dying in silence.

Today, I am back in my field, back in my church. I can feed my family. I used to harvest only about two bags of maize for the past years because I was too weak. Now I am talking of harvesting 10 bags of maize just this year alone. I feel I have a future. My neighbours started coming to see me again like before.

I myself feel I have changed. I can now talk to my children, telling them that I have AIDS.

I tell my neighbors, too. When there is a meeting in the district, I go and speak about my disease because I am not ashamed anymore and I know I will not be rejected because I am just like everybody else. Somebody with AIDS, who is very sick, makes everybody afraid because you see your own death in his eyes. But when you are under treatment, you feel and look better, and then people do not reject you. I would like to say to all the people here that treatment is the best tool against stigma. I used to think that there was no hope for those of us living with HIV, but treatment has changed this.

There are some other people who also said that treatment is too expensive and therefore only prevention should be done in a country like Malawi. First of all, would that mean that I would have to die with all my friends of the PLWA group as well as with the 1 million people who are already infected in Malawi? For us, talking about prevention is too late, but I believe we still have the right to be treated. I would like to say that I consider my life important not only for me but for my family and my country as a whole. I feel whatever little I contribute for my family, friends, relatives, neighbors and to the Malawi nation, could be just a complement of the work of all those who are feeding Malawi.

Secondly, I think having affordable treatment available will encourage people to know about their serostatus rather than when there is nothing to be offered. It will help break the silence. My friends and I from the PLWA group have been trained as counselors and we are convincing a lot of people to go for testing because they know that if they are already positive they will be treated. They know they will have a real benefit for themselves. They see what the treatment has done for us.

Thirdly, we are using generic medicines from India in the program in Malawi, which keeps the price as low as possible. The less expensive the drugs, the less expensive the program, and the more people can be treated.

I would like to say that prevention and treatment are linked together. I am sure that today I can convince more people to change their behavior, to use a condom, to go for testing. Why? Because I am treated myself. I can also tell them that treatment is still difficult and that they have to try hard to stay negative.

I would also like to ask those people who say we should only do prevention: if this epidemic were claiming so many lives in your community, would you really accept letting all of us already living with HIV die?

I hope I will be able to meet a lot of people who are also HIV positive and who could help me become more active. We are the ones who need these drugs and we are the ones who need to fight for it. I hope that our voice will convince the people who decide to give money for drugs to do it and not only for prevention.

In closing, I would like to appeal to all the pharmaceutical companies manufacturing ARVs to reduce the prices of their drugs for all low income countries, to our national governments to support treatment and ask for help, and to rich governments to give us the money to be able to do it.

The war against HIV/AIDS can only be won if we all come together.

Source: Statement of Fred Minandi of Chiradzulu, Malawi, Delivered July 7, 2002 by Fred Minandi in Barcelona, Spain, at a satellite meeting co-sponsored by MSF and Health Gap of the XIV International AIDS Conference

NO ⤶

Alan F. Holmer

The Case for Innovation: The Role of Intellectual Property Protection

[The pharmaceutical industry is] America's greatest, most creative industry—an industry that is the unquestioned leader in creating value for humankind.

How? By staying committed to discovering cures for illnesses, preventing disease, responding to worldwide health crises, and working to expand access to prescription medicines.

In order to create this value, pharmaceutical companies need the enabling conditions of innovation. . . .

I will focus . . . on what may be the single most critical but also least understood enabling condition of pharmaceutical innovation: intellectual property incentives. Pharmaceutical intellectual property [patents] is an issue because of the widely-held view that prescription medicines are making health care unaffordable. Therefore, I'll start by addressing pharmaceutical costs head-on.

If you read the press and listen to the debate in Congress, you constantly hear claims that prescription drug costs are unsustainable, unaffordable, and harming the competitiveness of American businesses. But if you look at the facts, you'll learn that these claims are wrong.

Prescription medicines make up about ten cents of each health care dollar. And that ten cents pays for a lot more than buying brand name drugs from manufacturers. It also pays for generic drugs, and the costs of pharmacies, and prescription benefit managers, and so on.

When you look at the medical progress we're buying for ten cents out of the health care dollar, prescription medicines are an extraordinary value. Consider just a few of the new types of medicines invented by pharmaceutical research companies in recent years:

- antipsychotic medicines that have revolutionized treatment for mental illnesses,
- protease inhibitors that dramatically improve outcomes for patients with HIV/AIDS,
- cholesterol-lowering medicines that prevent heart attacks,

- the first treatments for Alzheimer's disease, and
- medicines that allow children with asthma to breathe freely and go to school, rather than the emergency room.

We could spend less than ten cents of the health care dollar on prescription medicines if pharmaceutical research companies hadn't created these and many other advances. But imagine the cost, first in human terms, and then in economic terms, such as higher hospital and nursing home expenditures and lost productivity. . . .

Turning now to intellectual property. Intellectual property [IP] is the essential foundation for all of the wonderful progress I've already mentioned—and for the progress yet to come. Patents allow companies to invest the $800 million it takes, on average, to discover and bring to patients just one new medicine—with seven out of ten that do reach the market failing to recoup average R&D [research and development] costs.

Today, some policymakers are proposing to further weaken pharmaceutical patents. Weakening intellectual property would constrain progress in treating and curing disease, precisely when the aging of our society suggests we must be working the hardest and investing the most on making medical care better and more affordable.

When we discuss the importance of patents to promoting pharmaceutical research and innovation, we're sometimes met with skepticism. But consider this—copying most pharmaceuticals costs a small fraction of the amount invested in the research and development of that medicine. Without patents, competitors who have not invested in research could easily undercut the price charged by the company that did make the investment. No company could afford to invest hundreds of millions of dollars per drug in research on such terms.

While the role of intellectual property is not always well understood in the pharmaceutical context, it frequently is recognized as essential to other industries. For example, media companies understand that they could not remain in business without copyright protection for their intellectual property. Notably, the duration of pharmaceutical patents is far shorter than copyright. . . .

In the developing world [many are] attacking patent protection. Two articles that appeared in [*The Economist*] in September [2002] talked about "the wrongs of intellectual property rights" in the developing world.

I respectfully disagree with [that] position on this issue. So do the leaders of developing countries around the world. Rather than roll back IP protection, these governments have expedited adoption of WTO TRIPS [World Trade Organization Agreement on Trade-Related Aspects of Intellectual Property Rights] disciplines in the realization that development follows IP protection, and not the reverse. Egypt has recently adopted comprehensive IP legislation for the purpose of meeting its WTO obligations. China has followed through on its WTO commitments to put into place effective protection for confidential test data, known as data exclusivity, and Colombia has also implemented a Data Exclusivity Decree. Even India has recently passed legislation intended to meet its current WTO obligations and is considering steps needed to provide protection for clinical data. Singapore and Jordan have agreed to enhance their protection of intellectual property relating to pharmaceuticals through Free Trade Agreements with the U.S. In Jordan, the local pharmaceutical industry has increased exports by 25% since WTO accession in 1999, contrary to fears that it would be harmed by adoption of pharmaceutical patents. PhRMA [Pharmaceutical Research and Manufacturers of America] members have also turned to Jordan as a promising venue for clinical trial and other research activities, given the improved intellectual property environment under King Abdullah II.

In order to provide medicines to poor people in developing countries, we first have to make sure the medicines exist. But new medicines won't continue to be developed absent strong intellectual property incentives, since investors will shy away from risky, expensive research it takes to come up with cures.

Where, then, will the next generation of AIDS medicines—or the vaccine against AIDS that developing countries so desperately need—come from? Just as researchers in Thailand, a county with strong intellectual property, recently developed a new, patented vaccine for Dengue Fever, with more and more developing countries recognizing the value of IP to spur research, perhaps new cures in the next century will also emerge from research in the developing world.

While we rejoice that our HIV/AIDS medicines have helped reduce U.S. AIDS death rate dramatically over the last decade, we share the world's horror that millions continue to die of AIDS in the poorest countries, leaving millions of orphaned children behind. What is the pharmaceutical industry's response?

First, the pharmaceutical companies that discover and develop new medicines have donated nearly $2 billion in assistance—medicines, funds, expertise —over the last four years alone. Public health experts and, more importantly, patients, are familiar with our companies' extraordinary efforts. We cannot alone solve the public health crises that result primarily from poverty, inadequate health infrastructure, inadequate international financial help and, in some cases, local strife or corrupt governance. But we are contributing partners, working with national and international organizations as well as local authorities to make a difference for patients in the poorest countries.

Researchers have shown that not many AIDS medicines are patented in the poorest countries, and that those limited patents are not a significant, much less leading, barrier to more effectively treating disease in the developing world. The vast majority of medicines judged "essential" by the World Health Organization [WHO] for the developing world are not patented and may be copied freely. But we in the pharmaceutical industry want to be part of a solution to improve access to needed medicines for patients of grave epidemics like HIV/AIDS, malaria and TB in the poorest countries. That's why, second, we support the [WHO's] Doha Declaration on TRIPS and Public Health, which gives flexibility for developing countries through TRIPS provisions on what is called "compulsory licensing" to improve their access to medicines. In plain English, that means in certain circumstances that a patented medicine may be copied even without the patent owner' s consent. The Doha Declaration underscores that patients around the world need us to remain true to two goals: both improving access to medicines already invented at huge risk and cost, and also continuing the high-risk, costly search for new medicines and cures. That's why [Gro Harlem] Brundtland, head of the WHO, has said that patent protection is essential to public health: by encouraging investment in the search for more medicines to treat HIV/AIDS and other dread diseases, and vaccines to prevent them.

Third, we support negotiations to find a "solution" for the very poorest countries battered by the unprecedented pandemic of HIV/AIDS and epidemics of similar gravity and scope like malaria and TB, that lack manufacturing capacity in the pharmaceutical sector. The flexibility in TRIPS to manufacture a patented medicine locally without the patent owner's consent doesn't help patients in a country that lacks manufacturing capacity. Therefore, in certain circumstances, we support the ongoing efforts to find a "solution" to help these particular countries by allowing what is otherwise prohibited by TRIPS: compulsory licensing for exports. Again in English, that means in certain circumstances allowing a country to import patented medicines without the patent owner's consent, if that country lacks capacity to manufacture the medicine locally.

But any wholesale evisceration of TRIPS would harm public health, for the reasons Mrs. Brundtland has proclaimed: because strong patent protection is the incentive for innovation that benefits patients. The CDC [Centers for Disease Control and Prevention] has documented that drug-resistant strains of HIV/AIDS are growing in the U.S. and Canada; the current medicines are not enough for patients with these strains. And we lack the vaccine that most believe will be most helpful and cost-effective in the developing world. That's why we do *not* support the efforts of leading copy countries like Brazil, India and China—which have companies that thrive by copying our innovative medicines without the patent owner's consent—to open up compulsory licensing for export in broader circumstances. They propose essentially to gut TRIPS for medicines, so that any country (including developed countries) could make and export any medicine (including for chronic diseases like diabetes, asthma and heart disease) without the patent owner's consent. They propose that any country (including developed countries) could import such medicines (including for chronic diseases) without the patent owner's consent. And they propose that none of the decisions to take these actions would be subject to any WTO review. Instead, each country would make its own decision and simply notify the rest of the world of its unilateral decision.

If we wanted to send patients with unmet medical needs around the globe a message of hopelessness, this would be constructive. But we in the pharmaceutical industry want to continue to try to serve *both* groups of patients: those who need better access to the medicines already discovered and developed, *and* those whose medical needs are still unmet, whose hopes for better health rely primarily on our drug R&D, which in turns relies heavily on strong patent protection.

We can, and we must, find a reasonable balance between access and patent protection. Both goals are essential to help patients. We will continue to support the efforts underway in the WTO to essentially disregard patent rights for the patients in the poorest countries battered by HIV/AIDS, malaria, TB and other epidemics of similar gravity and scope, where the country cannot make its own medicines because it lacks pharmaceutical manufacturing capacity. But to protect the patients in America and around the world who depend on our industry to continue the search for new medicines and cures, we will resist broadening it to apply beyond the bounds for which it was intended. . . .

One of the negative things you hear about the drug industry is that we fight hard to protect intellectual property. Our critics say this is just about money. They don't understand that money is the means to an end. The end is to create new medicines that give value, quality of life, and life itself to patients and their families.

POSTSCRIPT

Are Patents on HIV/AIDS Drugs Unfair to Poor Countries?

It is easy to jump to an emotional conclusion on the issue of whether or not the high costs of patent-protected HIV/AIDS drugs (and many other medicines) are unfair to poor countries facing health crises. After all, millions are dying every year in countries where paying for drugs is not just difficult, it is impossible. It is also easy to dismiss the pharmaceutical companies' discounting of prices as mere window dressing and to condemn the CEOs of those companies as capitalist monsters who care more for their ledger sheets and stock prices than for human life. Certainly there are many who take this approach.

There is, however, another side of the debate that, while less appealing emotionally, cannot be easily dismissed. Pharmaceutical companies spend vast sums of money doing research to develop new drugs. When a company makes an important breakthrough, it can earn huge profits because the cost of producing that drug will be much lower than what it sells for. Focusing on a single such item can make the drug companies seem greedy beyond redemption. However, for every successful medication that is developed, there is an enormous amount of research that leads nowhere. The process to produce a drug takes many years and is expensive. In 1999 the U.S. pharmaceutical industry association estimated that the average research and development (R&D) cost for each drug that successfully came to market was $500 million and that only 1 in every 1,500 experimental drug programs proved worthy of licensing and production. Many analysts contend that those figures are substantially padded, but even discounting the cost by 50 percent would still leave substantial R&D costs. Thus, while siding with the drug companies might not feel as uplifting as criticizing them, there is something to Holmer's argument that undercutting the profits of the firms that produce drugs would dry up the supply of new, innovative medicines being developed. More on the views of the Pharmaceutical Research and Manufacturers of America can be found on its Web site at http://www.phrma.org.

Some people might argue that since the poor cannot afford the drug anyway, what is the loss if the industry just gives it to them or allows firms in less developed countries to make a generic version cheaply? The answer that Holmer would probably give is that there is a strong possibility that at least some of the free or steeply discounted medicine would not reach the patients it was intended for but instead be siphoned off by corrupt officials in the recipient countries and sold on the black market in wealthy countries, thereby diminishing sales of the legitimate drug. Similarly, companies fear that generic versions of drugs that are meant for poor countries would begin to legally or illegally be imported into wealthier countries. On the general topic of patents, copyrights, and other types of intellectual property protection in the international political economy, see Keith E. Maskus and C. Fred Bergsten, *Intellectual Property Rights in the Global Economy* (Institute for International Economics, 2000). An analysis of the TRIPS Agreement can be found in Carlos Maria Correa, *Intellectual Property Rights, the WTO and Developing Countries: The TRIPS Agreement and Policy Options* (Zed Books, 2000). A more recent report published by Médecins Sans Frontières is "Drug Patents Under the Spotlight: Sharing Practical Knowledge About Pharmaceutical Patents," by Pascale Boulet, Christopher Garrison, and Ellen 't Hoen (May 2003), which can be found online at http://www.accessmed-msf.org/documents/patents_2003.pdf.

In the end, the solution has to provide medicines at affordable prices to poor people in poor countries (and poor people in wealthy countries) while not constraining the profits of the pharmaceutical industry so much that the R&D ends. At least one initiative was taken by the United States in January 2003, when President George W. Bush announced that he would triple U.S. contributions to international HIV/AIDS programs to $15 billion over five years. A significant part of these funds will go to providing drugs to countries that cannot afford them. Information about the global HIV/AIDS crisis and the international effort to ease it can be found at the UNAIDS Web site at http://www.unaids.org.

U.S.-Russia Developments

This site, maintained by the Acronym Institute for Disarmament Diplomacy, provides up-to-date news and analysis of U.S. and Russian disarmament activity.

http://www.acronym.org.uk/start/

The Center for Security Policy

The Web site of this Washington, D.C.–centered "think tank" provides a wide range of links to sites dealing with national and international security issues.

http://www.centerforsecuritypolicy.org/index.jsp?section=today

Selected Papers on National Security

A supportive view of the military can be found at this Heritage Foundation site.

http://www.nationalsecurity.org/terrorism.html

Office of the Coordinator for Counterterrorism

This worthwhile site explores the range of terrorist threats and activities, albeit from the U.S. point of view, and is maintained by the U.S. State Department's Office of Counterterrorism.

http://www.state.gov/www/global/terrorism/

Centre for the Study of Terrorism and Political Violence

The primary aims of the Centre for the Study of Terrorism and Political Violence are to investigate the roots of political violence; to develop a body of theory spanning the various and disparate elements of terrorism; and to recommend policy and organizational initiatives that governments and private sectors might adopt to better predict, detect, and respond to terrorism and terrorist threats.

http://www.st-and.ac.uk/academic/intrel/research/cstpv/

Worldwide Guide to Women in Leadership

This site is an excellent compilation of women who have served as party, legislative, and executive political leaders.

http://www.guide2womenleaders.com

PART 4

Issues About Violence

*W*hatever we may wish, war, terrorism, and other forms of physical coercion are still important elements of international politics. Countries calculate both how to use the instruments of force and how to implement national security. There can be little doubt, however, that significant changes are under way in this realm as part of the changing world system. Strong pressures exist to expand the mission and strengthen the security capabilities of international organizations and to gauge the threat of terrorism. This section examines how countries in the international system are addressing these issues.

- Does the Moscow Treaty Advance Nuclear Arms Reductions?

- Are Military Means the Best Way to Defeat Terrorism?

- Is Government-Ordered Assassination Sometimes Acceptable?

- Would World Affairs Be More Peaceful If Women Dominated Politics?

ISSUE 12

Does the Moscow Treaty Advance Nuclear Arms Reductions?

YES: Donald H. Rumsfeld, from Statement Before the Committee on Foreign Relations, U.S. Senate (July 17, 2002)

NO: Christopher E. Paine, from Statement Before the Committee on Foreign Relations, U.S. Senate (July 23, 2002)

ISSUE SUMMARY

YES: Donald H. Rumsfeld, U.S. secretary of defense, argues that the Moscow Treaty represents a decision by the United States and Russia to move toward historic reductions in their deployed offensive nuclear arsenals.

NO: Christopher E. Paine, codirector of the Nuclear Warhead Elimination and Nonproliferation Project, Natural Resources Defense Council, argues that the Moscow Treaty is too vague to be legally binding on either the United States or Russia and, worse, that the treaty is actually a step backward in the long-standing effort to reduce, even eliminate nuclear weapons.

Almost as soon as the United States leveled Hiroshima and Nagasaki with atomic bombs in 1945, the countries of the world began devising ways to restrain the number of nuclear weapons and who possessed them. As early as 1946, the organization that is now called the International Atomic Energy Agency (IAEA) was created with the goal of limiting the use of nuclear technology to peaceful purposes.

During the 1950s the cold war kept arms control efforts in a deep freeze, but by the early 1960s worries about nuclear weapons began to overcome frosty U.S.-Soviet relations. The first major step toward nuclear arms control came in 1963, when most countries agreed to cease testing nuclear weapons in the atmosphere.

The 1970s witnessed an easing of cold war tensions, and arms control was able to advance. The Anti-Ballistic Missile (ABM) Treaty (1972) constrained U.S.

and Soviet efforts to build a ballistic missile defense (BMD) system, which many analysts believed could destabilize nuclear deterrence. During that decade, Washington and Moscow also concluded the Strategic Arms Limitation Talks I (SALT I) Treaty (1972) and the Strategic Arms Limitation Talks II (SALT II) Treaty (1979). The two treaties put important limits on the future growth of nuclear weapons.

As the cold war thawed even more during the 1980s and then ended in late 1991 with the collapse of the Soviet Union, the pace of arms control sped up even more. The effort now was not to limit the future growth of nuclear arsenals but to actually reduce them. Moscow and Washington agreed to the Intermediate-Range Nuclear Forces (INF) Treaty (1987), which banished an entire class of nuclear delivery vehicles (missiles with ranges between 500 and 5,500 kilometers). Then, after years of bargaining, the two superpowers agreed in 1991 to the Strategic Arms Reduction Talks (START I) Treaty. Each country was required to reduce their nuclear inventories to 1,600 delivery vehicles (missiles and bombers) and 6,000 strategic explosive nuclear devices (warheads and bombs).

Two years later, Presidents Boris Yeltsin and George Bush took a further step toward cutting their respective arsenals when they signed the second Strategic Arms Reduction Talks (START II) Treaty. The provisions of START II required the two countries to slash their arsenals of nuclear warheads and bombs to 3,500 for the United States and 2,997 for Russia by 2007. The treaty also has a number of clauses relating to specific weapons, the most important of which is the elimination of all multiple independently targetable reentry vehicle (MIRV) intercontinental ballistic missiles (ICBMs).

After START II, progress became less straightforward. On the positive track, Presidents Bill Clinton and Boris Yeltsin agreed in 1997 on principles for a third START Treaty, which would reduce the number of each side's nuclear warheads and bombs to between 2,000 and 2,500. This goal was seemingly advanced in May 2002 when President George W. Bush met with President Vladimir Putin in Moscow and the two signed what is officially entitled the Treaty Between the United States of America and the Russian Federation on Strategic Offensive Reductions and is commonly referred to as the Moscow Treaty. Under the treaty, the two countries agree to cut their nuclear arsenals of nuclear warheads and bombs to no more than 2,200 by 2012. Unlike START I and START II, there are no provisions relating to delivery vehicles.

Many analysts and policymakers have hailed the Moscow Treaty as an important advance that, by 2012, will reduce the U.S. and Russian arsenals to 20 percent of what they were a quarter of a century earlier. This viewpoint is taken by Donald H. Rumsfeld in the first of the following selections. Others see the treaty as deeply flawed. Critics argue that since the treaty deals with only deployed warheads and bombs, both sides can store massive arsenals and quickly reintroduce them. Also, the lack of stepped reductions means that neither side has to do anything more, including meeting the defunct START II levels, until 2012. These and other worries are voiced by Christopher E. Paine in the second selection.

Donald H. Rumsfeld

 **YES**

Treaty on Strategic Offensive Reduction: The Moscow Treaty

When President [George W.] Bush took office [in 2002], he made clear his determination to transform the Russian-American relationship, to put hostility and distrust that has been built-up over so many decades behind us, and to set our two nations on a course toward greater cooperation. Some naysayers insisted that it really could not be done. They looked at his agenda, his promise to withdraw from the ABM [Anti-Ballistic Missile] treaty, his commitment to build defenses for friends and allies, to protect the friends and allies and ourselves from ballistic missile attack, his determination to strengthen the NATO [North Atlantic Treaty Organization] alliance by making new allies of old adversaries, and the prediction was that the U.S. and Russia were really on a collision course.

The past year suggests what a difference a year can make. None of these dire predictions came to pass. To the contrary, the U.S.-Russian relationship is stronger today than perhaps at any time in my adult lifetime. Far from a clash over NATO expansion, the President has submitted a new NATO-Russian relationship that will permit increasing cooperation between Russia and the members of the Atlantic Alliance. Far from causing a deep chill in relations, the U.S. withdrawal from the ABM treaty was greeted in Russia with something approximating a yawn. Indeed, President [Vladimir] Putin declared the decision to not pose a threat to Russia, which of course it does not.

Far from launching a new arms race, the U.S. and Russia have both decided to move toward historic reductions in their deployed offensive nuclear arsenals, reductions to be codified in the Moscow Treaty before [the Committee on Foreign Relations].

Indeed, President Putin chose to announce the Russian reductions on the very same day that President Bush announced his intention to withdraw from the ABM treaty. In little over a year, President Bush has defied the critics and set in motion a fundamental transformation in the U.S.-Russian relationship, one that is designed to benefit the people of both of our nations and indeed the world. As the record shows, it is a transformation that began before the terrible events of September 11.

From U.S. Senate. Committee on Foreign Relations. *Treaty on Strategic Offensive Reduction: The Moscow Treaty.* Hearing, July 17, 2002. Washington, D.C.: U.S. Government Printing Office, 2002. (S.Hrg 107-622.)

In the last 12 months, the Presidents of the United States and of Russia have had probably more interaction and forged more areas of cooperation across a broader range of political, economic, and security issues than at any time. Today the U.S. and Russia are working together to develop new avenues for trade and economic cooperation. We are working together to fight terrorism and deal with new and emerging threats that will face both of our countries in this new and dangerous century. We are working together to reduce deployed offensive nuclear weapons, weapons that are a legacy of the past and which are no longer needed when Russia and the U.S. are basing our relationship on one of increasing friendship and cooperation, rather than a fear of mutual annihilation.

So these are historic changes, changes of a breadth and scale that few imagined and many openly doubted could be achieved in so short a period of time. Of course there is a good deal of work ahead and challenges to overcome, let there be no doubt. Our success is by no means assured, but we have an opportunity to build a new relationship for our peoples, a relationship that can contribute to peace and stability and prosperity for generations of Russians and Americans.

It will require a change in our thinking, thinking in the bureaucracy of both countries, thinking in the Congress and in the Duma [Russia's parliament], thinking in the press and in academic institutions. We have decades of momentum going in the opposite direction and we need to recalibrate our thinking and our approaches with respect to this relationship.

In both of our countries, there are those who are still struggling with the transition. Habits built up over many decades become ingrained and hard to break. Here in the U.S., there are some what would have preferred to see us continue the adversarial arms control negotiations of the Soviet era, where teams of lawyers drafted hundreds of pages of treaty text and each side worked to gain the upper hand, while focusing on ways to preserve a balance of nuclear terror.

That's an approach that President Bush rejected, insisting instead that we deal with Russia as we deal with other normal countries, in a spirit of friendship and cooperation.

Similarly, in Russia today there are those who are stuck in the past, who look warily at American offers of greater cooperation and friendship, preferring to keep us at arm's length. I have had many, many hours of meetings with them, as has General [Dick] Myers [Chairman of the Joint Chiefs of Staff].

There are others in Russia who want to see her embrace the future and take her rightful place in Europe through increased integration with the western industrialized democracies and by embracing political and economic freedom and the prosperity and improved standard of living, democracy peace and thriving culture that are the products of free societies. Sometimes these divergent impulses can even be found in the same people.

Russia and the United States entered this new century saddled with two legacies of the cold war, the adversarial relationship to which we had both grown accustomed and the physical manifestation of that adversarial relationship, the massive arsenals of weapons that we built up to destroy each other. In the past year we have made progress in dealing with both.

Last November [2001] at the Crawford Summit President Bush announced his intention to reduce the U.S. operationally deployed strategic nuclear warheads by some two-thirds, to between 1,700 and 2,200 weapons. Soon after that, President Putin made a similar commitment. These reductions, these proposed reductions, are a reflection of our new relationship.

When President [Ronald] Reagan spoke to the students at Moscow State University in 1988, he told them nations do not distrust each other because they are armed; they are armed because they distrust each other. Clearly, we do not distrust each other the way the U.S. and the Soviet Union once did.

But what is remarkable is not simply the fact of these planned reduction, but how they have happened. After a careful review, President Bush simply announced his intention to cut our stocks of operationally deployed nuclear warheads. This was the result of the nuclear posture review that we spent many months on. . . .

President Putin shortly thereafter did exactly the same thing. When they met in Moscow [in May 2002], they recorded these unilaterally announced changes in a treaty that will survive their two presidencies, the Moscow Treaty which the Senate will now consider. But it is significant that, while we consulted closely and we engaged in a process that has been open and transparent, we did not engage in the lengthy, adversarial negotiations in which the U.S. kept thousands foreign weapons it did not need as a bargaining chip and Russia did the same. We did not establish standing negotiating teams in Geneva, with armies of arms control aficionados ready to do battle over every colon and every comma.

If we had done so, we would still be negotiating today. Instead, we are moving directly toward dramatic reductions in the ready nuclear weapons of our two countries and in clearing the way for a new relationship between our countries based on increasing trust and friendship.

If you want an illustration of how far we have come in that regard, consider: There [indicating] is the START [Strategic Arms Reductions Talks] treaty, if I can lift it. It is massive. There [indicating] is the Moscow Treaty; it is three pages. The START thank you between President [George] Bush and Michael Gorbachev is 700 pages long and took 9 years to negotiate. The Moscow Treaty was concluded in the summer, some 6 months to negotiate, and it is 3 pages long.

. . . [W]e are working toward the day when the relationship between our two countries is such that no arms control treaties will be necessary. That is how normal countries deal with each other. The U.S. and Great Britain both have nuclear weapons, yet we do not spend hundreds of hours negotiating with each other the fine details of mutual reductions on offensive weapons. We do not feel the need to preserve a balance of terror between us. We would like the relationship with Russia to move in that direction, and indeed it is.

We would have made these cuts regardless of what Russia did with its arsenal. We are making them not because we signed the treaty in Moscow, but because the fundamental transformation in the relationship with Russia means that we do not need so many deployed weapons. Russia has made a similar calculation, and the agreement we reached in Moscow was the result of those judgments, those determinations, not the cause of those judgments or determinations.

That is also why we saw no need to include detailed verification measures in the treaty. First, there simply is not any way on Earth to verify what Russia is doing with all their warheads and their weapons. Second, we do not need to. Neither side has an interest in evading the terms of the treaty since it simply codifies unilateral announced intentions and reductions, and it gives both sides broad flexibility in implementing those decisions.

Third, we saw no benefit in creating a new forum for bitter debates over compliance and enforcement. Today the last place in the world where U.S. and Russian officials still sit across a table arguing with each other is in Geneva. Our goal is to move beyond that kind of cold war animosity, not to find new ways to extend it into the twenty first century.

Similarly flawed in my view is the complaint that, because the Moscow Treaty does not contain a requirement to destroy warheads removed from the missiles and the bombers, that the cuts are reversible and therefore they are not real. Put aside for the moment the fact that no previous arms control agreement, not SALT [Strategic Arms Limitations Talks Treaty], not START, not the INF [Intermediate-Range Nuclear Forces Treaty], has required the destruction of warheads, and no one offered objections to those treaties on the basis that they did not require the reductions in warheads, the destruction of warheads.

This charge is based in my view on a flawed premise, that irreversible reductions in nuclear weapons are possible. In point of fact, there is no such thing in my view as irreversible reductions in nuclear weapons. The knowledge of how to build nuclear weapons exists. There is no possibility that that knowledge is going to disappear from the face of the Earth. Every reduction is reversible given enough time and enough money.

Indeed, when it comes to building nuclear weapons, Russia has a distinct advantage over the United States. Today Russia can and does produce both nuclear weapons and strategic nuclear delivery vehicles. They have open, warm production lines. The U.S. does not produce either ICBM's [intercontinental ballistic missiles] or nuclear warheads. It has been a decade since we produced a nuclear weapon and it would likely take us the better part of a decade to begin producing some capabilities again.

In the time it would take us to redeploy decommissioned nuclear warheads, Russia could easily produce a larger number of new ones because they have an open, warm production line. But the question is why would we want to do so? Barring some unforeseen or dramatic change in the global security environment, like the sudden emergence of a hostile peer competitor on a par with the old Soviet Union, there is no reason why we would want to redeploy the warheads we are reducing.

The reason to keep rather than destroy some of those decommissioned warheads is to have them available in the event of a problem with safety or reliability in our arsenal. Since we do not have an open production line, it would be in my view simply mindless for us to destroy all of those warheads and then not having them for the backup in the event that we run into safety or reliability problems or indeed a sudden, unexpected change in the global security environment. Russia, by contrast, has less need to maintain reserve warheads since it has an active production capability.

. . . [I]f we had pursued the path of traditional arms control as some suggested, we would not be proceeding with the reductions outlined in this treaty before you. Rather, we would be still at the negotiating table arguing over how to reconcile these and a dozen other asymmetries that exist between how Russia is arranged and how we are arranged. They have different geography, they have a different technical base, they have a different GNP [gross national product], they have different currently deployed capabilities from the United States of America. So it ought not to be surprising that trying to make an agreement with countries that have those numerous asymmetries would be an enormously difficult thing had we not done what we did, namely to each look at our own circumstance and make the best judgment we could as to what was in the interest of our respective national securities.

For example, if we had said that we are going to pursue the traditional approach, we would have had to address those asymmetries. We would have had to try to balance Russia's active production capacity against the U.S. lack of a production capacity. Russia might have insisted that any agreement take into account the size of our economy and our ability to mobilize resources quickly to develop new production facilities.

We might have argued that Russia's proximity to rogue nations allows them to deter these regimes with tactical systems because they are many thousands of miles away from us; the United States' distance from them requires more intercontinental systems, possibly, than theater systems. This could have resulted in a mind-numbing debate over how many non-strategic systems, which was raised I believe by Senator [Richard] Lugar [R-Indiana], should equal an intercontinental system, or open the door to a discussion of whether an agreement should include all nuclear warheads regardless of whether they are strategic or tactical.

Russian negotiators might have countered that the U.S. advantage in advanced high-tech conventional weapons should also be taken into account, and so forth ad infinitum.

But the point is this. We do not need to reconcile all those asymmetries because neither Russia nor the U.S. has an interest in taking advantage of the other by increasing its respective deployed nuclear forces. The approach we have taken is to treat Russia not as an adversary, but as a friendly power. In so doing, we have been able to preserve the benefits attributed to arms control—the dialog, the consultations, lower force levels, predictability, stability, we hope greater transparency—but we have done so without all the drawbacks—the protracted negotiations, the withholding of bargaining chips, the legalistic and adversarial process, that more often than not becomes a source of bitterness between the participants, and the extended, embittered debates over compliance and enforcement agreements.

Because Russia and the U.S. are no longer adversaries, our interests have changed. As enemies we had an interest in each other's failure. As friends we ought to have an interest in each other's success. As enemies we had an interest in keeping each other off balance. As friends we have an interest in promoting stability.

When Russia and the U.S. were adversaries, our principal focus was trying to maintain and freeze into place the balance of nuclear terror. With the recently completed nuclear posture review, the U.S. has declared that we are not interested in preserving that balance of terror with Russia. Today the threats we both face are no longer from each other. They come from new sources, and as our adversaries change our deterrence calculus must change as well.

That is why we are working to transform our nuclear posture from one aimed at deterring a Soviet Union that no longer exists to one designed to deter new adversaries, adversaries who may not be discouraged from attacking us by the threat of U.S. nuclear retaliation, just as the terrorists who struck us on September 11 were certainly not deterred by the United States' massive nuclear arsenal.

With the nuclear posture review, President Bush is taking a new approach to strategic deterrence, one that combines deep reductions in offensive nuclear forces with new conventional offensive and defensive systems more appropriate for deterring the potential adversaries we face in the twenty first century.

Some have asked why in the post–cold war world we need to maintain as many as 1,700 to 2,200 operationally deployed warheads. The fact that the Soviet threat has receded does not mean that we no longer need nuclear weapons. To the contrary, the U.S. nuclear arsenal remains an important part of our deterrent strategy and helps us to dissuade the emergence of potential or would-be peer competitors by underscoring the futility of trying to sprint toward parity with us.

Indeed, . . . our decision to proceed with reductions as deep as the ones outlined in the Moscow Treaty is premised on decisions to invest in a number of other critical areas, such as intelligence, ballistic and cruise missile defense, and a variety of conventional weapons programs funded in our 2003 budget request.

Others have asked why there is no reduction schedule in the treaty. The answer is quite simple: flexibility. Our approach to the nuclear posture review was to recognize that we're entering a period of surprise and uncertainty, when the sudden emergence of unexpected threats will be an increasingly common feature of our security environment.

We were surprised on September 11, and let there be no doubt we will be surprised again. I was interested to note that we ought to have a healthy respect for all we do not know. . . .

If one thinks back to the rapidity with which Iran went from being a regional power in close and intimate relationship with the United States to being led by the Ayatollah [Khomeini of Iran] and hostile to the United States, we have to recognize that it is an uncertain world. It is not only an uncertain world, it is a world that, besides promising surprise and promising little or no warning, is a world that has weapons of mass destruction. So the penalty for not being able to cope with surprise or cope with little or no warning could be enormous.

Our intelligence has repeatedly underestimated the capabilities of different countries of concern to us. I say that not to be critical of the intelligence community. It stretches back over decades. But the fact is that it's a big world, there are a lot of closed societies, and we have historically had significant gaps in our knowledge, gaps where some significant event occurred in a country and we did not know about it, an important country that we were looking at, a significant event, and we did not know it for 2 years, 4 years, 6, 8, 10, in one case 12 years or 13 years before we became aware of that event.

Indeed, the only surprise is that so many among us are still surprised when we find that there were things happening in the world that we did not know. We have to accept that. This problem is certainly more acute in an age when the spread of weapons of mass destruction in the hands of terrorist states and potentially terrorist networks means that our margin of error is significantly less than it had been. The cost of a mistake could be not thousands of lives, but tens of thousands of lives.

Because of that smaller margin for error and the uncertainty of the future security environment, the U.S. will need flexibility. Through the Nuclear Posture Review, we determined the force levels and the flexibility that we will need to deal with that new world, and then we negotiated a treaty that allows both deep reductions in offensive weapons and the flexibility to be able to respond to sudden changes in our strategic environment should that be necessary.

We are working to develop the right mix of offensive and defensive capabilities. If we do so, we believe the result will be that nations are less likely to acquire or use nuclear weapons.

None of these changes in any way is a threat to Russia. Far from it, this new approach to deterrence should help us better contribute to peace and stability and address the new threats and challenges that we both will face in this century.

In this century, Russia and the United States both face new and different security challenges, not exactly the same, but certainly the threats of terrorism and fundamentalism and the spread of weapons of mass destruction to rogue states are common. The difference is that these are threats our two nations have in common and that we can face together, rather than threats from each other.

It means that we have entered a period when cooperation will be increasingly important to our security and our prosperity. We can work together to stop the spread of weapons of mass destruction into the hands of terrorists and we can work together to support Russia's economic transformation and deeper integration into the Euro-Atlantic community, because a prosperous Russia will not face the same pressures to sell to rogue states the tools of mass destruction. And we can work together to help Russia's transformation into a stable free market democracy.

If one were to look down from Mars, one would see that the world pretty neatly divides countries into those that are doing pretty well by their people and those that are not doing very well by their people. The countries that are doing well are the ones that have freer political systems, freer economic systems, the rule of law, transparency, predictability, and are integrated into the world economy. These are nations where there is growth and opportunity.

If Russia hopes to attract foreign capital or retain her most gifted citizens, she must provide them with a climate of economic opportunity and political freedom, a climate that is the critical foundation upon which prosperity, cultural creativity, and national greatness is built.

We in the United States can encourage Russia by working together to put the past behind us and establish bonds and friendships between our people. In the end, of course, the choice and the struggle belong to the Russian people. This treaty is by no means the foundation of that new relationship. It is just one element of a growing multi-faceted relationship between our two countries that involves not just security, but also increasing political, economic, diplomatic, and cultural and other forms of cooperation.

These reductions in the nuclear arsenal of our two countries are an important step in that process. The reductions characterized in the Moscow Treaty will help eliminate the debris of past hostility that has been blocking our way as we build a new relationship. The treaty President Bush has fashioned and the process by which he fashioned it are a model for future cooperation between our two countries. We will achieve deep reductions and enhance security of both of our countries and do so without perpetuating a cold war way of thinking that hinders our desire for better relations.

Christopher E. Paine  **NO**

Treaty on Strategic Offensive Reduction: The Moscow Treaty

[I will] present the views and concerns of the Natural Resources Defense Council regarding the proposed Moscow Treaty. I also chair the board of the Los Alamos Study Group, which is a small non-profit think tank in Santa Fe, New Mexico, that is an independent source of information and analysis about nuclear weapons, research and radioactive contamination at Los Alamos, and the study group has asked me to state they concur in the testimony I am going to present today.

We are not persuaded that the document which lies before you rises to the level of what the legal profession would call a legally cognizable treaty obligation. That is, an agreement with binding, self-evident reciprocal obligations such that an impartial authority or the parties themselves can reasonably ascertain their mutual obligations and ensure compliance.

Whatever the treaty's immediate political value to Presidents [Vladimir] Putin and [George W.] Bush, as a legally binding agreement, it is a sham, a mere memorandum of conversation masquerading as a treaty, and that raises the question of whether the Senate can or should ratify the treaty in its present form without the addition of significant binding conditions to cure the most serious defects.

Lest we forget, a treaty once ratified becomes part of the law of the land, and no Senator or committee of Senators should knowingly vote for a law that is so flawed in its construction that its essential mandates cannot be divined, adjudicated, or enforced. . . . [T]he present agreement brings to mind that old nostrum that was once very popular with some of our Republican friends, a bad treaty is worse than having no treaty at all. It may actually be quite apt in this case.

Unfortunately, as we have discussed, the legal deficiencies in this agreement are just the surface manifestation of a much deeper problem. This administration is abandoning binding, verified nuclear arms control agreements as a tool of American diplomacy. It is systematically replacing cooperative approaches to security based on verified mutual or multilateral arms prohibitions and constraints with unilateral military preparedness and preemptive strike

From U.S. Senate. Committee on Foreign Relations. *Treaty on Strategic Offensive Reduction: The Moscow Treaty.* Hearing, July 23, 2002. Washington, D.C.: U.S. Government Printing Office, 2002. (S.Hrg 107-622.)

planning. We saw it first with the President's rejection of the Comprehensive Test Ban Treaty, an agreement that Russia and all our allies have ratified.

Then came the unilateral withdrawal from the ABM [Anti-Ballistic Missile] treaty, and now this, the jettisoning of predictability, verifiability, irreversibility, and mutual accountability as objectives in our nuclear relationship with Russia in favor of obtaining increased unilateral flexibility for the U.S. nuclear force posture, which of course translates into increased nuclear flexibility for Russia's nuclear forces as well. It is a stunningly bad tradeoff.

As a consequence of these misplaced priorities, the Moscow Treaty imposes no limitation whatsoever on the current or future size of U.S. and Russian nuclear forces and warhead stockpiles; nor does it require improvements in cooperative monitoring and storage for tens of thousands of nondeployed Russian warheads, warhead components, and stocks of nuclear weapon usable materials. All categories of nuclear warheads and delivery systems are left uncontrolled, including tactical nuclear systems.

Even the treaty's sole purported limit on operationally deployed strategic warheads turns out to be hollow, a public relations stunt that actually expires the very moment it enters into force.

. . . [T]his agreement does not require the destruction of a single Russian or U.S. missile silo, strategic bomber, submarine, missile, warhead, or nuclear warhead component. It does nothing to move Russia or the United States down the road toward deep verified nuclear force reductions or toward verified warhead elimination and eventual nuclear disarmament. It therefore very clearly does not fulfill the U.S. obligation under Article 6 of the Nonproliferation Treaty [NPT] to engage in good faith negotiations on effective measures relating to nuclear disarmament.

This treaty is clearly not an effective measure as described by its own proponents, and certainly it is not effective within the meaning of the NPT. Even a cursory reading . . . of the administration's Nuclear Posture Review will convince you that this agreement was not undertaken as a good faith step toward nuclear disarmament.

One would have thought after September 11 that reducing nuclear proliferation risk from Russia would have leapt to the forefront of the Bush administration's nuclear arms control agenda, but achieving meaningful verified controls on Russia's nuclear arsenal requires extensive American reciprocity and, despite all the talk about a new U.S.-Russian relationship, the administration remains transfixed with the possibilities inherent in the unilateral use of U.S. global nuclear strike capabilities to deter and combat proliferation.

. . . [T]his new element of our nuclear posture—planning the use of U.S. weapons of mass destruction [WMD] to preempt the possible use of WMD by other nations—wreaks havoc with our nuclear nonproliferation obligations and assurances, and only validates and encourages other nations in their quest for similarly destructive deterrent and war-fighting capabilities.

Let me outline briefly the major difficulties with this treaty.

The effective date of the treaty's only constraint, a reduction to 2,200 10 years hence, in what President Bush calls operationally deployed strategic warheads, precisely coincides with the treaty's expiration, and in my prepared statement I go into some detail on this. I have concluded that this one portion of the treaty, the only effective part of it, never enters into force. It never becomes binding. I think this is a matter for the attention of your legislative counsel.

The question you should pose is this: Does the Moscow Treaty actually contain any legally binding obligations on the parties to do anything? My own personal view is that it does not, and that the intent of the treaty expressed in the preamble, "to implement significant reductions in strategic arms," is not borne out in the binding articles. Obviously, as a supporter of deep nuclear arms reductions I am eager to be proven wrong, but I am not betting on it.

. . . [T]he treaty lacks verification and inspection provisions of any kind. A decade ago this month [Senator Joseph Biden (D-Delaware)] offered condition 8 to the START I [Strategic Arms Reduction Talks] treaty, the START I treaty resolution of ratification which was adopted and remains binding on the President. This condition reads as follows:

> Inasmuch as the prospect of loss of control of nuclear weapons or fissile material in the former Soviet Union could pose a serious threat to the United States and to international peace and security, in connection with any further agreement reducing strategic offensive arms the President shall seek an appropriate arrangement, including the use of reciprocal inspections, data exchanges, and other cooperative measures to monitor the number of nuclear stockpile weapons on the territory of the parties to this treaty, and the location and inventory of facilities on the territory of the parties capable of producing or processing significant quantities of fissile material.

I wish we had done that in the last 10 years. [Senator Biden was] very right to offer that provision, and we would have been far ahead of the game now in dealing with the threat of nuclear terrorism if those controls were in effect.

Now, it is obvious that the Moscow Treaty makes no provision for the measures outlined in the Biden condition. It is unclear from the testimony received to date whether the administration even sought to establish such reciprocal arrangement in connection with this treaty, as the condition required.

The committee has received a variety of conflicting responses on the verification question. Secretary [of Defense Donald] Rumsfeld testified he saw no need to include detailed verification procedures in this treaty, because there simply is not any way on earth to verify what Russia is doing with all their warheads and their weapons.

But [the Joint Chiefs of Staff chairman General Richard B.] Myers testified that we pushed hard on the verification regime trying to get some action there, and it just never materialized. On the other hand, the General's prepared statement suggested that a virtue of the Moscow Treaty is that it will not subject the U.S. to intrusive inspections in some of our most sensitive military areas.

Secretary [of State Colin] Powell testified that START I verification procedures would give us transparency as they go below the START level, but he did not explain how this would apply, if, in fact, at all, to the problem of monitoring Russia's nuclear stockpile.

At another point in the transcript, Secretary Powell stated that this treaty has nothing to do with the problem of loose nukes and materials flowing to Iraq, which he claimed was being dealt with under . . . debt reduction efforts.

. . . I would point out that the [debt reduction] program did not originate in a void but, rather, came into being to help Russia and the other new states of the Former Soviet Union carry out their START I and NPT commitments. There obviously has to be further U.S.-Russian agreement of some kind on comprehensive threat reduction in order to expand the scope and effectiveness of the [debt reduction] program.

The Moscow Treaty provides no real basis for expanding this effort, despite all the good words that have been spoken today. I doubt that verifying the future force loadings on Russian strategic bombers and missiles was what [Congress] had in mind when [it] began this program.

There are sufficient inconsistencies in the hearing record for the committee to be very concerned. To sort this out, I think you need to obtain the negotiating record of the discussions concerning verification.

By radically condensing or, more accurately, skipping the treaty negotiation process, the administration's approach clearly has the short-term political benefit of appearing to achieve more rapid progress than the previous formal START negotiating process, but I wonder if the President understands that the Moscow Treaty is actually START III Lite, that is, the 2,500 warhead limit from the 1997 Clinton-Yeltsin Helsinki Accord, modified by some accounting gimmicks to exclude several hundred warheads associated with strategic systems in overhaul.

Third, the treaty lacks any interim reduction milestones for assessing compliance. The treaty permits either side to do nothing, or even to increase its operational deployed strategic theater and tactical nuclear forces for a period of 10 years prior to this evanescent culminating moment of simultaneous compliance combined with dissolution of the treaty.

The administration officials who have come before the committee have all said that this President does not care and will not care how many warheads Russia deploys. Even if Russia were to stick its 6,000 at the START I level, Secretary Powell testified that the President would say, fine, I am safe with 1,700 to 2,200, so do what you think you have to do.

But . . . a future President might not be comfortable with that. Members of this body might not be comfortable with that. Depending on when and where Russia rolled out these additional weapons, some of our European allies, Japan or China might not be comfortable with that, and their responses could easily rebound on our own and on global security.

President Bush may think that his own comfort level, buttressed by thousands of potentially deployable nuclear weapons that he plans to keep in reserve, would remain undisturbed, but this Nation's and the world's comfort level might not. They may not look at it that way, and that is precisely why the Constitution requires the advice and consent process, so that treaties reflect more than the personal predilections of the President.

Furthermore, I simply do not believe Secretary Powell's testimony on this point, however well-intentioned it might be. Certainly no modern Republican President, indeed no President of either party has ever withstood the political pressures that would ensue from the development of such a nuclear disparity between the U.S. and a foreign power.

. . . [E]ven the whiff of nuclear parity America experienced at the hands of the Soviets in the late 1970's caused a near meltdown in Republican circles, and Secretary Rumsfeld has attested to the opposite position, which is that if any nation tried to sprint toward parity or superiority in nuclear capabilities with the U.S., that the response would come from our reserve warheads and uploading either for deterrence or defense. So in other words, the Joint Chiefs might not feel safe with 1,700 to 2,200 warheads.

There is no agreed definition or common understanding of what is being reduced. The treaty simply cites two statements which are disparate and do not refer to the same force structures or categories of weapons.

. . . So President Putin talked about further drastic irreversible and verifiable reductions, and we know what the President talked about, reductions in operationally deployed strategic warheads. Those are not the same things.

The Moscow Treaty does not require the elimination of a single missile silo or submarine, nuclear warhead, or bomb. The treaty has an exceedingly permissive withdrawal clause, and I think that needs to be attended to. To fix the treaty, . . . we recommend a number of conditions, and let me just summarize those.

I think the first condition would be to direct the President to achieve a peacetime ceiling of 1,700 operationally deployed strategic weapons within 5 years. The current reduction schedule, or lack of it, is way too loose.

A second condition would direct the President to bring down the active U.S. nuclear reserve stockpile to 1,000 weapons or less within 5 years, and either verifiably store or retire and dismantle inactive stockpile weapons.

A third condition would direct the President to reduce the total stockpile of nuclear weapons in the custody of the Department of Defense to 3,500 weapons within 5 years.

Condition 4 would force a choice, in Secretary Powell's term, "pressurize the system" to choose between verified component storage or demilitarization and disposal.

And finally, require the Senate's advice and consent to exercise the withdrawal clause.

With the addition of these conditions, the administration's memo treaty could be made minimally acceptable, and we then could support its ratification. Without significant conditions like those I have outlined, we believe the treaty is for all practical purposes meaningless. In that event, in order to avoid further damage to the integrity of the U.S. treaty process, we could not support ratification of the Moscow Treaty.

POSTSCRIPT

Does the Moscow Treaty Advance Nuclear Arms Reductions?

As the introduction to this debate indicates, the efforts to constrain nuclear arms that began seriously in the 1960s made progress, albeit slowly, for more than 30 years. The number of new countries (apart from the United States, Soviet Union/Russia, Great Britain, France, and China) with nuclear weapons grew by only one: Israel. The Soviets and Americans restrained testing, put caps on nuclear expansion, and began to reduce their massive arsenals.

Then, in the late 1990s, the seeming progress began to falter, or at least became less clear. India, Pakistan, and perhaps North Korea joined the ranks of the acknowledged and unacknowledged nuclear powers. Other countries, such as Iran, seemed intent on acquiring nuclear weapons. The U.S. Senate refused to ratify the Comprehensive Test Ban Treaty. The United States withdrew from the ABM Treaty. There were persistent reports that the George W. Bush administration was exploring new types of nuclear weapons, such as "mini-nukes," which could be used to attack individual targets such as underground bunkers but which might also make it easier for a country to step over the heretofore absolute line of not using nuclear weapons.

The current debate over the Moscow Treaty belongs in this context. From the most dire point of view, both sides are free to do nothing, not even observe the limits in the abortive START II agreement, until 2012. Moreover, even the uncertain cut does not mean that the two sides cannot simply store the warheads and bombs for rapid redeployment. From a more optimistic viewpoint, in 2012 there will be far fewer nuclear weapons than there would have been without the Moscow Treaty and earlier arms agreements. One way to begin to explore the treaty further is to read it. The entire text is on the U.S. Department of State's Web site http://www.state.gov/t/ac/trt/18016.htm. For a look at the wisdom of disarmament, including the somewhat unusual argument that more nuclear weapons may be better, see Scott D. Sagan and Kenneth N. Waltz, *The Spread of Nuclear Weapons: A Debate Renewed* (W. W. Norton, 2003).

According to *Bulletin of the Atomic Scientists*, in mid-2002, the United States had about 6,500 and Russia approximately 5,000 nuclear devices (warheads, bombs) deployed on strategic-range delivery systems (missiles, bombers). Beyond these, the Russians have 3,000 shorter-range nuclear weapons, and the Americans have about 1,000, which can be delivered by cruise missiles or tactical aircraft. Each country has many other nuclear devices in storage. To learn more about the nuclear weapons of the United States, Russia, and other countries, see the *Bulletin of the Atomic Scientists* Web site at http://www.thebulletin.org, particularly the archives. Toward the end of most issues is a section entitled "Nuclear Notebook," which contains valuable data.

ISSUE 13

Are Military Means the Best Way to Defeat Terrorism?

YES: Benjamin Netanyahu, from Statement Before the Committee on Government Reform, U.S. House of Representatives (September 20, 2001)

NO: Bill Christison, from "Why the 'War on Terror' Won't Work," *CounterPunch* (March 4, 2002)

ISSUE SUMMARY

YES: Benjamin Netanyahu, former prime minister of Israel, argues that there would be little or no organized terrorism if it were not supported by the governments of various countries and that the only way to reduce or eliminate terrorism is to make it clear to countries that support it that they face military retaliation.

NO: Bill Christison, a former member of the Central Intelligence Agency, identifies what he describes as the six root causes of terrorism and argues that using force to address them may not only be ineffective but also counterproductive.

Terrorism is a form of covert attack directed at targets that are outside a certain range of clearly military targets. The line between military action and terrorism is not precise; some people argue that actions conducted by uniformed military forces can sometimes fall into the category of terrorism. Whatever the truth of that controversy, this debate is about the best way to deter and react to covert attacks launched by individuals or groups, rather than by military forces as part of an overt military campaign.

The focus here on individuals and groups does not necessarily mean that countries *cannot* be behind such attacks. In the first of the following selections, for example, Benjamin Netanyahu stresses the importance of what is called "state terrorism." This involves terrorist attacks carried out by covert units of an established government or by private individuals or groups financed and perhaps given terrorist weapons by such a government. For instance, the U.S.

Department of State designates Cuba, Iran, Iraq, Libya, North Korea, Sudan, and Syria as countries that are guilty of state terrorism. Terrorism can also be conducted by one or more individuals (such as Timothy McVeigh, who was convicted for the Oklahoma City bombing in 1995) or organized groups independent of any established government. Al Qaeda, led by Osama bin Laden, is probably the most notorious of these groups, but there are many others, such as Red Hand Defenders in Northern Ireland, Hizballah (Party of God) in the Middle East, and Sendero Luminoso (Shining Path) in Peru.

Although the use of terrorism extends far back into history, recent decades have seen a rise in the practice for several reasons. One is the overwhelming advantage in weapons that governments usually have over dissident groups. Because many governments are armed with aircraft and other high-tech weapons that are unavailable to opposition forces, it has often become nearly suicidal for armed dissidents to use conventional tactics. Second, terrorists' targets are now more readily available than in the past: people are more concentrated in urban areas and even in large buildings; there are countless airline flights; and more and more people travel abroad. Third, the mass availability of instant visual news through television and satellite communications makes it easy for terrorists to gain an audience. This is important because terrorism is not usually directed at its victims as such; rather, it is intended to frighten others. Fourth, technology has led to the creation of increasingly lethal weapons that terrorists can use to kill and injure large numbers of people. These technological "advances" include biological, chemical, nuclear, and radiological weapons.

Terrorist attacks are a relatively regular event. In 2000 there were 423 international terrorist attacks—those that are carried out across national borders—and many other incidents of domestic terrorism. However, through this time, Americans worried little about terrorism. For example, in a survey conducted in 1999 that asked Americans to name two or three top foreign policy concerns, only 12 percent of the respondents mentioned terrorism as a worry.

America's sense of security was shattered by the September 11, 2001, terrorist attacks, which included the destruction of the World Trade Center, major damage to the Pentagon, the crash of a hijacked airliner in Pennsylvania, and the deaths of over 3,000 people. Soon thereafter, President George W. Bush responded by announcing a war on terrorism. An American-led coalition of forces intervened in Afghanistan, toppling the Taliban government that had supported Al Qaeda and attacking Al Qaeda forces in the country. Later, President Bush charged that Iraq, Iran, and North Korea constituted an "axis of evil" that, among other things, was guilty of state terrorism. In March 2003 the United States, in alliance with Great Britain, attacked Iraq, arguing in part that Iraq's support of terrorism made it an international threat and an outlaw nation.

In the following selection, Netanyahu supports this military, "fight fire with fire" approach to deterring and defeating terrorism. In the second selection, Bill Christison argues that military retaliation addresses only the symptoms, not the causes, of terrorism. Indeed, Christison contends that military responses exacerbate the tensions that lead to terrorism.

 **YES**

Preparing for the War on Terrorism

I feel a profound responsibility addressing you in this hour of peril [nine days after the September 11, 2001, terrorist attacks] in the capital of liberty. What is at stake today is nothing less than the survival of our civilization. Now, it might have been some who would have thought a week ago that to talk in these apocalyptic terms about the battle against international terrorism was to engage in reckless exaggeration or wild hyperbole. That is no longer the case. I think each one of us today understands that we are all targets, that our cities are vulnerable and that our values are hated with an unmatched fanaticism that seeks to destroy our societies and our way of life.

I am certain that I speak today on behalf of my entire nation [Israel] when I say, today we are all Americans, in grief and in defiance. In grief, because my people have faced the agonizing horrors of terror for many decades, and we feel an instant kinship, an instant sympathy with both the victims of this tragedy and the great Nation that mourns its fallen brothers and sisters. In defiance, because just as my country continues to fight terrorism in our battle for survival, I know that America will not cower before this challenge.

I have absolute confidence that if we, the citizens of the free world, led by President [George W.] Bush, marshal the enormous reserves of power at our disposal, if we harness the steely resolve of free peoples, and if we mobilize our collective will, we'll succeed at eradicating this evil from the face of the Earth.

But to achieve this goal, we must first answer several questions. First, who is responsible for this terrorist onslaught? Second, why? What is the motivation behind these attacks? And, third and most importantly, what must be done to defeat these evil forces?

The first and most crucial thing to understand is this: There is no international terrorism without the support of sovereign states. International terrorism simply cannot be sustained for any length of time without the regimes that aid and abet it, because, as you well know, terrorists are not suspended in midair. They train, arm, indoctrinate their killers from within safe havens in the territory or territories provided by terrorist states. Often these regimes provide the terrorists with money, with operational assistance, with intelligence, dispatching them to serve as deadly proxies to wage a hidden war against more powerful enemies, which are very often, by the way, democracies, and these regimes

From U.S. House of Representatives. Committee on Government Reform. *Preparing for the War on Terrorism.* Hearing, September 20, 2001. Washington, D.C.: U.S. Government Printing Office, 2001.

mount a worldwide propaganda campaign to legitimize terror, besmirching its victims, exculpating its practitioners. . . .

I think that to see Iran, Libya and Syria call the United States and Israel racist countries that abuse human rights, I think even [the author George] Orwell could not have imagined such a grotesque world.

Take away all the state support, and the entire scaffolding of international terrorism will collapse into the dust. The international terrorist network is thus based on regimes, in Iraq, in Iran, in Syria, in Taliban Afghanistan, Yasser Arafat's Palestinian Authority, and several other Arab regimes such as the Sudan. These regimes are the ones that harbor the terrorist groups; Osama bin Laden in Afghanistan, Hezbollah and others in Syria-controlled Lebanon, Hamas, Islamic Jihad and the recently mobilized Fatah and Tanzim factions in the Palestinian territories, and sundry other terror organizations based in such capitals as Damascus, Baghdad and Khartoum.

These terrorist states and terror organizations together constitute a terror network whose constituent parts support each other operationally as well as politically. For example, the Palestinian groups cooperate closely with Hezbollah, which in turn links them to Iran and Syria, and to [Osama] bin Laden. These offshoots of terror also have affiliates in other states that have not yet uprooted their presence, such as Egypt, Yemen, Saudi Arabia.

Now, the question is, how did this come about? How did this terror network come into being? The growth of this terror network is the result of several crucial developments in the last two decades. Chief among them is the Khomeini revolution, which established a clerical Islamic state in Iran. This created a sovereign spiritual base for fomenting a strident Islamic militancy, a militancy that was often backed by terror.

Equally important was the victory in the Afghan war of the international mujaheedin brotherhood. I suppose that the only way I can compare it is to say that the international mujaheedin is to Islam what the International Brigade was for international communism in the Spanish Civil War. It created an international band of zealots. In this case, the ranks include Osama bin Laden, who saw their victory over the Soviet Union as providential proof of the innate superiority of faithful Muslims over the weak infidel powers. They believed that even the superior weapons of a superpower could not withstand their superior will.

To this should be added Saddam Hussein's escape from destruction at the end of the Gulf war, his dismissal of U.N. monitors, and his growing confidence that he can soon develop unconventional weapons to match those of the West.

And finally, the creation of Yasser Arafat's terror enclave centered in Gaza gave a safe haven to militant Islamic terrorist groups, such as Islamic Jihad and Hamas. Like their mujaheedin cousins, they and their colleagues drew inspiration from Israel's hasty withdrawal from Lebanon, glorified as a great Moslem victory by the Syrian-backed Hezbollah.

Now, under Arafat's rule, the Palestinian Islamic terrorist groups made repeated use of the technique of suicide bombing, going so far, by the way, as to organize summer camps, for Palestinian children, beginning in kindergarten, to teach them how to become suicide martyrs.

Here is what Arafat's government-controlled newspaper—he controls every word that appears there. Here is what his newspaper, his mouthpiece, Al Hayat Al Jadida, said on September 11th, the very day of the suicide bombing in the Twin Towers and the Pentagon, "The suicide bombers of today are the noble successors of the Lebanese suicide bombers, who taught the U.S. Marines a tough lesson in Lebanon. These suicide bombers are the salt of the Earth, the engines of history. They are the most honorable people among us."

Suicide bombers, so says Arafat's mouthpiece, are the salt of the Earth, the engines of history, the most honorable people among us.

. . . [A] simple rule prevails here. The success of terrorists in one part of the terror network emboldens terrorists throughout the network.

This, then, is the who. Now, then, for the why. Though its separate constituent parts may have local objectives and take part in local conflicts, the main motivation driving the terror network is an anti-Western militancy that seeks to achieve nothing less than the reversal of history. It seeks to roll back the West and install an extreme form of Islam as the dominant power in the world, and it seeks to do this not by means of its own advancement and progress, but by destroying the enemy. This hatred is the product of a seething resentment that has simmered for centuries in a certain part of the Arab and Islamic world.

Now, mind you, most Moslems in the world, including the vast majority of Moslems in the growing Moslem communities in the West, are not guided by this interpretation of history, nor are they moved by its call for a holy war against the West. But some are, and though their numbers are small compared to the peaceable majority, they nonetheless constitute a growing hinterland for this militancy.

Militant Islamists resented the West for pushing back the triumphant march of Islam into the heart of Europe many centuries ago. Its adherents, believing in the innate superiority of Islam, then suffered a series of shocks when in the last two centuries, beginning with Napoleon's invasion in Egypt, by the way, that same hated, supposedly inferior West came back and penetrated Islamic realms in north Africa, the Middle East and the Persian Gulf. For them, the mission was clear and defined. The West had to be first pushed out of these areas. So pro-Western Middle Eastern regimes in Egypt and Iraq, these monarchies in Libya, were toppled in rapid succession, including in Iran. And indeed Israel, the Middle East's only democracy and its purest manifestation of Western progress and freedom, must be wiped off the face of the Earth.

Thus, the soldiers of militant Islam do not hate the West because of Israel. They hate Israel because of the West, because they see it as an island, an alien island of Western democratic values in a Moslem-Arab sea; a sea of despotism, of course. That is why they call Israel the Little Satan, to distinguish it clearly from the country that has always been and will always be the Great Satan, the United States of America.

I know that this is not part of normal discourse on TV, where people think that Israel is guiding Osama bin Laden. Well, nothing better illustrates the true order of priorities of the militant Islamic terror than Osama bin Laden's call for Jihad [holy war] against the United States in 1998. He gave as his primary reason for this Jihad not Israel, not the Palestinians, not the peace process, but, rather, the very presence of the United States, "occupying the land of Islam in the holiest of places." What do you think that is? Jerusalem? Temple Mount? No. "The Arabian Peninsula," says bin Laden, where America is, "plundering its riches, dictating to its rulers and humiliating its people." Israel, by the way, comes a distant third, after the, "continuing aggression against the Iraqi people."

So for the bin Ladens of the world, Israel is merely a sideshow. America is the target. But reestablishing a resurgent Islam requires not just rolling back the West, it requires destroying its main engine, the United States. And if the United States cannot be destroyed just now, it can be first humiliated, as in the Tehran hostage crisis 20 years ago, and then ferociously attacked again and again until it is brought to its knees. But the ultimate goal remains the same: Destroy America, win eternity.

Now, some of you may find it hard to believe that Islamic militants truly cling to this mad fantasy of destroying America. Make no mistake about it. They do. And unless they are stopped now, their attacks will continue and become even more lethal in the future.

The only way I can explain the true dangers of Islamic militancy is to compare it to another ideology bent on world domination: communism. Both movements pursued irrational goals, but the Communists at least pursued theirs in a rational way. Any time they had to choose between ideology and their own survival, as in Cuba or in Berlin, they always backed off and chose survival.

Not so for the Islamic militants. They pursue an irrational ideology irrationally with no apparent regard for human life, neither their own lives nor the lives of their enemies. The Communists seldom, if ever, produced suicide bombers, while Islamic militancy produces hordes of them, glorifying them, promising them for their dastardly deeds a reward in a glorious afterlife.

This highly pathological aspect—I can use no other words—this highly pathological aspect of Islamic militancy is what makes it so deadly for mankind. But in 1996, I wrote in my book about fighting terrorism, I warned about the militant Islamic groups operating in the West with the support of foreign powers, serving as a new breed of what I called domestic international terrorists; that is, basing themselves in America to wage Jihad against America. Such groups, I wrote then, nullify in large measure the need to have air power or intercontinental missiles as delivery systems for an Islamic nuclear payload. They, the terrorists, will be the delivery system. In the worst of such scenarios, I wrote, the consequences could be not a car bomb, but a nuclear bomb in the basement of the World Trade Center.

Well, . . . they didn't use a nuclear bomb. They used two 150-ton, fully loaded jetliners to wipe out the Twin Towers. But does anyone doubt that given the chance, they will throw atom bombs at America and its allies; and perhaps long before that, they'd employ chemical and biological weapons?

This is the greatest danger facing our common future. Some states of the terror network already possess chemical and biological capabilities, and some are feverishly developing nuclear weapons. Can one rule out the possibility that they will be tempted to use such weapons openly or secretly through their terror proxies, seemingly with impunity, or that their weapons might fall into the hands of the terrorist groups they harbor?

We have received a wake-up call from hell. Now the question is simple: Do we rally to defeat this evil while there is still time, or do we press a collective snooze button and go back to business as usual? The time for action is now. Today the terrorists have the will to destroy us, but they do not have the power. There is no doubt that we have the power to crush them. Now we must also show that we have the will to do so, because once any part of the terror network acquires nuclear weapons, this equation will fundamentally and irrevocably change, and with it the course of human affairs. This is the historical imperative that now confronts us all.

And now to my third point. What do we do about it? First, as President Bush said, we must make no distinction between the terrorists and the states that support them. It is not enough to root out the terrorists who committed this horrific act of war. We must dismantle the entire terrorist network. If any part of it remains intact, it will rebuild itself, and the specter of terrorism will reemerge and strike again. Bin Laden, for example, has shuttled over the last decades from Saudi Arabia to Afghanistan to the Sudan and back again. So we cannot leave any base of this terror network intact.

To achieve this goal we must first have moral clarity. We must fight terror wherever and whenever it appears. We must make all states play by the same rules. We must declare terrorism a crime against humanity, and we must consider the terrorists enemies of mankind, to be given no quarter and no consideration for their purported grievances. If we begin to distinguish between acts of terror, justifying some and repudiating others based on sympathy with this or that cause, we will lose the world clarity that is so essential for victory. This clarity is what enabled America and Britain to wipe out piracy in the 19th century. This is how the allies rooted out Nazis in the 20th century. They didn't look for the root cause of piracy, nor for the root cause of nazism, because they knew that some acts are evil in and of themselves and do not deserve any consideration or any "understanding." They didn't ask if Hitler was right about the alleged wrong done to Germany at Versailles. They left that to the historians. The leaders of the Western Alliance said something entirely different. They said, nothing justifies nazism, nothing.

Well, we must be equally clear-cut today. Nothing justifies terrorism, nothing. Terrorism is defined not by the identity of its perpetrators nor by the cause they espouse. Rather, it is defined by the nature of the act. Terrorism is the deliberate attack on innocent civilians. In this it must be distinguished from legitimate acts of war that target combatants and may unintentionally harm civilians.

When the British Royal Air Force bombed the Gestapo headquarters in Copenhagen in 1944 and one of their bombs unintentionally struck a children's hospital nearby, that was a tragedy, but it was not terrorism. When Israel [recently] fired a missile that killed two Hamas archterrorists, and two Palestinian children who were playing nearby were tragically struck down, that is not terrorism, because terrorists do not unintentionally harm civilians. They deliberately murder, maim and menace civilians, as many as possible.

No cause, no grievance, no apology can ever justify terrorism. Terrorism against Americans, against Israelis, against Spaniards, against Britons, against Russians or anyone else is all part of the same evil and must be treated as such. It is time to establish a fixed principle for the international community. Any cause that uses terrorism to advance its aims will not be rewarded. On the contrary, it will be punished, severely punished, and placed beyond the pale.

. . . [A]rmed with this moral clarity in defining terrorism, we must possess an equal clarity in fighting it. If we include Iran, Syria and the Palestinian Authority in the coalition to fight terror, even though they currently harbor, sponsor and dispatch terrorism—as we speak, terrorists struck innocent people, [having just] murdered a woman . . . from Yasser Arafat's domain against Israel. If we include these terrorist regimes in the coalition, then the alliance against terror will be defeated from within. We might, perhaps, achieve a short-term objective of destroying one terrorist fiefdom, but it will preclude the possibility of overall victory. Such a coalition will necessarily melt down because of its own internal contradictions. We might win a battle, but we will certainly lose the war.

These regimes, like all terrorist states, must be given a forthright demand: Stop terrorism, not temporarily for tactical gains, stop terrorism permanently, or you will face the wrath of the free world through harsh and sustained political, economic and military sanctions.

Now, obviously, some of these regimes today will scramble in fear and issue platitudes about their opposition to terror, just as Arafat, Iran and Syria did, while they keep their terror apparatus intact. Well, we shouldn't be fooled. These regimes are already on the U.S. list of states supporting terrorism; and if they're not, they should be.

The price of admission for any state into the coalition against terror must be first to completely dismantle the terrorist infrastructures within their realm. Iran will have to dismantle the worldwide network of terrorism and incitement based in Tehran. Syria will have to shut down Hezbollah and a dozen other terrorist organizations that operate freely in Damascus and in Lebanon. Arafat will have to crush Hamas and Islamic Jihad, close down their suicide factories and training grounds, rein in his own Fatah and Tanzim terrorists and cease the endless incitement of violence.

To win this war, we have to fight on many fronts. Well, the most obvious one is direct military action against the terrorists themselves. Israel's policy of preemptively striking at those who seek to murder its people is, I believe, better understood today and requires no further elaboration.

But there's no substitute for the key action that we must take: imposing the most punishing diplomatic, economic and military sanctions on all terrorist states. To this must be added these measures: Freeze financial assets in the West of terrorist regimes and organizations. Revise legislation, subject to periodic renewal, to enable better surveillance against organizations inciting violence. Keep convicted terrorists behind bars. Do not negotiate with terrorists.

And train special forces to fight terror. And, not least important, impose sanctions, heavy sanctions, on suppliers of nuclear technology to terrorist states.

. . . I've had some experience in pursuing all of these courses of action in Israel's battle against terrorism. . . . But I have to be clear: Victory over terrorism is not at its most fundamental level a matter either of law enforcement or intelligence. However important these functions are, they could only reduce the dangers, not eliminate them. The immediate objective is to end all state support for and complicity with terror.

If vigorously and continuously challenged, most of these regimes can be deterred from sponsoring terrorism, but there is a possibility that some will not be deterred, and those may be the ones that possess weapons of mass destruction. Again, we cannot dismiss the possibility that a militant terrorist state will use its proxies to threaten or launch a nuclear attack with a hope of apparent immunity and impunity. Nor can we completely dismiss the possibility that a militant regime, like its terrorist proxies, will commit collective suicide for the sake of its fanatical ideology. In this case, we might face not thousands of dead, but hundreds of thousands and possibly millions.

This is why the United States must do everything in its power to prevent regimes like Iran and Iraq from developing nuclear weapons and to disarm them of their weapons of mass destruction. This is the great mission that now stands before the free world. That mission must not be watered down to allow certain states to participate in the coalition that is now being organized. Rather, the coalition must be built around this mission.

It may be that some will shy away from adopting such an uncompromising stance against terrorism. If some free states choose to remain on the sidelines, America must be prepared to march forward without them, for there is no substitute for moral and strategic clarity. I believe that if the United States stands on principle, all democracies will eventually join the war on terrorism. The easy route may be tempting, but it will not win the day.

On September 11th, I, like everyone else, was glued to a television set, watching the savagery that struck America, but amid the smoking ruins of the Twin Towers, one could make out the Statue of Liberty holding high the torch of freedom. It is freedom's flame that the terrorists sought to extinguish, but it is that same torch so proudly held by the United States that can lead the free world to crush the forces of terror and to secure our tomorrow. It is within our power. Let us now make sure that it is within our will.

Why the "War on Terror" Won't Work

On January 15 [2002] the Attorney General of the United States, John Ashcroft, held a press conference in order to describe the initial criminal charges that the government would make against John Walker, the 20-year-old American citizen who had joined the Taliban military forces. In his talk, Ashcroft said this, and I quote: "The United States does not casually or capriciously charge one of its own citizens with providing support to terrorists. We are impelled to do so today by the inescapable fact of September the 11th, a day that reminded us in no uncertain terms that we have enemies in the world and that these enemies seek to destroy us. We learned on September 11 that our way of life is not immune from attack, and even from destruction."

The guts of what Ashcroft said is—and I quote again—"We have enemies in the world and these enemies seek to destroy us." Unquote. I submit to you that this is simply not a true statement. The evidence I've seen shows that the real objective of the Muslim extremists led by Osama bin Laden was to rid the Muslim world itself of American domination and influence. They wanted NOT to destroy the United States; rather they wanted the U.S. out of their own land. Bin Laden and his supporters also wanted, and those yet alive still want, to unite Muslim nations behind an extreme version of Islam, believing that the Islamic world can thereby better control its own future. I think they realize full well there is no possibility they can "destroy" the United States, and their objective, while still pretty grandiose, is considerably more limited. Their aim, according to one recent analysis that appeared in the *New York Review of Books*—and I quote again—"is to create one Islamic world . . . This is a call to purify the Islamic world of the idolatrous West, exemplified by America. The aim is to strike at American heathen shrines, and show, in the most spectacular fashion, that the U.S. is vulnerable, a paper tiger." Unquote.

These Islamic extremists are not nice people. Those still alive, and other future adherents to their cause, will continue to try to kill innocent people in the U.S. and elsewhere. But what the extremists see themselves as trying to do is to stop the United States from continuing its drive for global hegemony, including hegemony over the Islamic world. I think it's important to understand this, because if people in the United States believe that some enemy is trying to "destroy" the U.S. and actually has some possibility of doing so, then waging an all-out war against that enemy can be more easily justified. But what if the U.S. is not trying to prevent its own destruction, but instead is trying to preserve and extend its global hegemony? In that case, I think we should all step back and start demanding of our government a serious public debate over future U.S. foreign policies....

President George W. Bush's references in his own speeches to America's enemies as "the evil ones" tend in the same direction [as Ashcroft's views]. Although acts of terrorism—which I'm defining here as killings of, or other violence against, innocent noncombatants—are always inexcusable, simply labeling perpetrators as "the evil ones" makes it easier for the U.S. government to avoid any inconvenient discussion of ways in which the U.S. might modify its foreign policies to reduce the likelihood of future terrorist acts. But are all Afghans "evil ones?" Or all members of the Taliban? Or did only a few Taliban leaders know about the planned terrorist attacks before September 11? In any case, is it clear that all Taliban members were accomplices of Al Qaeda and Osama bin Laden? And if they were accomplices, is it not true that the better legal systems of the world do not punish accomplices to a crime as severely as the criminals themselves? Is it right that in this war the U.S. is punishing the accomplices just as much as the criminals themselves? It seems to me that the use of the term "evil ones" is intended to avoid discussion of a lot of nuances.

... The U.S. government, by the way, seems uninterested in even estimating how many innocent noncombatants have in fact been killed, but it is possible that the number is as large as or larger than the 3,000 killed in the U.S. on September 11. Whatever the military success of the U.S., however, a couple of years hence new extremists just as clever as bin Laden, and hating the U.S. even more, will almost certainly arise somewhere else in the world. That's why we need to understand the root causes behind the terrorism. If I am right that military action will not prevent future terrorism, but only delay it, we should start working on these root causes right away. We should not wait until the military actions are finished before looking at root causes, as some people would urge us to do.

So let's go. I'm going to list six major root causes of the terrorism that I think are important....

> ONE. *My number one root cause is the support by the U.S. over recent years for the policies of Israel with respect to the Palestinians, and the belief among Arabs and Muslims that the United States is as much to blame as Israel itself for the continuing, almost 35-year-long Israeli occupation of the West Bank and the Gaza Strip.*

My first comment on this issue is that it is a more controversial root cause than any of the others on our list. The government of Israel, and many supporters of Israel in the United States, really did not want to talk about any root causes immediately after September 11. Top leaders in the United States, most of whom strongly support Israel, preferred to talk only in general terms about how the terrorists were mad and irrational, and how they had attacked "freedom itself," out of mindless hatred. More recently, when pressured to talk about root causes at all, the Israelis and their supporters have gone to great lengths to reject arguments that Israel's behavior toward the Palestinians, or U.S. support for Israel, are in any way even a partial cause of the terrorism. When forced to say something positive about root causes, they tend to allege a broader Islamic religious hatred of the West and its modern technology than I think exists. They also emphasize the internal tensions within the Arab world, the lack of democracy and the dictatorial rulers of Arab nations, who are depicted as trying to distract their people from their own internal grievances by whipping up hatred of Israel.

. . . In the last couple of months, a sizable propaganda campaign has been launched suggesting that Saudi Arabia is the most important root cause of the September 11 terrorism. I certainly agree that the dictatorial and decrepit Saudi government and its support throughout the Muslim world for a harsh and immoderate version of Islam can be seen as one—but only one—of the root causes behind the recent terrorism. I'll have more to say about this later.

What I want to point out here is that I suspect supporters of Israel are aggressively pressing this campaign against Saudi Arabia, in the hope of persuading other world leaders that the issue of Palestine is NOT a significant root cause. . . .

TWO. My number two root cause is the present drive of the United States to spread its hegemony and its version of big-corporation, free enterprise globalization around the world. At the same time, the massive poverty of average people, not only in Arab and Muslim nations but also in the whole third world, has become more important as a global political issue. The gap between rich and poor nations, and rich and poor people within most of the nations, has grown wider during the last 20 years of globalization or, more precisely, the U.S. version of globalization. Animosities against the United States have grown among the poor of the world, who have watched as the U.S. has expanded both its hegemony and a type of globalization based on its own economic system, while they themselves have seen no or very little benefit from these changes.

This problem of poverty around the world is so immense that it's almost impossible to grasp. Global statistics are far from perfect, buy they show that the world's population hit 6 billion [in 2001]. 2.8 billion people, almost half of the world's total, have incomes of less than two dollars a day. Here's another statistic: the richest one percent of the world's people receive as much income as the poorest 57 percent. And here's a final statistic: The richest 25 million people in the United States receive more income than the 2 billion poorest people of the world—one third of the world's total population. Can we . . . even comprehend the magnitude of the injustice that these figures represent? . . .

The catalog of reasons for animosity toward the U.S. throughout the world includes a number of things in addition to our overbearing assertion of both economic and political hegemony: our arrogance in insisting that whatever we say goes, our penchant for abrogating or ignoring international treaties that we don't happen to like, as well as the influence of U.S. corporations that exploit cheap labor in third world countries to make consumer goods for Americans.

Take all these things together and you have a wide sense among the poor people of the world of being oppressed by the United States. This in turn made it possible for Osama bin Laden and the fundamentalists around him to instill and spread intense hatred of us, just as a sense of being oppressed by the Allies after World War I made it possible for Hitler to arouse the kind of fear and hatred among Germans that led both to the slaughter of Jews and to World War II. . . .

> THREE. *The number three root cause I want to discuss is the continuing sanctions and lack of food and medicines for the people of Iraq, deaths of Iraqi children, and the almost daily bombing of Iraq by the U.S. and Great Britain. Right or wrong, the Arab and Muslim "street" blames this on the U.S., not on Saddam Hussein.*

I don't have much to comment about on this one. The sanctions and the bombings have been in effect for ten years, and have neither brought about the ouster of Saddam Hussein nor significantly weakened him. And they have caused the deaths of children variously estimated at up to or over a million. The U.S. government's position is that Saddam himself is to blame for the troubles of the Iraqi people, but the fact remains that after all these years, the Iraqi people are the ones hurt by U.S. actions, not Saddam.

My view is that simple justice argues for an end to both the sanctions and the bombings. My proposal is that we do precisely that.

> FOUR. *My number four root cause is the continued presence of U.S. troops in Saudi Arabia.*

Ten years ago this was the principal cause of Osama bin Laden's hostility toward the United States. (His hostility on account of U.S. actions against Iraq and then the massive U.S. support for Israel came later and in both cases may be tactical—an effort to broaden his own popularity in the Arab world.) Today the thousands of U.S. military personnel in Saudi Arabia are a constant irritant in Saudi-U.S. relations. The Saudi people clearly do not want them there. Unless we plan to invade Iraq again, I doubt there is any longer a vital reason to keep men and U.S. ground-based military facilities there.

My proposal? The obvious one—that we remove the troops. I understand, of course—you'd have to be blind and deaf not to know this—that some people at high levels in the U.S. government do want to invade Iraq again. All I can say is, I hope such people do not carry the day. I can't think of a thing that would do more to broaden this "war on terrorism" into a Judeo-Christian war against Islam—despite any U.S. governmental protestations to the contrary.

> FIVE. *The fifth root cause on my list is the dissatisfaction and anger of many average and even elite Arabs and Muslims over their own authoritarian, undemocratic, and often corrupt governments, which are supported by the United States.*

My first comment here is that Osama bin Laden is a good example of this particular root cause. His extremist wrath was directed as much against the

Saudi government, for example, as it was against the United States. His opposition to what used to be his own government was probably the main reason why he had the support of a majority of the young men under 25 in Saudi Arabia. He received similar support from many young men in other Arab and Muslim states as well. Right now these groups of angry young men obviously no longer have a viable leader in Osama bin Laden, but other extremist leaders are almost sure to arise. In addition, the next generation of leaders in at least some of these states may well emerge from among these young men. If any of them do come into power, their future governments will likely be more anti-American than the present governments, which Washington likes to call "moderate," but which are really nothing of the sort. If we have not reduced our energy dependence on oil in the meantime, we may face serious trouble. . . .

I think we should . . . gradually reduce the closeness of our ties with the present authoritarian governments in Arab and Muslim states, and try to develop a better understanding of and improved relations with groups in these states that oppose their own present governments. We should seek out groups that appear to be democratically inclined and "moderate" in the true meaning of the word. Difficult? Of course it will be. But it is the best shot we've got, in my opinion, to have a decent relationship with many Muslim states in the future. It's also the best shot we've got if we wish to diminish, over time, the support for future Osama bin Ladens that arises from the anger of Arabs and Muslims with their own governments.

SIX. The sixth and last root cause on my list arises directly from the U.S. "war on terrorism." It has to do with the kind of war the U.S. is now able to fight. On three recent occasions—the Gulf War of 1990–1991, the Kosovo war of 1999 against Yugoslavia, and the current war against Afghanistan—the United States has easily achieved victories by relying almost exclusively on air power, on missiles launched from a great distance, and now even on drone aircraft with no humans on board. The U.S. has won these wars with practically no casualties among its own forces. But while few Americans get killed, sizable numbers of other nationalities do.

Most people in the United States are proud both of these victories and of the low U.S. casualties in these three wars. From the viewpoint of anyone who supports the wars, this prowess of U.S. armed forces deserves to be honored. But elsewhere in much of the world, especially the underdeveloped world, this overwhelming invincibility of the U.S. military intensifies the frustrations about and hatred of the United States. This in turn makes future terrorist acts against the U.S.—or what is now called by U.S. strategic thinkers asymmetrical warfare—even more likely. Those in underdeveloped lands who oppose the U.S. drive for worldwide hegemony are increasingly coming to see no means other than terrorism as an effective method of opposing the United States.

This is an issue that demands a lot more discussion than it's been getting, and it goes to the heart of our future foreign policies. For the immediate future, perhaps the next five or ten years, it's going to be tempting for any government of the United States to implement and enforce whatever foreign policies it chooses by going to war, because it will be confident—even overconfident—that it won't lose a military confrontation and won't suffer many casualties. The U.S. government in fact has already started moving in this direction, by threatening to launch preemptive wars against nations that are trying to develop nuclear weapons or other weapons of mass destruction. Another thing the U.S. is already doing is to militarize the United States to an unprecedented, and wholly unnecessary, degree in comparison with other nations. An editorial in the March 3 [2002] *New York Times* puts it bluntly. "If Congress cranks up the Pentagon's budget as much as President Bush would like, the United States will soon be spending more on defense than all the other countries of the world combined." To me, this is absurd—but there you are. These military expenditures will clearly lead to cuts in spending on domestic U.S. problems such as poverty and healthcare, and make it harder to do anything about solving the problems of global poverty and income inequality that I've already discussed. In this same five to ten year period, the readily available military option will also encourage the U.S. to avoid facing up to the hard decisions necessary for a peaceful resolution of our more intractable foreign policy problems.

This leads me to a very important conclusion. Since the greater willingness to initiate and fight wars intensifies hatred of the U.S., it is in the U.S. interest to show restraint and voluntarily stop employing warfare based on bombing in order to combat future acts of terrorism. The fact that U.S. bombs and missiles have already killed innocent civilians is tragic and puts us on a par with the extremists who committed the September 11 acts. The U.S. should stop, right now, all further military action that risks killing more civilians.

At the same time, I want to emphasize that I am quite sure there is enough evidence of Osama bin Laden's complicity in the September 11 terrorist actions to arrest and indict him. Assuming he is still alive, I would therefore support covert or Green-Beret-type operations to capture, but not assassinate, him. Maximum precautions should be taken, however, to prevent such operations from killing or injuring any more innocent civilians. Once captured, bin Laden should be prosecuted and tried in an international court.

I fully understand that compared to most views you hear concerning the U.S. "war on terrorism," my views are RADICAL. But I believe that unless the U.S. moves in the directions I've been suggesting throughout this [selection], in five or ten years the terrorism against the United States will become so intense that our global relationships with other nations will be in shambles. On the other hand, if the U.S. government voluntarily moves toward the kind of foreign policy changes I've been talking about, I think that its actions might start a trend toward a considerably more peaceful, and stable, 21st Century than now seems likely.

POSTSCRIPT

Are Military Means the Best Way to Defeat Terrorism?

Probably the best way to think about this debate is in terms of finding the proper balance between two approaches. One is the use of military means to defeat and deter terrorism. The other is an attempt to address some of the socioeconomic causes of terrorism by seeking a diplomatic solution to the persistent problems in the Middle East, using foreign aid and other economic tools to relieve poverty in the less developed countries, and utilizing other nonmilitary techniques. Very few people beyond those who are true pacifists would reject any use of the military to counter terrorism. It would be hard to make the case that the world will not be a safer place when Osama bin Laden is no longer able to direct terrorism and that the use of force is not necessary to end his threat and that of Al Qaeda.

Yet it can also be said that counterviolence is not the complete answer. In a sense, the September 11 attacks served as a wake-up call to the connection between terrorism and the horrific conditions in many parts of the world. When world leaders met in Monterrey, Mexico, in March 2002 to discuss conditions in the world's poor countries, many made the connection between poverty and violence.

The first step toward coming to a fully informed opinion is to learn more about terrorism. One good place to start is *A History of Terrorism* by Walter Laqueur (Transaction Press, 2001). An even more recent volume, and one that looks at many aspects of terrorism, is David J. Whittaker, ed., *Terrorism Reader* (Routledge, 2003). To learn more about continuing terrorist activity, see the Naval Postgraduate School's Terrorist Group Profiles Web page at http://library.nps.navy.mil/home/tgp/tgp2.htm.

From that point you can turn to studies of the various possible policy responses to terrorism. A helpful book on that topic is Lawrence Freedman, ed., *Superterrorism: Policy Responses* (Blackwell, 2002). It is also good to understand the morality and perceived morality of terrorism and counterterrorism, a topic that can be pursued in Tony Coady and Michael O'Keefe, eds., *Terrorism and Justice: Moral Argument in a Threatened World* (Melbourne University Press, 2003). A volume that specifically addresses America's post–September 11 response is Jean Bethke Elshtain's *Just War Against Terror: The Burden of American Power in a Violent World* (Basic Books, 2003). Finally, you should know more about the potential of the use of weapons of mass destruction by terrorists. Two helpful volumes are Jonathan B. Tucker, ed., *Toxic Terror: Assessing Terrorist Use of Chemical and Biological Weapons* (MIT Press, 2000) and Nadine Gurr and Benjamin Cole, *The New Face of Terrorism: Threats From Weapons of Mass Destruction* (St. Martin's Press, 2000).

ISSUE 14

Is Government-Ordered Assassination Sometimes Acceptable?

YES: Bruce Berkowitz, from "Is Assassination an Option?" *Hoover Digest* (Winter 2002)

NO: Margot Patterson, from "Assassination as a Weapon," *National Catholic Reporter* (September 6, 2002)

ISSUE SUMMARY

YES: Bruce Berkowitz, a research fellow at the Hoover Institution at Stanford University, argues that while government-directed political assassinations are hard to accomplish and are not a reliably effective political tool, there are instances where targeting and killing an individual is both prudent and legitimate.

NO: Margot Patterson, a senior writer for *National Catholic Reporter*, contends that assassinations are morally troubling, often counterproductive, and have a range of other drawbacks.

Political assassination by internal plotters has existed seemingly as long as politics have. Ancient kings and other rulers were poisoned, stabbed to death, or otherwise killed with some regularity. Assassinations of heads of state and government have continued into modern times. In the United States, Presidents Abraham Lincoln, James Garfield, William McKinley, and John F. Kennedy were all felled by an assassin's bullet.

Much less common throughout history has been the assassination or attempted assassination of the leader of a country by a national government. Certainly this has occurred, especially in times of particular stress. In the United States, for example, extraordinary hostility and a sense of threat occasioned by the cold war promoted assassination. A report issued in 1975 by the U.S. Senate's Select Committee on Intelligence chaired by Senator Frank Church (D-Idaho)—and thus known as the Church Committee—disclosed that the Central Intelligence Agency had allegedly attempted to assassinate at least five foreign leaders during the previous 15 years: President Salvador Allende of Chile, Cuban president Fidel Castro, Congolese prime minister Patrice Lumumba, Iraqi dictator Abdul Karim Qassem, and the Dominican Republic's Rafael Trujillo. The Soviet Union and other countries also reportedly engaged in attempted assassinations.

The Church Committee report was issued during a period of American post–Vietnam War reaction against the excesses of the cold war. Reflecting this changed mood, President Gerald R. Ford issued Executive Order 11905 in 1976. The directive ordered that "no employee of the United States Government shall engage in, or conspire to engage in, political assassination." Since then, every president through President Bill Clinton reaffirmed the U.S. stance against government-directed assassinations, and two presidents (Jimmy Carter and Ronald Reagan) issued supplementary executive orders banning the practice.

According to some analysts, this stance does not mean that the United States did not try to assassinate foreign leaders during this time. Perhaps the clearest example occurred when President Reagan authorized a military attack using warplanes and conventional bombs on the home of Libyan leader Muammar Qaddafi in 1986. He was not home that evening, but his infant daughter was killed.

One of the difficulties of governments' authorizing assassinations is where to draw the line. Consider, for example, the reasonably neutral definition of assassination offered by Bruce Berkowitz in the first of the following readings. He characterizes assassination as "deliberately killing a particular person to achieve a military or political objective, using the element of surprise to gain an advantage." That would cover a range of activities, from bribing the cook of a head of state during peacetime to specifically targeting leaders such as Qaddafi in conventional military attacks. The Reagan administration argued that the attack on Qaddafi's residence was part of a larger strike in retaliation for Libyan activities that had led to a terrorist bombing of a popular nightspot in Germany, killing many off-duty American soldiers. In this and similar cases, governments have also made the case that the "targeted killing" of a leader using conventional military means is acceptable.

Attitudes in the United States toward assassinations changed in the aftermath of the September 11, 2001, terrorist attacks. For example, before the attacks, the White House reprimanded Israel for having bombed an apartment building in Gaza in an attempt to kill a leader of the terrorist organization Hamas. Not long after the September 11 attacks, however, President George W. Bush signed a so-called intelligence finding authorizing the CIA to engage in "lethal covert operations" to kill Osama bin Laden and to destroy Al Qaeda. The following November, Bush also authorized the CIA to kill Qaed Salim Sinan al-Harethi, who was suspected of masterminding the attack on the USS *Cole* in Yemen. He was killed when a hellfire missile from a CIA-controlled Predator drone destroyed his car in Yemen.

The issue here is whether or not such tactics are wise and moral. In the first selection, Berkowitz endorses, albeit cautiously, governments' carrying out political assassinations. In the second selection, Margot Patterson argues that whether referred to as "assassinations" or "targeted killing" (or some other euphemism), specifically setting out with forethought to kill an individual is neither practical nor moral.

Bruce Berkowitz  **YES**

Is Assassination an Option?

Soon after the September 11 terrorist attacks on New York and Washington, U.S. officials announced that they had evidence linking Osama bin Laden to the attacks. As Americans began to recover from their initial shock, many of them asked, "Why don't we just get rid of the guy?"

The terrorist tragedy reopened one of the most controversial issues in national security policy: assassination. Few topics raise more passion. Yet, despite the intense emotions assassination raises, assassination rarely gets the kind of dispassionate analysis that we routinely devote to other national security issues. That is what I will do here. When it comes to assassination, four questions are key: What is it? Is it legal? Does it work? And when, if ever, is assassination acceptable?

What Is It?

One reason assassination—or, for that matter, banning assassination—provokes so much disagreement is that people often use the term without a precise definition and thus are really arguing about different things. One needs to be clear. Depending on the definition, one can be arguing about activities that are really quite different.

For example, is killing during wartime assassination? Does assassination refer to killing people of high rank, or can anyone be the target of assassination? Does it matter if a member of the armed forces, a civilian government official, or a hired hand does the killing? Depending on the definition, killing a military leader during a bombing raid might be "assassination" but killing a low-level civilian official with a sniper might not.

For what it is worth, the *Merriam-Webster Dictionary* defines *assassination* by referring to the verb *assassinate,* which is defined as "to injure or destroy unexpectedly and treacherously" or "murder by sudden or secret attack usually for impersonal reasons." In other words, assassination is murder—killing a person—using secrecy or surprise. Assassination stands in contrast to murder without surprise (e.g., a duel). Also, assassination is not murder for personal gain or vengeance; assassinations support the goals of a government, organization, group, or cause.

Although people associate assassinations with prominent people, strictly speaking, assassination knows no rank. Leaders are often the targets of state-sponsored assassination, but history shows that generals, common soldiers, big-time crime bosses, and low-level terrorists have all been targets, too. Also, it does not seem to matter how you kill the target. It does not matter if you use a bomb or a booby trap; as long as you target a particular person, it's assassination.

For our purposes, assume that assassination is "deliberately killing a particular person to achieve a military or political objective, using the element of surprise to gain an advantage." We can call such a killing "sanctioned assassination" when a government has someone carry out such an action—as opposed to, say, "simple assassination," killing by an individual acting on his own. Then the question is, should we allow the United States to sanction such activities? And, if we allow the government to sanction assassination, when and how should do it?

Is It Legal?

You might be surprised to learn that there are no international laws banning assassination. The closest thing to a prohibition is the 1973 Convention on the Prevention and Punishment of Crimes Against Internationally Protected Persons, Including Diplomatic Agents. This treaty (which the United States signed) bans attacks against heads of state while they conduct formal functions, heads of government while they travel abroad, and diplomats while they perform their duties.

The Protected Persons Convention was intended to ensure that governments could function and negotiate even during war. Without it, countries might start a war (or get drawn into one) and then find themselves unable to stop because there was no leader at home to make the decision to do so and because their representatives were getting picked off on their way to cease-fire negotiations.

But other than these narrow cases, the Protected Persons Convention says nothing about prohibiting assassination. Even then it applies only to officials representing bona fide governments and "international organizations of an intergovernmental character." So presumably the convention shields the representatives of the United Nations, the World Trade Organization, the International Red Cross, and, probably, the PLO [Palestine Liberation Organization]. It does not protect bosses of international crime syndicates or the heads of terrorist groups such as Al Qaeda.

Another treaty that some might construe as an assassination ban is the Hague Convention on the "laws and customs" of war. The Hague Convention states that "the right of belligerents to adopt means of injuring the enemy is not unlimited." (This was a bold statement in 1907, when the convention was signed.)

The Hague Convention tried to draw a sharp line between combatants and noncombatants; combatants were entitled to the convention's protections but were also obliged to obey its rules. For example, the Hague Convention tried to distinguish combatants by requiring them to wear a "fixed distinctive emblem recognizable at a distance." Wear the emblem while fighting, and you are entitled to be treated as a POW [prisoner of war] if captured; fail to follow the dress code, and you might be hanged as a mere bandit.

Alas, maintaining this definition of a "combatant" proved a losing battle throughout the twentieth century. Guerrilla warfare transformed civilians into soldiers. Strategic bombing transformed civilians into targets. Headquarters staff, defense ministers, and civilian commanders in chief today are all more likely to wear suits than uniforms. Teenage paramilitary soldiers in Liberia are lucky to have a pair of Levis to go along with their AK-47s, let alone fatigues or insignia. That is why, practically speaking, a "combatant" today is anyone who is part of a military chain of command.

Yet the Hague Convention may be more interesting not for what it prohibits but for what it permits. The closest the convention comes to banning assassination is when it prohibits signatories from killing or wounding "treacherously individuals belonging to the hostile nation or army." But when it refers to "treachery," it is referring to fighting under false pretenses (e.g., flying the enemy's flag or wearing his uniform to lure him to death). The Hague Convention specifically permits "ruses of war." Snipers, land mines, deception, camouflage, and other sneaky tactics are okay. In fact, one might even argue that, since the convention prohibits *indiscriminate* killing, state-sanctioned assassination—the most precise and deliberate killing of all—during war is exactly what the treaty calls for.

The third international agreement that is relevant to assassination is the Charter of the United Nations, which allows countries to use military force in the name of self-defense. If a country can justify a war as "defensive," it can kill any person in the enemy's military chain of command that it can shoot, bomb, burn, or otherwise eliminate. And it can use whatever "ruses of war" it needs to get the job done. As a result, the main legal constraints on sanctioned assassination other than domestic law, which makes murder a crime in almost all countries, are rules that nations impose on themselves.

The U.S. government adopted such a ban in 1976, when President Ford—responding to the scandal that resulted when the press revealed CIA involvement in several assassinations—issued Executive Order 11905. This order prohibited what it called "political assassination" and essentially reaffirmed an often-overlooked ban that Director of Central Intelligence Richard Helms had adopted for the CIA four years earlier. Jimmy Carter reaffirmed the ban in 1978 with his own Executive Order 12036. Ronald Reagan went even further in 1981; his Executive Order 12333 banned assassination in toto. This ban on assassination remains in effect today.

Even so, there has been a disconnect between our policy and practice. The United States has tried to kill foreign leaders on several occasions since 1976, usually as part of a larger military operation.

For example, in 1986, U.S. Air Force and Navy planes bombed Libya after a Libyan terrorist attack against a nightclub frequented by American soldiers in Berlin. One of the targets was Muammar Qaddafi's tent. During Desert Storm in 1991, we bombed Saddam Hussein's official residences and command bunkers.

After the United States linked Osama bin Laden to terrorist bombings of U.S. embassies in Kenya and Tanzania in 1998, we launched a cruise missile attack at one of his bases in Afghanistan.

In each case, U.S. officials insisted that our forces were merely aiming at "command and control" nodes or at a building linked to military operations or terrorist activities. In each case, however, the same officials admitted off the record that they would not have been upset if Qaddafi, Saddam, or bin Laden had been killed in the process.

More recently, according to press reports, presidents have also approved so-called lethal covert operations—operations in which there is a good chance that an unfriendly foreign official might be killed. For example, the press reported a CIA-backed covert operation to topple Saddam in 1996 that probably would have killed him in the process, given the record of Iraqi leadership successions (no one has left office alive). After the September 11 terrorist strikes on New York and Washington, former Clinton officials leaked word to reporters that the CIA had trained Pakistani commandos in 1999 to snatch bin Laden. Given the record of such operations, bin Laden would likely not have survived.

In short, the unintended result of banning assassinations has been to make U.S. leaders perform verbal acrobatics to explain how they have tried to kill someone in a military operation without really trying to kill him. One has to wonder about the wisdom of any policy that allows officials to do something but requires them to deny that they are doing it. We would be better off simply doing away with the prohibition, at least as it applies to U.S. military operations.

Does It Work?

The effectiveness of assassination has depended much on its objectives. Most (but not all) attempts to change the course of large-scale political and diplomatic trends have failed. Assassination has been more effective in achieving small, specific goals.

Indeed, past U.S. assassination attempts have had great difficulty in even achieving the minimal level of success: killing the intended target. According to the available information, *every* U.S. effort to kill a high-ranking official since World War II outside a full-scale war has failed. This record is so poor that it would be hard to find an instrument of national policy that has been less successful in achieving its objectives than assassination. . . .

According to the [Senator Frank] Church Committee investigations of the 1970s, the CIA supported assassins trying to kill Patrice Lumumba of the Congo in 1961 and repeatedly tried to assassinate Fidel Castro between 1961 and 1963. In addition, American officials were either privy to plots or encouraged coups that caused the death of a leader (Rafael Trujillo of the Dominican Republic in 1961, Ngo Dinh Diem of South Vietnam in 1963, General René Schneider of Chile in 1970, and, later, President Salvador Allende in 1973). And, as noted, in recent years the United States has tried to do away with Qaddafi, Saddam, and bin Laden.

What is notable about this record is that it is remarkably free of success. Castro, Qaddafi, Saddam, and (at least at this writing) bin Laden all survived. (As this is being written, U.S. forces are hunting bin Laden as part of the larger war against the Taliban in Afghanistan.) What is more, Qaddafi continued to support terrorism (e.g., the bombing of Pan Am flight 103). Saddam has managed to outlast the terms of two presidents who wanted to eliminate him (George Bush and Bill Clinton), while continuing to support terrorism—and developing weapons of mass destruction.

One might have predicted this dismal record just by considering why American leaders have resorted to the assassination option. More often than not, assassination is the option when nothing seems to work but officials think that they need to do *something.* When diplomacy is ineffective and war seems too costly, assassination becomes the fallback—but without anyone asking whether it will accomplish anything.

This seems to have been the thinking behind the reported U.S. covert operation to eliminate Saddam in the mid-1990s. Despite a series of provocations—an assassination attempt against former president George Bush, violence against Shi'ite Muslims and Kurds, and violations of U.N. inspection requirements—the Clinton administration was unwilling to wage a sustained, full-scale war against him. Diplomacy was also failing, as the United States was unable to hold together the coalition that won Desert Storm. Covert support to Saddam's opponents in the military was the alternative. It was an utter failure.

True, some other countries have been more successful in that they have killed their target. For example, after the terrorist attack on Israeli athletes in the 1972 Munich Olympics, Israeli special services tracked down and killed each of the Palestinian guerrillas who took part (they also killed an innocent Palestinian in a case of mistaken identity). In 1988 Israeli commandos killed Khalil Al-Wazir, a lieutenant of Yasser Arafat's, in a raid on PLO headquarters in Tunisia. More recently, Israel has killed specifically targeted Palestinian terrorist leaders—for example, Yechya Ayyash, who was killed with a booby-trapped cell phone.

Other countries have also attempted assassinations with some degree of tactical success. During the Cold War, the KGB [Soviet political police] was linked to several assassinations. Most recently, the Taliban regime in Afghanistan was suspected of being involved in the assassination of Ahmed Shah Massoud, the leader of the Northern Alliance opposition.

But even "successful" assassinations have often left the sponsor worse off, not better. The murder of Diem sucked the United States deeper into a misconceived policy. The assassination of Abraham Lincoln (carried out by a conspiracy some believe to have links to the Confederate secret service) resulted in Reconstruction. German retribution against Czech civilians after the 1942 assassination of Nazi prefect Reinhard Heydrich by British-sponsored resistance fighters was especially brutal. The 1948 assassination of Mohandas Gandhi by Hindu extremists led to violence that resulted in the partition of India.

In short, assassination has usually been unreliable in shaping large-scale political trends the way the perpetrators intended (though the assassination of Yitzhak Rabin by a Zionist extremist in 1995 may be the exception). When it accomplishes anything beyond simply killing the target, it is usually by depriving an enemy of the talents of some uniquely skilled individual. For example, in 1943 U.S. warplanes shot down an aircraft known to be carrying Admiral

Isoroku Yamamoto—the architect of Japan's early victories in the Pacific. His loss hurt the Japanese war effort. The same could be said of the loss of Massoud to the Northern Alliance.

The problem is, picking off a talented individual is almost always harder than it looks. One paradox of modern warfare is that, although it is not that hard to kill many people, it can be very difficult to kill a particular person. One has to know exactly where the target will be at a precise moment. This is almost always hard, especially in wartime.

Should We Do It, and If So, How?

This is the most complex issue, of course. The morality of sanctioned assassination depends mainly on whether and when one can justify murder. Most religions and agnostic philosophies agree that individuals have the right to kill in self-defense when faced with immediate mortal danger. This principle is codified in American law. And, as we have seen, even international law seems to allow killing—even killing specific individuals—when it can be justified as armed self-defense.

Although most Americans do not like the idea of deliberate killing, they do not completely reject it, either. Most would agree that their government should be allowed to kill (or, more precisely, allow people to kill in its behalf) in at least two situations.

One situation is when a police officer must eliminate an immediate threat to public safety—for example, shooting an armed robber or apprehending a suspect who has proven dangerous in the past and who resists arrest. The other situation is when soldiers go to war to defend the country from attack. In addition, many—but not all—Americans believe that the government should be allowed to kill in the case of capital crimes.

It is probably not a coincidence that the U.S. Constitution also envisions these three—and only these three—situations in which the federal government might take a life: policing, going to war, and imposing capital punishment. Logically, then, assassination must fit into one of these three tracks. Assassination can be considered a police act, in which case it must follow the rules for protecting accused criminals. Or it can be considered a military act, in which case it must follow the rules that control how the United States wages war. Or it can be considered capital punishment, in which case it must follow the rules of due process.

Given this, when would we want to allow government to kill a particular foreign national? Clearly we should not use assassination as a form of de facto capital punishment. Unless the intended target presents a clear and immediate threat, there is always time to bring a suspect to justice, where we could guarantee due process. Similarly, although police should be able to protect themselves and others while making an arrest, we would not want police to pursue their targets with the expectation that they would routinely kill them.

The only time we should consider assassination is when we need to eliminate a clear, immediate, lethal threat from abroad. In other words, assassination is a military option. We need to understand it as such because the United States will face more situations in which it must decide whether it is willing, in effect, to go to war to kill a particular individual and how it will target specific individuals during wartime. Two factors make this scenario likely.

First, technology often makes it hard for one *not* to target specific people. Weapons are so accurate today that, when one programs their guidance systems, you aim not just for a neighborhood, or a building in the neighborhood, but for a particular *room* in a particular building. In effect, even bombing and long-range missile attacks have become analogous to sniping. You cannot always be sure you will hit your target—just as snipers often miss and sometimes hit the wrong target—but you still must aim at specific people.

Second, the nature of the threats we face today will likely require us to target specific individuals. Terrorist organizations today use modern communications to organize themselves as worldwide networks. These networks consist of small cells that can group and regroup as needed to prepare for a strike. This is how the bin Laden organization has operated. Seeing how successful these tactics have been, many armies will likely often adopt a similar approach. To defeat such networked organizations, our military forces will need to move quickly, find the critical cells in a network, and destroy them. This inevitably will mean identifying specific individuals and killing them—in other words, assassination.

But when we do so, we should be clear in our own minds that, when the United States tries to assassinate someone, we are going to war—with all the risks and costs that war brings. These include, for example, diplomatic consequences, the danger of escalation, the threat of retaliation against our own leaders, the threat of retaliation against American civilians, and so on.

Because assassination is an act of war, such activities should always be considered a military operation. American leaders need to resist the temptation to use intelligence organizations for this mission. Intelligence organizations are outside the military chain of command. Intelligence operatives are not expected to obey the rules of war and thus are not protected by those rules. At the same time, intelligence organizations are also not law enforcement organizations. In many situations, having intelligence organizations kill specific individuals looks too much like a death sentence without due process.

Indeed, there is reason to question whether intelligence organizations are even technically qualified for assassination. In every publicly known case in which the CIA has considered killing a foreign leader, the agency has outsourced the job. In most cases, it has recruited a foreign intelligence service or military officials with better access. In some of the attempts to kill Castro, the CIA recruited Mafia hit men. Even in the more recent reported cases of lethal covert actions, foreigners would have done the actual killing. It is hard to maintain control and quality when you subcontract assassination services—as the record shows.

The United States did not ask for the threats we currently face, and killing on behalf of the state will always be the most controversial, most distasteful policy issue of all. That is why we need to use blunt language and appreciate exactly what we are proposing. Sugarcoating the topic only hides the tough issues we need to decide as a country. But if we do need to target specific people for military attack, it is important that we get it right.

NO

Margot Patterson

Assassination as a Weapon

Osama bin Laden Dead or Alive. Those were the words President [George W.] Bush used to describe U.S. policy toward the man believed to have masterminded the Sept. 11 attacks on the World Trade Center and Pentagon. So far Osama bin Laden appears to have eluded U.S. forces and their allies, but the war on terror that the United States unleashed after Sept. 11 has triggered far-reaching changes in the United States, ushering in an era of growing police powers at home and greater bellicosity abroad.

Increasingly, that bellicosity is triggering alarm, as the United States comes to be perceived as fighting fire with fire. A recent front-page story in *USA Today,* headlined "Global warmth for U.S. after 9/11 turns to frost," describes how the United States is coming to be seen by its allies as a rogue state. Internally, too, the United States faces criticism that it is becoming what it deplores: a society of men, not laws, operating without either internal or external brakes.

"Bush and his administration are pushing the edge on all moral fronts right now. We are dangerously treading on civil liberties in this country today," said Robert Ashmore, a professor emeritus of philosophy at Marquette University.

Civil rights advocates have numerous concerns, not least of which are the approximately 1,000 people picked up after Sept. 11 and held for months without charges being filed against them. The American Civil Liberties Union, which has filed a Freedom of Information Act request for information about the detainees, says it doesn't know how many people are still being held. Plans for an unprovoked attack on Iraq, the refusal of the United States to join the International Criminal Court after earlier recognizing its establishment, the embrace of pre-emptive strikes as new U.S. military doctrine, and greater discussion of assassination as a legitimate tactic in the war against terrorism indicate that the moral boundaries of U.S. policy are far different from what they were [in 2001].

For Robert Johansen, a professor of government and international studies at the University of Notre Dame, many of the tactics the United States has adopted in its war against terrorism are counterproductive.

"When we talk about terrorism, the single most important thing the United States can do in combating terrorism is to clarify the difference between the United States and terrorism," Johansen said. "Those engaging in terrorism want to deny that there is any difference. We do not intend to kill innocent people, and the moment we begin to move closer to killing innocent people the differentiation between us and terrorists begins to diminish."

For Johansen, therefore, assassination is unacceptable policy. "To kill people without some assessment of guilt is morally inappropriate, and that would mean some kind of trial," he said.

Fundamentally Different

Johansen said assassination is fundamentally different from the conduct of war. Murder is not legitimate killing; killing in war is considered legitimate. One problem with declaring a war on terrorism and treating the terrorists as if they were conducting a war is that it gives the other side some legitimacy, Johansen said. In his view, the attacks of Sept. 11 would be better considered a crime, not an act of war.

Johansen said the most fundamental principle in international relations is reciprocity. "We will not do anything ourselves that we're unwilling to have done against us. If we don't want U.S. leaders to be targets of assassins, I don't think the U.S. can legitimately use a policy of assassination against others. Inevitably this will come back to haunt us," he said.

Is the United States trying to assassinate Osama bin Laden? Some would say undoubtedly; others argue that attempts on the life of Osama bin Laden can be viewed as part of an effort to disable the enemy's command and control center and are covered by the rules of wartime engagement.

"It is not against the laws of war in attacking an enemy force to look especially for a particular individual or individuals, who if they resist at all you're entitled to kill them," said Col. Dan Smith, an analyst at the Center for Defense Information. What you can't do, said Smith, is capture an enemy and then tie him up and execute him.

The Confines of Warfare

"The pursuit of Osama bin Laden and Mullah Omar are supposedly undertaken within the confines of warfare, within the definition of warfare, so the fact that we are after them in the same way we were after the high command of the Nazis or Japanese imperial forces . . . keeps this from becoming a legal problem," said Smith.

If this strikes some as a legalistic and linguistic nicety, it also underscores the fact that to decision-makers the word "assassination" can be as flexible and open to interpretation as the word "is" famously was to President [Bill] Clinton. Officially, Washington is still bound by President Gerald Ford's 1976 executive order prohibiting U.S. government employees from engaging in political assassination, but some analysts believe that the United States is moving closer to adopting policies that it previously condemned. In October, *The Washington Post* reported that President Bush had signed an intelligence finding authorizing preemptive covert lethal action against Osama bin Laden and Al Qaeda.

"Washington's previous position was that assassinations were basically wrong morally, and politically counterproductive," said Fawaz Gerges, a Mideast scholar at Sarah Lawrence College. According to Gerges, Sept. 11 represented a watershed in Washington's thinking, not just toward the Mideast but toward other conflicts as well.

"Now the new thinking in Washington is that the United States under certain conditions and in certain situations should be able to empower the CIA to assassinate terrorists or certain people who represent a threat to the United States.

"We are slowly and steadily injecting life into the policy of targeted or limited assassination of certain individuals," Gerges said.

The wisdom of such a policy can and is being debated. Israel, for instance, the only country that openly uses assassination, is the model and sometimes the inspiration for this debate. As such, Israel's strategy of what it calls "active defense" against suspected terrorists is a test case of the pros and, many would say, the cons of such a policy.

Since the start of the second intifada [Palestinian civil uprising] in the occupied territories in the fall of 2000, Israel has slain close to 100 Palestinians it alleges were involved in terrorism. Gerges said there are two narratives about Israel's targeting of Palestinian activists. According to the Israeli narrative, Israel does not target Palestinian leaders or leaders who do not use violence as part of struggle. According to the Palestinian narrative, the policy of assassination has no limits and no checks and balances. Scores of innocent people have been killed as collateral damage, and Palestinians assert that Israel's policy of preemptive strikes is intended to keep the two sides from ever sitting down at the negotiating table.

Counterproductive at Best

Gerges' own take on these different narratives is that Israel's assassination policy has been counterproductive at best. "Violence has not only not resolved the conflict, it has produced opposite results. It has led to more bloodshed and destruction on both sides."

Mark Regev, a spokesman for the Israeli Embassy in Washington, said Israel conducts surgical targeted strikes against a known terrorist operative. "We use the latest military technologies specifically to hit only the person we want to hit and to avoid any collateral damage," said Regev, who rejected the word "assassinate" to describe Israel's policy of targeting suspected terrorists.

But Israel's execution in Gaza of Hamas leader Saleh Shehada in a July missile strike that also killed 14 civilians shows that Israel's targeted killings are not always so tidy.

Richard Falk, a professor of international law at Princeton University who was one of three people on a United Nations commission appointed by U.N. Commissioner for Human Rights Mary Robinson to investigate human rights in Israel, said the commission concluded that Israel's execution policy was as much about terrorizing the Palestinian population as defending Israel against terrorism.

"Some of the individuals targeted were very inappropriate from a security perspective. They were people active in the peace movement, people with strong contacts with Israeli peace activists. Israel produced no evidence that validated their selection as dangerous terrorists," Falk said.

Proponents of adding assassination to the U.S. national security arsenal say assassination can be a kinder, gentler way of achieving certain objectives than, say, war. Jeffrey Richelson, a senior fellow at the National Security Archive, and the author of *The U.S. Intelligence Community* and *A Century of Spies: Intelligence in the Twentieth Century,* said it's absurd to argue that assassination is immoral in all circumstances. "It may be a solution to a particular problem and in some circumstances it may be morally justified," he said.

"Profound Failure"

Writing in a recent issue of *The Journal of Intelligence and CounterIntelligence,* Richelson examines the case for assassination in an essay titled "When Kindness Fails: Assassination as a National Security Option." While noting that the United States' experience in efforts at assassinating foreign leaders is one of "profound failure," Richelson said the United States should not preclude using assassination in certain circumstances. Assassination should be employed only, but not always, to deal with terrorists or with heads of rogue states who are developing weapons of mass destruction, he said.

"Assassination should not be used as attempted in the past—as a foreign policy tool to eliminate troublesome foreign leaders such as Fidel Castro or Rafael Trujillo," he wrote. Richelson argued that if the United States assassinates foreign leaders and terrorists, it should acknowledge it as the Israeli government does now both as a deterrent to similar individuals and as a way of making clear under what circumstances the United States will resort to lethal means. "If the president can order such an operation, he should be able to defend it publicly," Richelson concluded.

Others question just how efficacious assassination would be as a policy. "Even as a strategic tool, assassination isn't useful," said Ashmore, who said that there is no evidence that Israel's resort to extrajudicial executions has made life any more secure for Israelis. He pointed out that Israel has the ability to go into Palestinian villages and put Palestinian suspects on trial but refuses to do this. "It prefers to act in the manner of a terrorist," Ashmore said.

Ashmore said that too much attention is focused on the terrorism of insurgents, those who are opposing government for one reason or other, and not enough on the terrorist practices of states. "Throughout history the vast majority of the victims of terrorism are the victims of state terrorism. Our own American history, unfortunately, provides many examples of state terrorism on the part of the United States," Ashmore said.

"What's tragic about so much of our support for terrorism is that it's blown up in our face," Ashmore said. "We make enemies among the people when we support despotic powers. It's very short-range strategic thinking to support someone like Pinochet or Mobotu or the Shah. I think the same thing is going to happen to us in Saudi Arabia and Egypt. We are supporting repressive regimes."

According to one former CIA analyst who prefers not to be identified, not only terrorism but also technology is driving some of the new debate about assassination.

"The weapons are so accurate today that it's hard to often avoid putting a weapon in a very precise place," he said. "My problem with assassination is that usually it comes up in the context of covert action. People are unwilling to accept the risk of war and diplomacy doesn't seem to be working, so they kid themselves that they have this bad magic. They also fool themselves that they aren't committing an act of war. If you decide you need to take military action, you shouldn't fool yourself that you're not taking military action."

Not Easy to Accomplish

Greg Treverton, a staffer on the Church Committee, the Senate committee headed by Sen. Frank Church that investigated CIA assassination plots prior to 1970, is the author of *Reshaping National Intelligence for an Age of Information.*

Like many others, Treverton said he sees signs that the United States is now moving toward easing the ban on political assassination, which resulted in large part from the discoveries made by the Church Committee. The ban applies to assassination efforts by U.S. government employees, but not the military.

"The idea at the time was that a willful targeting of a foreign leader was not a very good idea—beneath us morally, and not always easy to accomplish. I think we are reconsidering whether the lesser of evils might not be targeting leaders," Treverton said. "If the alternative to killing Saddam Hussein is fighting a major war in which Iraqis die like flies but Hussein stays in power, wouldn't it be morally superior to kill him directly if we could? If our goal is regime change, then there must be [a] better way than fighting a major war in which some Americans die and an awful lot of Iraqis die."

But Treverton said as with any assassination effort, it's unclear whether Saddam Hussein's successor would be better or worse.

POSTSCRIPT

Is Government-Ordered Assassination Sometimes Acceptable?

After the articles by Berkowitz and Patterson were written, the United States went to war against Iraq. Virtually the first U.S. action was to attack a bunker in which intelligence placed Iraq's president. According to press reports, President George W. Bush mulled over the legality of such an attack and decided that since the compound where Saddam Hussein was located was a command-and-control facility, it was a legitimate target. At another point, an American commentator noted that Hussein was actively engaged in directing Iraqi forces and was therefore an acceptable—indeed desirable—target. The logic was that a so-called decapitation attack held the prospect of disrupting the ability and will of the Iraqi military to fight. The legality of such acts falls in significant part under the theory of just war and international law, a topic that is examined in Ingrid Detter, *The Law of War* (Cambridge University Press, 2000).

Even if one assumes that a tacit state of war existed between the United States and Iraq at the time and, therefore, that the attack on Hussein was a military action, not an assassination attempt, there are still some troubling questions. One is whether it would have been equally legitimate for the Iraqis to try to kill President Bush. The logic of a decapitation attack could also be applied to an Iraqi attempt on the life of the U.S. commander in chief.

The controversy over government-sanctioned assassination also raises the troubling issue of means. Would it make a difference if Hussein had been killed by explosives delivered by U.S. warplanes while President Bush had been poisoned to death by a cook paid by Iraqi intelligence? Would the former have been legitimate military action and the latter a terrorist assassination? Does it make any difference that the CIA carries out some of the U.S. "targeted" attacks? Is that taking the country back to the practices condemned by the Church Committee's report? That 1975 report may be found in its original form in some government depository libraries, and it has also been published by several presses, including under the title *Alleged Assassination Plots Involving Foreign Leaders* (Fredonia Books, 2001).

In many ways, these questions touch on the larger question of what defines terrorism. The tendency for the United States and other militarily powerful countries has been to define terrorism in a way that, to a degree, allows killings that can be carried out using conventional military means and condemns attacks carried out by weak states and organizations using a sniper's bullet or some other nonmilitary means. From the point of view of some people, such parameters are nothing more than self-serving distinctions that are irrelevant to the basic issue of whether or not it is moral to specifically target and kill an individual.

Yet another controversy relates to whether or not it can be moral for a government to commit an act that would be immoral for an individual. This is hardly a new debate. In 1793, when France was at war with Great Britain, Paris asked for U.S. assistance, as called for in the treaty that had brought France to America's aid during the American Revolution. The problem is that war with Great Britain was dangerous. Secretary of State Thomas Jefferson argued that the United States was bound by its word, just as an individual would be. He said, "Between society and society the same moral duties exist as between the individuals composing them." Secretary of the Treasury Alexander Hamilton disagreed, contending that "the rule of morality . . . is not precisely the same between nations as between individuals. . . . Millions [of people] are [affected by] . . . matters of government; while the consequences of the private actions of an individual ordinarily terminate with himself."

A good source for more on government-directed assassinations is the March 2003 issue of *University of Richmond Law Review*, which contains a number of scholarly articles that resulted from the symposium "Terrorism and Assassination" held at the university in April 2002.

ISSUE 15

Would World Affairs Be More Peaceful If Women Dominated Politics?

YES: Francis Fukuyama, from "Women and the Evolution of World Politics," *Foreign Affairs* (September/October 1998)

NO: Mary Caprioli, from "The Myth of Women's Pacifism," An Original Essay Written for This Volume (August 1999)

ISSUE SUMMARY

YES: Francis Fukuyama, the Hirst Professor of Public Policy at George Mason University, contends that a truly matriarchal world would be less prone to conflict and more conciliatory and cooperative than the largely male-dominated world that we live in now.

NO: Assistant professor of political science Mary Caprioli contends that Fukuyama's argument is based on a number of unproven assumptions and that when women assume more political power and have a chance to act aggressively, they are as apt to do so as men are.

Political scientists are just beginning to examine whether or not gender makes a difference in political attitudes and the actions of specific policymakers and whether or not any gender differences that may exist as a result have a biological origin or are a product of the divergent ways in which males and females are socialized. The ultimate question is whether or not an equal representation of women among policymakers—or, even more radically, a reversal of tradition that would put women firmly in charge of foreign and defense policy—would make an appreciable difference in global affairs. That is what this debate is about.

Certainly there is good evidence that women in the mass public are less likely to countenance war than men. In the United States, for example, polls going back as far as World War II and extending to the present have indicated that women are less ready than men to resort to war or to continue war. Examining the difference between males and females in their opinions about the use of force against Iraq during the Persian Gulf crisis yields some fascinating results.

Polls of men and women in 11 countries found that in 10 of the countries (Belgium, France, Germany, Great Britain, Israel, Italy, Japan, Mexico, Nigeria, and Russia), men were more likely than women to favor using force against Iraq. Only in Turkey were women more bellicose than men. The pro-war average across the 11 societies was 55 percent for men and 47 percent for women. In the United States, 69 percent of the men thought that the benefits of war would be worth the cost in lives; only 49 percent of the women agreed.

Yet the attitudes of women who are in positions of authority may be much different from those of other women. So far only about two dozen women have been elected to lead countries, although that number is slowly growing. Yet the relative scarcity of female international leaders makes comparisons with their male counterparts difficult. There can be no doubt, though, that able, sometimes aggressive leadership has been evident in such modern female heads of government as Israel's Golda Meir and India's Indira Gandhi.

Moreover, it would be grossly inaccurate to say, "Men are apt to favor war, women are not." The reality is that favoring or not favoring war in each of the 11 countries in the study noted above was much more associated with the country itself. There were some countries in which both men and women strongly favored war. In Israel, which was threatened, then attacked, by Iraq, an overwhelming majority of both men and women favored using force. In other countries, a majority of both men and women opposed using force.

In a more extended sense, the following readings are also about equal political opportunity for men and women. There are two reasons to favor equal opportunity. The first is more philosophical and rests on justice. This argument holds that equity demands that women have the same ability as men to achieve political office. In most societies these days there are few people who would disagree, at least publicly, with this view.

The second argument in support of political gender equality is more controversial and at the heart of this debate. This is the disagreement over whether or not women would make a policy difference because they are inherently apt to have a different view of politics than men do. Many scholars, feminist and otherwise, agree that there is a deep-seated difference. Feminist scholar Betty Reardon suggests that "from the masculine perspective, peace for the most part has meant the absence of war and the prevention of armed conflict." She terms this "negative peace." By contrast, Reardon maintains that women think more in terms of "positive peace," which includes "conditions of social justice, economic equity and ecological balance." The implication of this is that in a world run much more by women than is now the case, military budgets would go down and budgets for social, educational, and environmental programs would go up. Even though he is not a feminist scholar, as such, Francis Fukuyama, author of the first of the following selections, would fit in with this school of thought.

There are other scholars, feminist and otherwise, who do not think that women and men have "hard-wired" political differences. For them, a matriarchal society would run generally the same as a patriarchal society, no matter what the issue. Mary Caprioli adheres to this line of thought in the second selection, and so the debate is joined.

Francis Fukuyama

Women and the Evolution
of World Politics

Chimpanzee Politics

In the world's largest captive chimp colony at the Burger's Zoo in Arnhem, Netherlands, a struggle worthy of Machiavelli unfolded during the late 1970s. As described by primatologist Frans de Waal, the aging alpha male of the colony, Yeroen, was gradually unseated from his position of power by a younger male, Luit. Luit could not have done this on the basis of his own physical strength, but had to enter into an alliance with Nikkie, a still younger male. No sooner was Luit on top, however, than Nikkie turned on him and formed a coalition with the deposed leader to achieve dominance himself. Luit remained in the background as a threat to his rule, so one day he was murdered by Nikkie and Yeroen, his toes and testicles littering the floor of the cage.

Jane Goodall became famous studying a group of about 30 chimps at the Gombe National Park in Tanzania in the 1960s, a group she found on the whole to be peaceful. In the 1970s, this group broke up into what could only be described as two rival gangs in the northern and southern parts of the range. The biological anthropologist Richard Wrangham with Dale Peterson in their 1996 book *Demonic Males* describes what happened next. Parties of four or five males from the northern group would go out, not simply defending their range, but often penetrating into the rival group's territory to pick off individuals caught alone or unprepared. The murders were often grisly, and they were celebrated by the attackers with hooting and feverish excitement. All the males and several of the females in the southern group were eventually killed, and the remaining females forced to join the northern group. The northern Gombe chimps had done, in effect, what Rome did to Carthage in 146 B.C.: extinguished its rival without a trace.

There are several notable aspects to these stories of chimp behavior. First, the violence. Violence within the same species is rare in the animal kingdom, usually restricted to infanticide by males who want to get rid of a rival's offspring and mate with the mother. Only chimps and humans seem to have a proclivity for routinely murdering peers. Second is the importance of coalitions

From Francis Fukuyama, "Women and the Evolution of World Politics," *Foreign Affairs*, vol. 77, no. 5 (September/October 1998). Copyright © 1998 by The Council on Foreign Relations, Inc. Reprinted by permission.

and the politics that goes with coalition-building. Chimps, like humans, are intensely social creatures whose lives are preoccupied with achieving and maintaining dominance in status hierarchies. They threaten, plead, cajole, and bribe their fellow chimps to join with them in alliances, and their dominance lasts only as long as they can maintain these social connections.

Finally and most significantly, the violence and the coalition-building is primarily the work of males. Female chimpanzees can be as violent and cruel as the males at times; females compete with one another in hierarchies and form coalitions to do so. But the most murderous violence is the province of males, and the nature of female alliances is different. According to de Waal, female chimps bond with females to whom they feel some emotional attachment; the males are much more likely to make alliances for purely instrumental, calculating reasons. In other words, female chimps have relationships; male chimps practice realpolitik.

Chimpanzees are man's closest evolutionary relative, having descended from a common chimp-like ancestor less than five million years ago. Not only are they very close on a genetic level, they show many behavioral similarities as well. As Wrangham and Peterson note, of the 4,000 mammal and 10 million or more other species, only chimps and humans live in male-bonded, patrilineal communities in which groups of males routinely engage in aggressive, often murderous raiding of their own species. Nearly 30 years ago, the anthropologist Lionel Tiger suggested that men had special psychological resources for bonding with one another, derived from their need to hunt cooperatively, that explained their dominance in group-oriented activities from politics to warfare. Tiger was roundly denounced by feminists at the time for suggesting that there were biologically based psychological differences between the sexes, but more recent research, including evidence from primatology, has confirmed that male bonding is in fact genetic and predates the human species.

The Not-So-Noble Savage

It is all too easy to make facile comparisons between animal and human behavior to prove a polemical point, as did the socialists who pointed to bees and ants to prove that nature endorsed collectivism. Skeptics point out that human beings have language, reason, law, culture, and moral values that make them fundamentally different from even their closest animal relative. In fact, for many years anthropologists endorsed what was in effect a modern version of Rousseau's story of the noble savage: people living in hunter-gatherer societies were pacific in nature. If chimps and modern man had a common proclivity for violence, the cause in the latter case had to be found in civilization and not in human nature.

A number of authors have extended the noble savage idea to argue that violence and patriarchy were late inventions, rooted in either the Western Judeo-Christian tradition or the capitalism to which the former gave birth. Friedrich Engels anticipated the work of later feminists by positing the existence of a primordial matriarchy, which was replaced by a violent and repressive patriarchy only with the transition to agricultural societies. The problem with this theory is, as Lawrence Keeley points out in his book *War Before Civilization,* that the most comprehensive recent studies of violence in hunter-gatherer societies suggest that for them war was actually more frequent, and rates of murder higher, than for modern ones.

Surveys of ethnographic data show that only 10–13 percent of primitive societies never or rarely engaged in war or raiding; the others engaged in conflict either continuously or at less than yearly intervals. Closer examination of the peaceful cases shows that they were frequently refugee populations driven into remote locations by prior warfare or groups protected by a more advanced society. Of the Yanomamö tribesmen studied by Napoleon Chagnon in Venezuela, some 30 percent of the men died by violence; the !Kung San of the Kalahari desert, once characterized as the "harmless people," have a higher murder rate than New York or Detroit. The sad archaeological evidence from sites like Jebel Sahaba in Egypt, Talheim in Germany, or Roaix in France indicates that systematic mass killings of men, women, and children occurred in Neolithic times. The Holocaust, Cambodia, and Bosnia have each been described as a unique, and often as a uniquely modern, form of horror. Exceptional and tragic they are indeed, but with precedents stretching back tens if not hundreds of thousands of years.

It is clear that this violence was largely perpetrated by men. While a small minority of human societies have been matrilineal, evidence of a primordial matriarchy in which women dominated men, or were even relatively equal to men, has been hard to find. There was no age of innocence. The line from chimp to modern man is continuous.

It would seem, then, that there is something to the contention of many feminists that phenomena like aggression, violence, war, and intense competition for dominance in a status hierarchy are more closely associated with men than women. Theories of international relations like realism that see international politics as a remorseless struggle for power are in fact what feminists call a gendered perspective, describing the behavior of states controlled by men rather than states per se. A world run by women would follow different rules, it would appear, and it is toward that sort of world that all postindustrial or Western societies are moving. As women gain power in these countries, the latter should become less aggressive, adventurous, competitive, and violent.

The problem with the feminist view is that it sees these attitudes toward violence, power, and status as wholly the products of a patriarchal culture, whereas in fact it appears they are rooted in biology. This makes these attitudes harder to change in men and consequently in societies. Despite the rise of women, men will continue to play a major, if not dominant, part in the governance of postindustrial countries, not to mention less-developed ones. The realms of war and international politics in particular will remain controlled by men for longer than many feminists would like. Most important, the task of resocializing men to be more like women—that is, less violent—will run into limits. What is bred in the bone cannot be altered easily by changes in culture and ideology.

The Return of Biology

We are living through a revolutionary period in the life sciences. Hardly a week goes by without the discovery of a gene linked to a disease, condition, or behavior, from cancer to obesity to depression, with the promise of genetic therapies and even the outright manipulation of the human genome just around the corner. But while developments in molecular biology have been receiving the lion's share of the headlines, much progress has been made at the behavioral level as well. The past generation has seen a revival in Darwinian thinking about human psychology, with profound implications for the social sciences.

For much of this century, the social sciences have been premised on Emile Durkheim's dictum that social facts can be explained only by prior social facts and not by biological causes. Revolutions and wars are caused by social facts such as economic change, class inequalities, and shifting alliances. The standard social science model assumes that the human mind is the terrain of ideas, customs, and norms that are the products of man-made culture. Social reality is, in other words, socially constructed: if young boys like to pretend to shoot each other more than young girls, it is only because they have been socialized at an early age to do so.

The social-constructionist view, long dominant in the social sciences, originated as a reaction to the early misuse of Darwinism. Social Darwinists like Herbert Spencer or outright racists like Madsen Grant in the late nineteenth and early twentieth centuries used biology, specifically the analogy of natural selection, to explain and justify everything from class stratification to the domination of much of the world by white Europeans. Then Franz Boas, a Columbia anthropologist, debunked many of these theories of European racial superiority by, among other things, carefully measuring the head sizes of immigrant children and noting that they tended to converge with those of native Americans when fed an American diet. Boas, as well as his well-known students Margaret Mead and Ruth Benedict, argued that apparent differences between human groups could be laid at the doorstep of culture rather than nature. There were, moreover, no cultural universals by which Europeans or Americans could judge other cultures. So-called primitive peoples were not inferior, just different. Hence was born both the social constructivism and the cultural relativism with which the social sciences have been imbued ever since.

But there has been a revolution in modern evolutionary thinking. It has multiple roots; one was ethology, the comparative study of animal behavior. Ethologists like Konrad Lorenz began to notice similarities in behavior across a wide variety of animal species suggesting common evolutionary origins. Contrary to the cultural relativists, they found that not only was it possible to make important generalizations across virtually all human cultures (for example, females are more selective than males in their choice of sexual partners) but even across broad ranges of animal species. Major breakthroughs were made by William Hamilton and Robert Trivers in the 1960s and 1970s in explaining instances of altruism in the animal world not by some sort of instinct towards species survival but rather in terms of "selfish genes" (to use Richard Dawkins' phrase) that made social behavior in an individual animal's interest. Finally, advances in neurophysiology have shown that the brain is not a Lockean tabula rasa waiting to be filled with cultural content, but rather a highly modular organ whose components have been adapted prior to birth to suit the needs of socially oriented primates. Humans are hard-wired to act in certain predictable ways.

The sociobiology that sprang from these theoretical sources tried to provide a deterministic Darwinian explanation for just about everything, so it was perhaps inevitable that a reaction would set in against it as well. But while the term sociobiology has gone into decline, the neo-Darwinian thinking that spawned it has blossomed under the rubric of evolutionary psychology or anthropology and is today an enormous arena of new research and discovery.

Unlike the pseudo-Darwininsts at the turn of the century, most contemporary biologists do not regard race or ethnicity as biologically significant categories. This stands to reason: the different human races have been around only for the past hundred thousand years or so, barely a blink of the eye in evolutionary time. As countless authors have pointed out, race is largely a socially constructed category: since all races can (and do) interbreed, the boundary lines between them are often quite fuzzy.

The same is not true, however, about sex. While some gender roles are indeed socially constructed, virtually all reputable evolutionary biologists today think there are profound differences between the sexes that are genetically rather than culturally rooted, and that these differences extend beyond the body into the realm of the mind. Again, this stands to reason from a Darwinian point of view: sexual reproduction has been going on not for thousands but hundreds of millions of years. Males and females compete not just against their environment but against one another in a process that Darwin labeled "sexual selection," whereby each sex seeks to maximize its own fitness by choosing certain kinds of mates. The psychological strategies that result from this never-ending arms race between men and women are different for each sex.

In no area is sex-related difference clearer than with respect to violence and aggression. A generation ago, two psychologists, Eleanor Maccoby and Carol Jacklin, produced an authoritative volume on what was then empirically known about differences between the sexes. They showed that certain stereotypes about gender, such as the assertion that girls were more suggestible or had lower self-esteem, were just that, while others, like the idea that girls were less competitive, could not be proven one way or another. On one issue, however, there was virtually no disagreement in the hundreds of studies on the subject: namely, that boys were more aggressive, both verbally and physically, in their dreams, words, and actions than girls. One comes to a similar conclusion by looking at crime statistics. In every known culture, and from what we know of virtually all historical time periods, the vast majority of crimes, particularly violent crimes, are committed by men. Here there is also apparently a genetically determined age specificity to violent aggression: crimes are overwhelmingly committed by young men between the ages of 15 and 30. Perhaps young men are everywhere socialized to behave violently, but this evidence, from different cultures and times, suggests that there is some deeper level of causation at work.

At this point in the discussion, many people become uncomfortable and charges of "biological determinism" arise. Don't we know countless women who are stronger, larger, more decisive, more violent, or more competitive than their male counterparts? Isn't the proportion of female criminals rising relative to males? Isn't work becoming less physical, making sexual differences unimportant? The answer to all of these questions is yes: again, no reputable evolutionary biologist would deny that culture also shapes behavior in countless critical ways and can often overwhelm genetic predispositions. To say that there is a genetic basis for sex difference is simply to make a statistical assertion that the bell curve describing the distribution of a certain characteristic is shifted over a little for men as compared with women. The two curves will overlap for the most part, and there will be countless individuals in each population who will have more of any given characteristic than those of the other sex. Biology is not destiny, as tough-minded female leaders like Margaret Thatcher, Indira Gandhi, and Golda Meir have proven. (It is worth pointing out, however, that in male-dominated societies, it is these kinds of unusual women who will rise to the top.) But the statistical assertion also suggests that broad populations of men and women, as opposed to exceptional individuals, will act in certain predictable ways. It also suggests that these populations are not infinitely plastic in the way that their behavior can be shaped by society.

Feminists and Power Politics

There is by now an extensive literature on gender and international politics and a vigorous feminist subdiscipline within the field of international relations theory based on the work of scholars like Ann Tickner, Sara Ruddick, Jean Bethke Elshtain, Judith Shapiro, and others. This literature is too diverse to describe succinctly, but it is safe to say that much of it was initially concerned with understanding how international politics is "gendered," that is, run by men to serve male interests and interpreted by other men, consciously and unconsciously, according to male perspectives. Thus, when a realist theorist like Hans Morganthau or Kenneth Waltz argues that states seek to maximize power, they think that they are describing a universal human characteristic when, as Tickner points out, they are portraying the behavior of states run by men.

Virtually all feminists who study international politics seek the laudable goal of greater female participation in all aspects of foreign relations, from executive mansions and foreign ministries to militaries and universities. They disagree as to whether women should get ahead in politics by demonstrating traditional masculine virtues of toughness, aggression, competitiveness, and the willingness to use force when necessary, or whether they should move the very agenda of politics away from male preoccupations with hierarchy and domination. This ambivalence was demonstrated in the feminist reaction to Margaret Thatcher, who by any account was far tougher and more determined than any of the male politicians she came up against. Needless to say, Thatcher's conservative politics did not endear her to most feminists, who much prefer a Mary Robinson [President of Ireland] or Gro Harlem Brundtland [first female prime minister of Norway] as their model of a female leader, despite—or because of—the fact that Thatcher had beaten men at their own game.

Both men and women participate in perpetuating the stereotypical gender identities that associate men with war and competition and women with peace and cooperation. As sophisticated feminists like Jean Bethke Elshtain have pointed out, the traditional dichotomy between the male "just warrior" marching to war and the female "beautiful soul" marching for peace is frequently transcended in practice by women intoxicated by war and by men repulsed by its cruelties. But like many stereotypes, it rests on a truth, amply confirmed by much of the new research in evolutionary biology. Wives and mothers can enthusiastically send their husbands and sons off to war; like Sioux women, they can question their manliness for failing to go into battle or themselves torture prisoners. But statistically speaking it is primarily men who enjoy the experience of aggression and the camaraderie it brings and who revel in the ritualization of war that is, as the anthropologist Robin Fox puts it, another way of understanding diplomacy.

A truly matriarchal world, then, would be less prone to conflict and more conciliatory and cooperative than the one we inhabit now. Where the new biology parts company with feminism is in the causal explanation it gives for this difference in sex roles. The ongoing revolution in the life sciences has almost totally escaped the notice of much of the social sciences and humanities, particularly the parts of the academy concerned with feminism, postmodernism, cultural studies, and the like. While there are some feminists who believe that sex differences have a natural basis, by far the majority are committed to the idea that men and women are psychologically identical, and that any differences in behavior, with regard to violence or any other characteristic, are the result of some prior social construction passed on by the prevailing culture.

The Democratic and Feminine Peace

Once one views international relations through the lens of sex and biology, it never again looks the same. It is very difficult to watch Muslims and Serbs in Bosnia, Hutus and Tutsis in Rwanda, or militias from Liberia and Sierra Leone to Georgia and Afghanistan divide themselves up into what seem like indistinguishable male-bonded groups in order to systematically slaughter one another, and not think of the chimps at Gombe.

The basic social problem that any society faces is to control the aggressive tendencies of its young men. In hunter-gatherer societies, the vast preponderance of violence is over sex, a situation that continues to characterize domestic violent crime in contemporary postindustrial societies. Older men in the community have generally been responsible for socializing younger ones by ritualizing their aggression, often by directing it toward enemies outside the community. Much of that external violence can also be over women. Modern historians assume that the Greeks and Trojans could not possibly have fought a war for ten years over Helen, but many primitive societies like the Yanomamö do exactly that. With the spread of agriculture 10,000 years ago, however, and the accumulation of wealth and land, war turned toward the acquisition of material goods. Channeling aggression outside the community may not lower societies' overall rate of violence, but it at least offers them the possibility of domestic peace between wars.

The core of the feminist agenda for international politics seems fundamentally correct: the violent and aggressive tendencies of men have to be controlled, not simply by redirecting them to external aggression but by constraining those impulses through a web of norms, laws, agreements, contracts, and the like. In addition, more women need to be brought into the domain of international politics as leaders, officials, soldiers, and voters. Only by participating fully in global politics can women both defend their own interests and shift the underlying male agenda.

The feminization of world politics has, of course, been taking place gradually over the past hundred years, with very positive effects. Women have won the right to vote and participate in politics in all developed countries, as well as in many developing countries, and have exercised that right with increasing energy. In the United States and other rich countries, a pronounced gender gap with regard to foreign policy and national security issues endures. American women have always been less supportive than American men of U.S. involvement in war, including World War II, Korea, Vietnam, and the Persian Gulf War, by an average margin of seven to nine percent. They are also consistently less supportive of defense spending and the use of force abroad. In a 1995 Roper survey conducted for the Chicago Council on Foreign Relations, men favored U.S. intervention in Korea in the event of a North Korean attack by a margin of 49 to 40 percent, while women were opposed by a margin of 30 to 54 percent. Similarly, U.S. military action against Iraq in the event it invaded Saudi Arabia was supported by men by a margin of 62 to 31 percent and opposed by women by 43 to 45 percent. While 54 percent of men felt it important to maintain superior world wide military power, only 45 percent of women agreed. Women, moreover, are less likely than men to see force as a legitimate tool for resolving conflicts.

It is difficult to know how to account for this gender gap; certainly, one cannot move from biology to voting behavior in a single step. Observers have suggested various reasons why women are less willing to use military force than men, including their role as mothers, the fact that many women are feminists (that is, committed to a left-of-center agenda that is generally hostile to U.S. intervention), and partisan affiliation (more women vote Democratic than men). It is unnecessary to know the reason for the correlation between gender and antimilitarism, however, to predict that increasing female political participation will probably make the United States and other democracies less inclined to use power around the world as freely as they have in the past.

Will this shift toward a less status- and military-power-oriented world be a good thing? For relations between states in the so-called democratic zone of peace, the answer is yes. Consideration of gender adds a great deal to the vigorous and interesting debate over the correlation between democracy and peace that has taken place in the past decade. The "democratic peace" argument, which underlies the foreign policy of the Clinton administration as well as its predecessors, is that democracies tend not to fight one another. While the empirical claim has been contested, the correlation between the degree of consolidation of liberal democratic institutions and interdemocratic peace would seem to be one of the few nontrivial generalizations one can make about world politics. Democratic peace theorists have been less persuasive about the reasons democracies are pacific toward one another. The reasons usually cited—the rule of law, respect for individual rights, the commercial nature of most democracies, and the like—are undoubtedly correct. But there is another factor that has generally not been taken into account: developed democracies also tend to be more feminized than authoritarian states, in terms of expansion of female franchise and participation in political decision-making. It should therefore surprise no one that the historically unprecedented shift in the sexual basis of politics should lead to a change in international relations.

The Reality of Aggressive Fantasies

On the other hand, if gender roles are not simply socially constructed but rooted in genetics, there will be limits to how much international politics can change. In anything but a totally feminized world, feminized policies could be a liability.

Some feminists talk as if gender identities can be discarded like an old sweater, perhaps by putting young men through mandatory gender studies courses when they are college freshmen. Male attitudes on a host of issues, from child-rearing and housework to "getting in touch with your feelings," have changed dramatically in the past couple of generations due to social pressure. But socialization can accomplish only so much, and efforts to fully feminize young men will probably be no more successful than the Soviet Union's efforts to persuade its people to work on Saturdays on behalf of the heroic Cuban and Vietnamese people. Male tendencies to band together for competitive purposes, seek to dominate status hierarchies, and act out aggressive fantasies toward one another can be rechanneled but never eliminated.

Even if we can assume peaceful relations between democracies, the broader world scene will still be populated by states led by the occasional [bloody dictator]. Machiavelli's critique of Aristotle was that the latter did not take foreign policy into account in building his model of a just city: in a system of competitive states, the best regimes adopt the practices of the worst in order to survive. So even if the democratic, feminized, postindustrial world has evolved into a zone of peace where struggles are more economic than military, it will still have to deal with those parts of the world run by young, ambitious, unconstrained men. If a future Saddam Hussein is not only sitting on the world's oil supplies but is armed to the hilt with chemical, biological, and nuclear weapons, we might be better off being led by women like Margaret Thatcher than, say, Gro Harlem Brundtland. Masculine policies will still be required, though not necessarily masculine leaders. . . .

Living Like Animals?

In Wrangham and Peterson's *Demonic Males* . . . , the authors come to the pessimistic conclusion that nothing much has changed since early hominids branched off from the primordial chimp ancestor five million years ago. Group solidarity is still based on aggression against other communities; social cooperation is undertaken to achieve higher levels of organized violence. Robin Fox has argued that military technology has developed much faster than man's ability to ritualize violence and direct it into safer channels. The Gombe chimps could kill only a handful of others; modern man can vaporize tens of millions.

While the history of the first half of the twentieth century does not give us great grounds for faith in the possibility of human progress, the situation is not nearly as bleak as these authors would have us believe. Biology, to repeat, is not destiny. Rates of violent homicide appear to be lower today than during mankind's long hunter-gatherer period, despite gas ovens and nuclear weapons. Contrary to the thrust of postmodernist thought, people cannot free themselves entirely from biological nature. But by accepting the fact that people have natures that are often evil, political, economic, and social systems can be designed to mitigate the effects of man's baser instincts.

. . . [To that end,] liberal democracy and market economies work well because, unlike socialism, radical feminism, and other utopian schemes, they do not try to change human nature. Rather, they accept biologically grounded nature as a given and seek to constrain it through institutions, laws, and norms. It does not always work, but it is better than living like animals.

Mary Caprioli

↩ NO

The Myth of Women's Pacifism

In a recent article in *Foreign Affairs,* Francis Fukuyama asserts that a more feminized world run by women would be more peaceful. Indeed, he concludes that biology is not destiny and that it is better not to live like animals. Presumably, this means that men can learn to be more peaceful in their behavior, and this new pacifism on the part of men would translate into a more peaceful world. What Fukuyama seems to eliminate from his analysis is that if men change their behavior, then so too would women. If a more egalitarian world would change men's behavior, then it would most likely change women's behavior, too. If men were to become more peaceful, women would potentially become more aggressive.

Fukuyama's argument is based on a number of assumptions. In order to conclude that a future world would be more peaceful if only men acted more like women, one must assume that a gender gap exists in support for the use of force. Furthermore, an acceptance of a gender gap assumes a dichotomy based on gender. Not only must there exist only two genders divided by sex, but there must also be a universality of experience for women and men across cultures. In order to understand Fukuyama's argument, it is necessary to have an understanding of gender and to explore the evidence for proclaiming the existence of a gender gap. Any prediction characterizing a future world must be made only after acquiring such understandings.

Gender

Gender is crucial to any study supporting and explaining the meaning of gender, in which gender is found to be a cause of state bellicosity. Is gender a function of genital-type, to each person's level of testosterone, or the extent to which each individual had been socialized into accepting standards of feminine and masculine attitudes, which would be a cultural measure of gender?

Gender is the crux of any argument based on the existence of a difference between men and women—the existence of a gender gap. We must, therefore, examine whether or not gender is a useful category of analysis. The literature on gender, especially within anthropology, is prolific yet surprisingly monolithic. All define gender as a culturally construed category. If a definition of gender is based on culture, then gender is not a universal category of analysis.

Indeed, the term "gender" is used to designate socially constructed roles attributed to women and men. These gender roles are learned, change over time, and vary both within and between cultures. Gender issues, therefore, have to do with differences in how women and men are supposed to act. In other words, gender definitions control and restrict behaviors. As Klaus Theweleit highlights in his 1993 piece "The Bomb's Womb and the Genders of War" in the edited volume *Gendering War Talk,* women traditionally have not had the power to act violently, which might explain the identified gender gap in support for war. The culturally proscribed role of women based on gender determines women's alternatives. In this instance, the use of violence is simply not an option for women. Women, therefore, are not necessarily choosing not to act violently but do not have the choice to act violently—a very important distinction.

Indeed, women who have obtained the power to act violently have done so. These leaders include prime ministers Margaret Thatcher (Great Britain), Indira Gandhi (India), Golda Meir (Israel), Khalida Zia (Bangladesh), Maria Liberia-Peters (Netherlands-Antilles), Gro Harlem Brundtland (Norway), and Benazir Bhutto (Pakistan). Using Stuart Bremer's 1996 Militarized Interstate Dispute (MID) data set, which covers the time period from 1816 until 1992, it is possible to examine the level of violence sanctioned by female leaders in comparison to male leaders.

Only twenty-four states have placed a female leader in office since 1900, with the first female leader obtaining power in 1960. In this instance, a female leader is defined as a president, prime minister, or any other decision-maker who is essentially the 'decision-maker of last resort' on decisions to use force and other high-level international decisions. Edith Cresson, who was premier of France in 1991–1992, is therefore not considered a leader in this discussion because that position is one of significantly lesser importance than that of the French president, who was a male.

We can compare the behavior of female and male leaders from 1960 until 1992 using MID. MID includes a variable for the level of hostility used during the crisis. Hostility level is divided into five levels coded as follows: 1) no militarized action, 2) threat to use force, 3) display of force, 4) use of force, and 5) war. Both female and male leaders rely on the fourth category, the use of force, most frequently. Furthermore, both female and male leaders' average use of violence is equal. According to this evidence, female leaders are no more peaceful than their male counterparts.

Of course, critics would argue that female leaders in a male world must act more aggressively in order to prove themselves. According to this argument women are still considered to be by nature more pacific than men are. Because women must operate in a social and political environment that has been defined, structured and dominated by men for centuries, they must recondition themselves to act more violently in order to gain power in a "man's world." On the other hand, women's assent to power and growing equality in relation to men may free women from gender stereotypes restricting their behavior. In this more equal world, women become free to exercise different alternatives, including the option of using violence.

The power and role of women varies across cultures because power is based on these culturally dependent gender roles. Because all people are born into, and socialized to a particular culture, the argument for biological determinism becomes impossible to prove. The only way to prove conclusively that women are more peaceful would be to raise a number of baby girls from birth in a cultural vacuum. Such an experiment would be unethical at best. Scholars, therefore, must speculate as to the 'nature' of women. For example, do women in power ignore and overcome inherent gender-based characteristics in order to gain and maintain power? Did Margaret Thatcher have to keep her natural tendency to be peaceful in constant check? Conversely, do people who are attracted to power share basic characteristics regardless of gender? Is Thatcher's biological tendency toward violence no different from a 'normal' tendency of any human's tendency toward violence?

The adoption of gender as a category of analysis implies a universality of experience across cultures and over time. This assumption, especially when espoused by educated and predominantly Western and European women, is as biased as some men's assumption of a universal human experience based on a male perspective. Placing women at the center of analysis is no less biased than focusing on men. Furthermore, the acceptance of this dichotomy between the only two recognized genders implies the acceptance that there do exist biological behavioral differences between the genders—an assumption that remains unproven.

Recognizing the exclusion of the female gender, as culturally defined, in most international relations literature is important to understanding the assumptions within existing research. To assume that including two genders or that taking the female-gender perspective will rectify the problems in current research and society is problematic, for both assume a universality of experience. As Third World feminists have clearly detailed, the experience of women is not universal. As scholars we must, therefore, recognize that the absence of a gendered analysis or conversely, the exclusive reliance on gendered analyses is a bias within research. We must understand that classifying an individual's behavior on the basis of gender is not a valid assumption.

Beyond the scope of this discussion, but important to note, is the argument that there are more than two genders. Arguments identifying a gender gap associate gender with sexuality. Furthermore, gender is assumed to include only two genders: the mythical male and female. If gender is equated with sexuality then children, the aged, especially post-menopausal women, gays, and lesbians would all constitute different genders. And, how would hermaphrodites—those 'sexless' people according to definitions of gender as male/female who are 'assigned' a sex at birth—be categorized?

Following a definition of gender based on sex or at least male and female sex organs, gender/sex differences should be the genesis of power inequality between women and men. These gender differences would necessarily have to be

Table 1

States With Female Leaders Since 1900–1994

State	Leader	Years in Office
Argentina	Isabel Perón	1974–1976
Bangladesh	Khalida Zia	1991–1996
Bolivia	Lidia Gueiler Tejada	1979–1980
Burundi	Sylvie Kingi	1993–1994
Canada	Kim Campbell	1993
Central African Republic	Elisabeth Domitien	1975–1976
Dominica	Mary Eugenia Charles	1980–1995
Haiti	Ertha Pascal-Trouillot	1990–1991
India	Indira Gandhi	1966–1977
		1980–1984
Israel	Golda Meir	1969–1974
Lithuania	Kazimiera Prunskiene	1990–1991
Malta	Agatha Barbara	1982–1987
Netherlands Antilles	Maria Liberia-Peters	1984–1985
		1988–1994
Nicaragua	Violeta Chamorro	1990–1997
Norway	Gro Harlem Brundtland	1981
		1986–1989
		1990–1996
Pakistan	Benazir Bhutto	1988–1990
		1993–1996
Philippines	Corazon C. Aquino	1986–1992
Poland	Hanna Suchocka	1993
Portugal	Maria de Lourdes Pintasilgo	1979
Rwanda	Agathe Uwilingiyimana	1993–1994
Sri Lanka	Sirimavo Bandaranaike	1960–1965
		1970–1977
Turkey	Tansu Ciller	1993–1996
United Kingdom	Margaret Thatcher	1979–1990
Yugoslavia	Milka Planinc	1982–1986

universal both between and within cultures. Yet, research conclusively demonstrates that gender changes over time within cultures and is not consistent among different cultures. Nonetheless, scholars continue to attempt to show a gender gap in the support for the use of force.

Often, this feminine pacifism is linked to women's maternal qualities and more specifically to their ability to have children. This argument is prejudiced in that it discounts men's 'maternal' instincts toward their children. As with the argument for two sexes, there exists a gray area in this argument about women's maternal instincts. For example, how would a single man who adopts a child be classified—or women who are infertile—or women who choose never to have children? Would the single dad not have any 'maternal' characteristics while assuming that women who choose not to have children possess 'maternal' characteristics by virtue of being female?

After examining the numerous gray areas surrounding much of the argument in support for women's pacifism, the argument seems rather thin. Yet, some studies do show that women seem to be less supportive of the use of violence than men. Of course, other groups of people including African Americans are also less supportive of the use of violence than are white men.

Gender Gap

All evidence in support for the existence of a gender gap comes from public opinion surveys of the Western world and in particular of the United States and the United Kingdom. The women of these countries can hardly be representative of women worldwide. In addition, the size of this identified gender gap varies from study to study. Another shortcoming of public opinion surveys is that women are more likely to indicate no opinion in public opinion surveys or merely fail to voice support for war. Little mention is made of political scientist John Mueller's admonition that not voicing support for a war is not the same as opposing it. In this instance, women might be constrained for voicing support for the use of force by gender stereotypes, so they offer no opinion.

Beyond the problems associated with public opinion surveys lie other challenges to a gender-gap theory for the support of violence. For instance, Nancy E. McGlen and Meredith Reid Sarkees in their 1993 book *Women in Foreign Policy: The Insiders* find varying degrees of a gender gap amongst the masses but none with women working within the State Department or the Defense Department. Some scholars might argue that women in the State and Defense Departments are not representative of women in general, or that these women might be forced to act violently to prove themselves in a predominantly male arena, or that they must adhere to traditional, institutional roles. Such arguments may be countered by the idea that these women are free to act violently—they have the opportunity to act violently and they do.

Not only do scholars challenge the existence of a gender gap but also the reasons for a gender gap. Some scholars suggest that the gender gap is created only by those women who identify with the women's movement. This argument suggests that it is not some inherent quality of women that creates a gender gap in support for the use of force but that women who happen to adhere to a feminist ideology of pacifism are more pacifist much in the same way that Democrats tend to be more pacific than Republicans. In keeping with Fukuyama's argument, it would be prudent, therefore, to encourage people to be Democrats in order to ensure world peace. This task should be less difficult than changing every culture of the world to have men change their behavior to act in the way women supposedly do—by not acting violently—and for women's supposed behavior to remain unchanged—to remain pacific.

A parallel discussion to the argument of women's nature would be asking whether or not people are born Democrat. Are party ideologies an inherent characteristic of some individuals, or are some individuals socialized into accepting the ideology of the Democrats? This is similar to questioning whether pacifism is an inherent quality of women or if women are socialized into being more pacific.

Others argue that any current gender gap associated with support for the commitment of armed forces or for war would be eradicated by the inclusion of more women in active duty within the armed forces. This argument follows the one outlined above in that women are not able to act violently and are traditionally excluded from activities and professions that include violence. Once women are included in active duty within the armed forces, they will not only achieve a certain level of equality but will also be free to act violently and support the use of violence.

Indeed, a group's relative position of power within society may determine its proclivity toward peaceful conflict resolution rather than some inherent quality. For example, John Mueller in his numerous research and public opinion surveys found that African Americans were more pacifistic in that as a group, African Americans were largely against escalation and for withdrawal in their general support for World War II, the Korean War, the Vietnam War, and the Persian Gulf War. According to this logic, one's placement in the hierarchy of power would best predict one's support for war as evidenced by women's and African Americans' more dovish nature.

Mark Tessler and Ina Warriner in an article in the January 1997 issue of *World Politics* argue that there is no evidence to show that women are less militaristic than men are. They do, however, find that individuals who are more supportive of equality between women and men are also less supportive of violence as a means of resolving conflict. This argument suggests that the relationship between more pacifist attitudes and international conflict rests upon the degree of gender equality that characterizes a society. Those who express greater concern for the status and role of women, and particularly for equality between women and men, are more likely than other individuals to believe that the international disputes in which their country is involved should be resolved through diplomacy and compromise. In other words, societies that have values that are less gender-based should be more pacific in their international behavior.

This argument that more gender-neutral societies are less internationally bellicose is similar to the theory outlined above that argues that people are more pacifist based on their position in society. In either situation, a society that is hierarchical in nature because it is based on prejudice against such classifications as gender and race will be more internationally bellicose. Once the social hierarchy is abolished in that all people gain the freedom to act as they choose, then differences among separate classifications of people will be eliminated. For instance, women and African Americans may become more aggressive as they gain more equality, more power. So too, may men become more pacific as they are freed from gender stereotypes that demand men to be aggressive.

Admittedly, I am no expert on chimpanzees. I can, however, speculate that female chimps might be more aggressive if they had the opportunity to act aggressively. Within the animal kingdom, power is often based on size. With humans, the base of power varies from physical size, to the size of one's bank account, to the size of one's intellect. Women, therefore, have an increasingly greater opportunity to act violently, an opportunity that her chimpanzee sisters may not enjoy.

The Future

The future may not be as rosy as Fukuyama suggests, at least not in terms of a global pacifism brought about by the increasing equality of women. Of course, equality in itself seems to be a positive force if only in freeing individuals to act according to their desires rather than by being restricted by cultural stereotypes related to, and defined by gender. This new freedom to act, however, does not necessarily translate into a global peace.

Fukuyama fleetingly mentions that the number of women incarcerated for violent crimes is increasing in proportion to that of men. The important question remains unanswered: Why are more women committing violent crimes? As American society becomes more egalitarian with regard to the sexes, women are gaining more power. This power may not be directed toward pacifist, nurturing ideals. Lord Acton wrote, "Power tends to corrupt, and absolute power corrupts absolutely." Perhaps he was correct.

POSTSCRIPT

Would World Affairs Be More Peaceful If Women Dominated Politics?

Studies of biopolitics, ethology, gender, genetics, and other related approaches are just beginning to probe the connection between biology and politics. Also, the so-called nature-versus-nurture debate continues and presents some fascinating questions. Bear in mind, though, that neither Fukuyama nor Caprioli (nor, for that matter, any other serious scholar) argues that "biology is destiny." Rather, all scholars recognize that human behavior is a mix of socialization and genetic coding. It is the ratio of that mix and its manifestations that are the points of controversy. Fukuyama clearly believes that genetics play a strong role. In contrast, Caprioli assigns a much greater role to socialization, at least as far as accounting for the fact that women seem less aggressive than men. More of her views can be found in Mary Caprioli, "Gendered Conflict," *Journal of Peace Research* (January 2000). A more general study is Inger Skjelsbaek and Dan Smith, eds., *Gender, Peace and Conflict* (Sage, 2001).

There are a number of good books on women and gender as they relate to world politics, including Christine Sylvester, *Feminist International Relations: An Unfinished Journey* (Cambridge University Press, 2002); J. Ann Tickner, *Gendering World Politics: Issues and Approaches in the Post–Cold War Era* (Columbia University Press, 2001); Vivienne Jabri and Eleanor O'Gorman, eds., *Women, Culture, and International Relations* (Lynne Rienner, 1998); and Jill Sterns, *Gender and International Relations: An Introduction* (Rutgers University Press, 1998).

Fukuyama contends that males tend to act like males and females like females and that a great deal of that is based on evolutionary biology. For a classic ethological view, read Desmond Morris, *The Naked Ape* (Dell, 1976) and Robert Ardrey, *The Territorial Imperative* (Atheneum, 1966). This line of thinking has many critics, some of whom directly dispute what Fukuyama has argued, including Barbara Ehrenreich and Katha Pollitt, in "Fukuyama's Follies," *Foreign Affairs* (January 1999).

Whatever the reality may be, its implications are important because women are increasingly playing leading roles. During the past few decades, women have led such important countries as Great Britain, India, and Israel. Just a few years ago a first was reached when, in Bangladesh, the choice for a country's prime minister came down to two women. Another breakthrough occurred when Madeleine Albright became the first female secretary of state for the United States. For studies of women in foreign policy leadership positions, read Nancy E. McGlen and Meredith Reid Sarkees's *The Status of Women in Foreign Policy* (Headline Series, 1995) and *Women in Foreign Policy* (Routledge, 1993).

United Nations Department of Peacekeeping Operations

This UN site provides access to descriptions of current and past peacekeeping operations, maps, lists of contributing countries, and other data related to UN military, police, and observer missions.

http://www.un.org/Depts/dpko/dpko/home.shtml

International Monetary Fund

The International Monetary Fund (IMF) was established to promote international monetary cooperation, exchange stability, and orderly exchange arrangements; to foster economic growth and high levels of employment; and to provide temporary financial assistance to countries to help ease balance of payments adjustment. Learn more about the IMF at its home page.

http://www.imf.org

The International Law Association

The International Law Association, which is currently headquartered in London, was founded in Brussels in 1873. Its objectives, under its constitution, include the "study, elucidation and advancement of international law, public and private, the study of comparative law, the making of proposals for the solution of conflicts of law and for the unification of law, and the furthering of international understanding and goodwill."

http://www.ila-hq.org

United Nations Treaty Collection

The United Nations Treaty Collection is a collection of 30,000 treaties, addenda, and other items related to treaties and international agreements that have been filed with the UN Secretariat since 1946. The collection includes the texts of treaties in their original language(s) and English and French translations.

http://untreaty.un.org

Public International Law

The faculties of Economics and Commerce, Education and Law at the University of Western Australia maintain this Web site, which has extensive links to a range of international law topics, ranging from institutions, such as the International Court of Justice, to topical links on crime, human rights, and other issues.

http://www.law.ecel.uwa.edu.au/intlaw/

PART 5

International Law and Organization Issues

*P*art of the process of globalization is the increase in scope and impor-tance of both international law and international organizations. The is-sues in this section represent some of the controversies involved with the expansion of international law and organizations into the realm of mili-tary security. Issues here relate to increasing international organizations' responsibility for security, the effectiveness of international financial or-ganizations, and the proposal to authorize international courts to judge those who are accused of war crimes.

- Should the United Nations Be Given Stronger Peacekeeping Capabilities?

- Do International Financial Organizations Require Radical Reform?

- Should the United States Ratify the International Criminal Court Treaty?

ISSUE 16

Should the United Nations Be Given Stronger Peacekeeping Capabilities?

YES: Lionel Rosenblatt and Larry Thompson, from "The Door of Opportunity: Creating a Permanent Peacekeeping Force," *World Policy Journal* (Spring 1998)

NO: John Hillen, from Statement Before the Subcommittee on International Operations, Committee on Foreign Relations, U.S. Senate (April 5, 2000)

ISSUE SUMMARY

YES: Lionel Rosenblatt and Larry Thompson, president and senior associate, respectively, of Refugees International in Washington, D.C., advocate the creation of a permanent UN peacekeeping force on the grounds that the present system of peacekeeping is too slow, too cumbersome, too inefficient, and too prone to failure.

NO: John Hillen, a policy analyst for defense and national security issues at the Heritage Foundation in Washington, D.C., contends that the United Nations was never intended to have, nor should it be augmented to have, the authority and capability to handle significant military operations in dangerous environments.

T he United Nations seeks to maintain and restore peace through a variety of methods. These include creating norms against violence, providing a forum to debate questions as an alternative to war, making efforts to prevent the proliferation of weapons, diplomatic intervention (such as mediation), and the placing of diplomatic and economic sanctions. Additionally, and at the heart of the issue here, the UN can dispatch troops under its banner or authorize member countries to use their forces to carry out UN mandates.

UN forces involving a substantial number of personnel have been used more than two dozen times since the organization came into existence (in 1945) and have involved troops and police from more than 75 countries. There is, therefore, a significant history of UN forces. Nevertheless, recent events and attitude changes have engendered renewed debate over the military role of the UN.

The increased number of UN operations is one factor contributing to the debate. Of all UN operations throughout history, about half are currently active. Furthermore, several of the recent missions have included large numbers of troops and, thus, have been very costly.

A second factor that has sparked controversy about UN forces is the successes and failures of their missions. Often UN forces have played an important part in the peace process; other times they have been unsuccessful. The limited mandate (role, instructions) and strength (personnel, armaments) of UN forces have frequently left them as helpless bystanders.

A third shift that has raised concerns about UN forces is the change in the international system. With the cold war ended, some people are trying to promote a new world order. This new world order would require countries to live up to the mandate of the UN charter that they only use force unilaterally for immediate self-defense or unless they are authorized to use force by the UN or a regional organization (such as the Organization of American States). This means that collective action under UN auspices is becoming more normal.

Two potential changes in the operation of UN forces apply to the issue here. The first is to increase the scope of the mission of UN forces. To date, UN forces have operated according to two concepts: *collective security* and *peacekeeping*. Collective security is the idea that aggression against anyone is a threat to everyone. Therefore, the collective body should cooperate to prevent and, if necessary, defeat aggression. The second, long-standing UN role of peacekeeping usually involves acting as a buffer between two sides to provide an atmosphere that will allow them to settle their differences, or at least to not fight. Neither collective security nor peacekeeping, however, precisely apply to situations such as domestic civil wars, in which there is no international aggressor or clearly identifiable aggressor. Some people consider this a gap in what the UN does to prevent the scourge of war and, therefore, would augment the UN's role to include *peacemaking*.

The second potential change for UN forces relates to proposals to create, at maximum, a standing UN army or, at least, a ready reserve of troops. These troops would remain with the forces of their home countries but would train for UN operations and be instantly available to the UN.

The immediate background to the issue debated here began with a January 1992 summit meeting of the leaders of the 15 countries with seats on the Security Council. The leaders called on the UN secretary-general Boutros Boutros-Ghali (1991–1996) to report on ways to enhance UN ability "for preventative diplomacy, for peacemaking, and for peacekeeping." In response, the secretary-general issued a report entitled *An Agenda for Peace,* in which he recommended a number of strategies to enhance UN peacekeeping.

That report set off a debate that continues today. In the following selections, Lionel Rosenblatt and Larry Thompson conclude that a permanent, rapid-reaction peacekeeping force should be created to respond at the beginnings of crises before they get out of control. John Hillen disagrees, arguing that if the UN is given difficult missions that it is not and cannot be ready to handle, then the result will be failure and damage to the UN.

**Lionel Rosenblatt and
Larry Thompson**

 **YES**

The Door of Opportunity: Creating a Permanent Peacekeeping Force

Somerset Maugham wrote a short story, "The Door of Opportunity," about a British colonial official who was dismissed in disgrace because he lacked the courage to face down a murderous crowd of rioters in some lonely, unimportant corner of the empire. "The utility of a government official depends very largely on his prestige," says the governor to the offender, "and I'm afraid his prestige is likely to be inconsiderable when he lies under the stigma of cowardice."

On the American scene, in John Ford's cavalry classic, *She Wore a Yellow Ribbon*, John Wayne and Ben Johnson, Jr. ride into a hostile Cheyenne village and avert an Indian uprising with a cultural sensitivity that would please a Berkeley don.

Are these two examples of fictional peacekeeping pure myth from a vanished time? Are the people in the age of the AK-47 [assault rifle] inherently more dangerous and less amenable to peaceful resolution of conflict than in the black powder era? Or is there some basis to believe that a few good men and women acting with "promptness and firmness" (Maugham's words) can avert some of the uncivil wars and ethnic slaughter that characterize our post–Cold War world?

Rwanda, 1994

Let us look at a contemporary example of a peacekeeping failure: Rwanda, April to August 1994.

On April 6, 1994, an airplane carrying Juvénal Habyarimana, the president of Rwanda, was shot down under mysterious circumstances, setting off a struggle for control of the country between the majority Hutu and minority Tutsi peoples, and the most horrific genocide of this decade. Some 800,000 people—mostly Tutsi and moderate Hutus—were killed by Hutu extremists in the space of three months.

Civil war and ethnic violence had been the norm in Rwanda for many years. A lightly armed U.N. peacekeeping force of 2,500 was stationed in the country on a "low-intensity" peacekeeping mission to monitor compliance with a prior agreement between Hutu and Tutsi, a "cakewalk" as one U.N. official said. In the spring of 1994, the flavor of the cake turned out [to] be devil's food. "They [Hutu extremists] were chopping off the breasts of women," said Gen. Romeo Dallaire, the Canadian commander of the U.N. peacekeepers. "They were slitting people open from their genitals right up to their sternum. They were chopping up the arms of two-year-old children just as if it [sic] was salami."

Among early victims were ten Belgian peacekeepers who were killed on April 7 by Hutu militia. With their deaths, the heart went out of the political masters of the peacekeeping mission in New York and other world capitals, especially Washington.

General Dallaire had minimal resources at his disposal to contend with the eruption of violence. According to the *Joint Evaluation of Emergency Assistance to Rwanda*, published in 1996, he had only "one working armored personnel carrier, a demoralized Belgian battalion, and an under-equipped, below-strength unit from Bangladesh." Moreover, the peacekeepers had no mandate from the U.N. Security Council to intervene to prevent the mass murders taking place before their eyes. But, in a sign of what might have been had the U.N. force been instructed to protect civilians, 12 "blue helmets" [U.N. soldiers who traditionally wear blue helmets with the U.N. emblem on them] at the Amahoro Stadium, armed only with bluff, hand weapons, and barbed wire, saved the lives of several thousand persons.

The response of Boutros Boutros-Ghali, the U.N. secretary general, and the Security Council to the events in Rwanda was to decide that there was no peace to keep, and on April 21, Boutros-Ghali ordered the withdrawal of most of the peacekeepers. A traumatized Belgium prompted the flight of the United Nations, but a jittery United States, which had recently lost 18 of its own army rangers in Somalia, backed the Belgians. (When the Belgian peacekeepers got back to Brussels, several of them shredded their U.N. blue berets for the television cameras.)

Later, as conscience crept back into the United Nations, Boutros-Ghali proposed a more forceful U.N. role, and the Security Council, including a reluctant and foot-dragging United States, finally approved sending a new peacekeeping force of 5,400 personnel to Rwanda. It was now May 17—six weeks and several hundred thousand lives into the genocide.

The U.N. Security Council, however, only mandates on paper. Several African countries came forward with offers of troops, but matching ill-equipped African troops with essential Western equipment (to be provided by the United States and other countries) proved to be too much of a bureaucratic obstacle to overcome quickly, given the lack of political will in Western capitals. On July 25, 1994, Boutros-Ghali reported sadly that only 500 of the 5,400 peacekeepers the Security Council had authorized were on the ground in Rwanda and that they were barely operational.

Disgracefully, the first American armored personnel carriers so vital to the success of a U.N. peacekeeping force arrived only on July 30. It was much too little, much too late. The civil war was over. The genocide was over. The United Nations and its members had fiddled for four months while Rwanda burned. The toll in a country of 8 million was astonishing: up to 800,000 people were dead, two million were refugees in neighboring Zaire and Tanzania, and three million were displaced within the country itself. The Rwandan conflict and genocide had a profound impact on Burundi, Uganda, Tanzania, and, especially, Zaire—now the Democratic Republic of Congo—the third largest country of Africa. President Laurent Kabila's rise to power in Congo began with the refugee flight from Rwanda and the spread of the Hutu-Tutsi conflict to eastern Zaire. Perhaps the least important of the consequences was that the international community would be saddled with a tab of half a billion dollars per year to feed the refugees and clean up the mess.

Fear and Loathing of the United Nations

The political commentator Michael Lind thinks the United Nations and its philosophy of collective security have been "finally and completely discredited" and the organization itself should be allowed to "wither away into irrelevance." Lind argues for a return to nineteenth-century statecraft, in which an ill-defined "great power concert" would run the world. But what, we would ask, will happen if the so-called great powers do not want to run the world? The abdication of responsibility by the great powers in the 1920s and 1930s led to the militaristic rise of Japan, Germany, and Italy. The great powers of the 1990s do not seem to have any more stomach for managing chaos than did France, Britain, or the United States back then.

The scholar David Rieff displays a spirit complementary to Lind's. Rieff discounts the ability of the world to have a positive effect on civil wars and ethnic conflicts. "I would go so far as to suggest that some of these conflicts are inevitable; that they have a certain political logic and also a certain political function; and that in the long run we're actually doing no one any favor by trying to intervene and prevent them."

A synthesis of the writings of the anti–United Nations, anti-intervention theorists might be that "the United Nations is no damn good but even if it were it would have no business sticking its nose into small wars in faraway places to save lives."

We beg to differ. It is precisely in these small wars in faraway places that the United Nations can make a difference. The United Nations is not going to be able to handle a conflict as big as the 1991 Gulf War for the foreseeable future. Even a Bosnia may be too big for the organization to handle. But the United Nations should be equipped to deal with the Liberias, Rwandas, and Macedonias of the world—with the chaos, ethnic conflicts, or threats of strife in countries that do not engage the urgent political concerns of the big powers: a few hundred thousand dead people here, a few hundred thousand there, add up.

It seems to us not too much to ask that the international community regard the preservation of lives as an important factor when it deliberates about intervention in crisis situations. We should be hardheaded; but we should not be coldhearted.

The Failure of a Peacekeeping Mission

International peacekeeping failed in Rwanda for several reasons. First, the U.N. peacekeeping force at the outbreak of the genocide was a mixed bag of soldiers from several countries, indifferently equipped and trained, inadequately financed, and barely coordinated because each national military unit ultimately looked to its own chain of command for orders rather than to the nominal commander, General Dallaire.

Second, the terms of reference for the peacekeeping force were all wrong. It was, to quote the *Joint Evaluation*, "a classic, minimalist peacekeeping operation," with no authority to protect innocent lives if the uneasy peace between Tutsi and Hutu it was supervising broke down. In fact, there was no option for action at all in case of trouble except withdrawal. With no peace to keep, the Security Council withdrew the peacekeepers.

Third, national considerations took precedence over the needs of the U.N. peacekeeping mission and the people of Rwanda. When the Belgian peacekeepers were brutally killed, the Belgian government decided immediately to withdraw its forces. The United States seconded Brussels, partially because of a penny-pinching attitude about peacekeeping worthy of Ebenezer Scrooge. (The U.N. peacekeeping mission to Rwanda cost the United States about $5 million per month; the failure of the peacekeeping mission would ultimately cost U.S. taxpayers $125 million in less than a month for an emergency military airlift to feed Rwandan refugees.) The French intervened unilaterally in southwestern Rwanda (Operation Turquoise). Although the French intervention undoubtedly saved some lives, it also reflected mixed political and humanitarian motives and helped the Hutu army and militia, the primary perpetrators of the bloodbath, remain intact.

Suppose a well-trained, well-equipped U.N. peacekeeping force had been in Rwanda when war broke out in April 1994. And suppose that it had had a mandate to protect civilian lives. Could such a force have stemmed the ensuing genocide and its many tragic aftershocks? Dallaire thinks so. "Had we been able to deploy the troops and equipment with a mandate to prevent a situation of crimes against humanity and to be more offensive in nature, I believe we could have curtailed a significant portion of it."

Dallaire goes on to pinpoint the U.N.'s number-one problem in its peacekeeping operations. "The United Nations is not a sovereign state with the resources of a sovereign state. It can take decisions but it has to go outside to get resources, particularly troops and equipment. When I left at the end of August 1994, I had barely 3,000 of the [promised] 5,500 troops."

Dallaire's words lead us to a proposal that is gaining currency in international circles: to create a standing, permanent, international rapid response force to deal with situations like the one in Rwanda.

A Rapid Reaction Force

The secretary general's "Supplement to an Agenda for Peace" in January 1995 called on the United Nations [to] consider the idea of a "rapid deployment force." Denmark and Belgium, mindful of the tragedy in Rwanda, are behind the idea; Canada lends support; and even the United States—reluctant though it may be—has come up with a variation on the theme: an African Crisis Response Initiative of 5,000–10,000 African troops, internationally trained and financed, to respond to crises on the continent.

The present system of international peacekeeping is too slow, too cumbersome, too inefficient, too prone to failure, too ad hoc to meet the necessities of the confusing, nameless era that has followed in the wake of the Cold War. Former U.N. under secretary general Brian Urquhart says the United Nations is at the "sheriff's posse" stage. "There are a lot of people who don't really agree with each other very much most of the time who suddenly are shocked by some horrendous human event into putting together some ad hoc and improvised posse to do something about it after the fact. It is better than doing nothing, of course. But we've got to move on from this stage." For one thing, the world body is too slow and unprofessional to create and mobilize a "sheriff's posse" whenever the need arises.

Stung by its failure in Rwanda, the United Nations has taken some steps toward improving its peacekeeping capabilities. U.N. headquarters in New York now has a 24-hour watch center to monitor crises around the world. A rapidly deployable headquarters team is being formed to assess crises early on, before peacekeeping forces are sent out. The Standby Arrangement System maintains a register of personnel and equipment volunteered by member states that can be utilized for peacekeeping missions. Planning, intelligence, and early warning functions have been enhanced.

All this may be superfluous if, when the necessity for deployment of peacekeepers arises, as it does suddenly, the need for troops and matériel cannot be met. "Unconditional commitments" of aid from member states have a way of evaporating when the United Nations calls in the chits. Domestic politics tends to govern the response of member states to an urgent demand for peacekeeping. National pride, particularly that of the French and the Americans, complicates command and control. The French do not work well with anybody, and the Americans will not work at all unless they are in charge.

Time is a critical factor. Opportunities to defuse a situation are lost in the time between a Security Council decision and the actual deployment of troops on the ground. Dallaire needed reinforcements in Rwanda in April [1998]—not in August.

The best solution to these problems would seem to be the creation of a standing, permanent international military and police force made up of volunteers and under the direct command of the U.N. Security Council: what has variously been called a rapid reaction force, a rapid deployment brigade, a U.N. foreign legion.

Proponents of such a rapid reaction force (RRF) differ in their estimates of the numbers needed—proposals range from 5,000 to 55,000 soldiers and police.

We are persuaded, however, by MIT professors Carl Kaysen and George W. Rathjens that the optimum size of an RRF would be about 15,000, of which 10,600 would be deployable troops and 4,400 headquarters, logistics, and training staff. An RRF of 15,000 would permit the simultaneous deployment of two peacekeeping forces, with one such force in reserve.

The annual U.N. peacekeeping budget now stands at $2–3 billion. The cost of a 15,000-person RRF, as calculated by Kaysen and Rathjens, would be on the order of $1.5 billion annually. But a good portion of the cost of the proposed RRF would be offset by savings realized from the revamping of the current peacekeeping system.

The most recent endorsement of the idea of a rapid reaction force comes from the prestigious Carnegie Commission on Preventing Deadly Conflict. The commission, cochaired by former U.S. secretary of state Cyrus Vance, recommended the "establishment of a rapid reaction force of some 5,000 to 10,000 troops. . . . The Commission offers two arguments for such a capability. First, the record of international crises points out the need in certain cases to respond rapidly and, if necessary, with force; and second, the operational integrity of such a force requires that it not be assembled in pieces or in haste. A standing force may well be a necessity to effective prevention."

What would a rapid reaction force do? First, it would be available to reinforce long-term U.N. peacekeeping missions on an emergency basis. For example, if events turned sour in Cyprus or Macedonia, a battalion or two from the RRF could be sent out to reinforce the existing U.N. peacekeeping forces in those countries. RRF deployments would be temporary, not to exceed six months in most cases. Either the crisis would be resolved during those six months or a long-term peacekeeping force would be assembled. The elite troops of the RRF should not end up walking the Green Line [that divides the Greek-Cypriot and Turkish-Cypriot communities] in Cyprus for 20 years.

Lost Opportunities

Equally important, the RRF would respond to fast-breaking crises. Going back to the example of Rwanda, there were three situations in three years in which such a force could have been deployed to assist in the resolution of a crisis that called for armed intervention.

The first was during the genocide of April to July 1994, when reinforcements of several thousand tough, well-trained, well-equipped soldiers and police could have saved hundreds of thousands of lives by protecting noncombatants. (One of the hazards of peacekeeping, however, is that if a U.N. peacekeeping force had succeeded in holding down the number of murdered civilians to, say, 100,000, the operation would likely have been regarded as a failure.)

A second opportunity for the deployment of a rapid reaction force came when the teeming refugee camps of Zaire—housing more than a million Rwandan refugees—were formed in July and August 1994. Armed Hutu militia took over the camps and imposed their will. The unarmed representatives of U.N. civilian agencies were helpless in the face of the armed militiamen.

As the U.N. High Commissioner for Refugees, Sadako Ogata, noted, "The Rwandan refugee camps in Zaire and Tanzania were controlled by armed men, many of whom were probably guilty of genocide. We asked for international help in getting these people out of the camps. No country offered to get involved. My staff had to continue feeding criminals as the price for feeding hundreds of thousands of innocent women and children."

Boutros-Ghali strongly recommended that the Security Council assemble a multinational force of up to 7,000 troops to disarm the militia and remove them from the camps. The Security Council members greeted his proposal with thundering silence. The United Nations, consequently, was unable to do anything to wrest control of the camps from the Hutu militia. Had a standing rapid reaction force existed, it could have been deployed to the camps to disarm the militia and free the refugees from their control.

A third opportunity came in November 1996, when Laurent Kabila's ADFL [Alliance of Democratic Forces for the Liberation of Congo/Zaire] rebel forces in Zaire broke up the refugee camps. Most of the refugees returned home to Rwanda, but hundreds of thousands were unwilling or unable to do so and were forced deep into the interior of the country where they have suffered excruciating hardships and thousands have been systematically murdered by ADFL forces. In November 1996, an RRF could have made a quick excursion into Zaire to set up a safe transport corridor, thereby permitting the United Nations and private humanitarian agencies to bring aid to and repatriate the refugees. An intervention of this limited scope and duration would not have affected the outcome of the civil war raging at that time in eastern Zaire/Congo between Kabila's forces and long-time dictator Sese Seko Mobutu.

A multinational rescue was proposed by Canada to aid the refugees lost in the forests of Congo. Initially, the United States went along with the plan, then quickly backtracked when some reports suggested and Washington chose to believe that there were few refugees in eastern Congo except Hutu militia who did not want to return to Rwanda anyway. Since, Washington believed, there were few refugees, there was no need for a multinational rescue force.

This bit of extraordinary self-deception by the United States and others cost tens of thousands of legitimate refugees their lives. The existence of a rapid reaction force would have permitted the quick dispatch of an early rescue mission, saved lives, and probably saved money. In the end, the U.N. High Commissioner for Refugees airlifted many of the surviving refugees back to Rwanda in a lengthy, high-cost operation.

The international community thus struck out three times in quick succession with respect to genocide and its aftershocks in Rwanda and Congo. (Moreover, if an intervention had been carried out successfully on the first occasion, the opportunity for interventions two and three would not have arisen. The lesson is do it right the first time.)

Getting Big Daddy on Board

Canada, Denmark, and the Netherlands have taken these failures to heart in proposing means by which the United Nations can deploy peacekeepers more rapidly and efficiently than at present. All the rhetoric in the world, however, is not going to result in the creation of a rapid reaction force unless the big daddy on the international scene, the United States of America, gets behind the idea. Penuriousness, neo-isolationism, and hostility toward the United Nations are holding it back.

Yet, a rapid reaction force has much to recommend itself to the United States. First, if a competent, international peacekeeping force existed, the need for U.S. bilateral deployment, as in Somalia and Haiti would decrease, thereby reducing America's costs and casualties. Second, the anguished political debate that arises at the prospect of U.S. soldiers being sent in harm's way would be stilled. (American volunteers, we foresee, might serve in an RRF, but they would be outside the jurisdiction and responsibility of the Department of Defense.) And, third, with such a force under the control of the Security Council, the United States would always be able to use its veto if it did not approve of a proposed deployment.

U.S. intelligence, logistics, and emergency airlift support would clearly be essential for a rapid reaction force to be feasible. Yet, the cost to the United States in providing these services to an RRF would not necessarily be high. The United States could negotiate with the United Nations to be reimbursed for any expenses exceeding the 25 percent share of peacekeeping costs that it sees as its fair share—and the repayment of U.S. arrears could give a strong boost to a new initiative in peacekeeping. Early-warning measures and fast deployment of a rapid reaction force might save the United States more money than they would cost. They would certainly save lives and contribute to America's national security and well-being.

This post–Cold War period has left America groping for a definition of its interests and role in the world. In 1992, in the euphoric aftermath of the Gulf War, President George Bush lauded the United Nations. "{[N]ever before has the United Nations been so ready and so compelled to step up to the task of peacekeeping." A year later—after Somalia—Americans were talking not about a New World Order but of a New World Disorder, and confidence in the United Nations had plummeted.

The occasional necessity for international peacekeeping forces is doubted by virtually nobody. With the states of Central Africa in turmoil and the Bosnian situation far from settled, to name just two areas of concern, the question we must ask ourselves is whether the present ad hoc system of international peacekeeping is the best possible mechanism. The answer is an emphatic no. The creation of a rapid reaction force would give the international community the means to prevent and contain conflicts and to protect noncombatants that it now sadly lacks.

The United States should promote the creation of a rapid reaction peacekeeping force. It is in America's own interest to quiet the brushfires of distant conflicts before they burn out of control.

John Hillen

← **NO**

United Nations Peacekeeping Missions and Their Proliferation

I will make some short remarks here on the strategy of UN military operations—that is, the level at which the political and military dimensions of peacekeeping meet. In the course of my work I studied some 50 UN and other multinational peacekeeping and peace enforcement operations. The lessons learned from those missions give us a fairly good idea of the challenges of these missions and the institutional competence and capabilities of the UN itself.

. . . [T]oday we sit on the cusp of a periodic upswing in the size, character, and ambitions of UN peacekeeping operations. [Recently] the UN has mandated three large and complex peacekeeping operations—in East Timor, Sierra Leone, and the Democratic Republic of the Congo—in which the UN itself will direct significant military forces operating in some difficult environments. In addition, of course, there is the fairly new UN mission to Kosovo, but in that mission NATO [North Atlantic Treaty Organization] is handling the military tasks while the UN restricts itself to policing, administrative, and other basic governmental functions.

I say periodic upswing because a survey of the 52-year history of UN peacekeeping shows that it goes in cycles. I'd like briefly to discuss these cycles in order to better understand where we might be headed now. My study shows that UN peacekeeping goes through recurrent phases—and the pattern has been repeated several times in the past half-century. In the first phase small peacekeeping successes lead an emboldened international community to give the UN larger, more complex, and ambitious military operations in more belligerent environments. In the second phase these sorts of operations quickly overwhelm the capabilities of the UN itself, which tries unsuccessfully to improvise in operations for which it has no institutional structure, authoritative management systems, or military competency. In the third phase, burned and discredited, the UN pulls back to a more traditional peacekeeping role that suits the institution. Finally, with time healing some of these wounds and challenges to the international community continuing to mount, short memories compel the international community to thrust the UN back onto the international

From U.S. Senate. Committee on Foreign Relations. Subcommittee on International Operations. *United Nations Peacekeeping Missions and Their Proliferation.* Hearing, April 5, 2000. Washington, D.C.: U.S. Government Printing Office, 2000. (S.Hrg 106-573.)

security stage in a more ambitious and central role than before. The lessons of each of these cycles are clear. The UN itself has never had, nor was it ever intended to have, the authority, institutions, and procedures needed to successfully manage complex military operations in dangerous environments. Conversely, the UN—the world's most accepted honest broker—has exactly the characteristics needed to manage some peacekeeping operations undertaken in supportive political environments. Even then, the UN has struggled to competently direct even small and innocuous operations. But the real problems for all involved have come when the international community puts the UN in a military role for which it is neither politically suited nor strategically structured. My book goes into great detail on exactly why the UN has shown—in almost 50 missions—that there are strict limits to its military role. Quite simply, the UN should not be in the business of running serious military operations. It has neither the legitimacy, authority, nor systems of accountability needed to build the means necessary to direct significant military forces.

Authoritative, specifically structured, and well-rehearsed military alliances or coalitions of the willing better manage multinational military operations of the sort we've recently seen in the former Yugoslavia, Somalia, and Africa led by a major military power. These sorts of organizations are specifically structured—legally, politically, and organizationally—to direct complex and coercive military operations in uncertain environments. The model we've seen in Kosovo and East Timor recently may work best. An alliance like NATO or a multinational coalition such as that Australia led in East Timor can do the heavy lifting before turning it over to the UN.

. . . [L]et me briefly summarize how these cycles have occurred and in particular the U.S. and UN role in them. In my full testimony I have the complete story of the most recent cycle—that of Somalia and Bosnia—and perhaps in questioning we can discern from those episodes lessons for these new missions on the horizon. In 1948/49, UN peacekeeping started with relatively innocuous missions to Palestine and India-Pakistan—missions which, we should note, are still in existence today. A largely successful peacekeeping mission in the Sinai in the 1950's encouraged the UN to mount a very ambitious mission to the Congo in 1960. That mission ended very badly, taking the life of some 234 Blue Helmets [UN soldiers who traditionally wear blue helmets with the UN emblem on them] and the Secretary-General [Dag Hammarskjöld]. It is still referred to by many as "the UN's Vietnam."

Chastened, the international community returned to what was emerging as a more tried and true formula for UN peacekeeping. Small, lightly armed, and relatively unambitious missions deployed after a peace was concluded. These Blue Helmets did best when they followed the so-called principles of peacekeeping: strict neutrality, passive military operations, and the use of force only in self-defense. Importantly, the UN recognized that the Blue Helmets were only supporting players, there to help belligerents that had agreed to the UN presence. UN peacekeeping was never intended to be a coercive military instrument—one that could force a solution on one side or another to a conflict. This role for the UN, which is not specifically referred to in the Charter (nor envisaged by the UN's founders) evolved over time, the nature of the technique (peacekeeping) uniquely suiting the character and management abilities of the institution (the UN). By the late 1980's, the UN's ability to manage a small number of peacekeeping operations was not in doubt. In fact, in 1988 the Blue Helmets were awarded the Nobel Peace Prize. We should remember that in 1988 UN peacekeeping represented a rather small and unambitious enterprise in the grand scheme of global security. In January of 1988 the UN was managing less than 10,000 troops in five long-running peacekeeping missions and on an annual peacekeeping budget of some $230 million. The U.S. then, as now, picked up about [one-third] the cost of those missions.

Things changed quickly, though, after the fall of the Berlin Wall. The thawing of the Cold War and the unprecedented cooperation shown by the Security Council during the Persian Gulf War presaged a new era of UN-sponsored collective security. The enthusiasm for more and newer forms of UN peacekeeping was quickly manifested in a series of ambitious, expensive, dangerous, and militarily complex missions.

By 1993, the UN was managing almost 80,000 peacekeepers in eighteen different operations, including large and heavily armed missions to Cambodia, Somalia, and the former Yugoslavia. The annual peacekeeping budget grew to $3.6 billion.

Less than two years on from that peak however, UN peacekeeping had been thoroughly discredited. The Blue Helmets' failure to halt political violence in Somalia, Rwanda, Haiti, and the former Yugoslavia was reinforced by images of peacekeepers held hostage in Bosnia, gunned down in Mogadishu, or butchered along with thousands in Kigali. The UN quickly retreated—turning a nascent peacekeeping mission in Haiti over to a U.S.-led coalition, passing Bosnia off to NATO, and leaving Somalia to slip back into chaos. By 1997, UN peacekeeping was down to a more manageable level of some 15,000 Blue Helmets operating in more mundane environments and on a budget of around $1.2 billion. All has been relatively quiet on the UN front until this past fall [1999], when Kosovo, East Timor, Sierra Leone, and the Congo sprang onto the scene. If those missions go forward as planned, they will add over 25,000 Blue Helmets and some $700 million–$1 billion in costs to the UN's plate. More important, several of these new missions, especially Sierra Leone and the Congo, look certain to take place in very uncertain and belligerent environments—the sort in which the UN rarely if ever succeeds. . . . [A] word on the U.S. role in this latest cycle in the rise and fall of UN peacekeeping in the six years after the end of the Cold War. This message I believe is critical for the U.S. policy community because our own actions drive these episodes as much as anything else. More coherence in U.S. policy could have prevented many of the recent disasters in places such as Somalia and Bosnia. While a broad range of observers drew the same basic conclusion from peacekeeping's recent past—that the UN should not be in the business of managing complex, dangerous, and ambitious military operations—most are split on how it happened and whom to blame. Conservatives in the United States charge the UN itself and especially a fiendishly ambitious Boutros Boutros-Ghali who tried openly to accrue more and more military legitimacy and power for the UN itself. Liberal internationalists blame a parochial U.S. Congress that pulled the U.S. out of Somalia at the first sign of trouble, and is now holding America's UN dues hostage to its provincial agenda.

Both views are off base. Ironically, those who put UN peacekeeping through the wringer and hung the organization and its last Secretary-General out to dry were those American internationalists most likely to promote a larger collective security role for the United Nations. Over the past seven years, American officials sought for the UN a much greater role in international security affairs. But even though they were philosophically amenable to that goal, they choose to propel the UN into uncharted waters more out of political expediency rather than as a carefully crafted manifestation of their predisposition towards collective security. In many cases a new role for the UN was not so much a matter of policy, but a way of avoiding hard policy decisions such as those concerning the former Yugoslavia and Somalia. In essence, we used the UN as an excuse, not a strategy.

Either way, American officials, especially in the first Clinton administration, pushed a reluctant UN into much greater military roles than it could hope to handle. Once its failures were manifest, the same officials joined in the conventional wisdom that the UN itself "tried to do too much." Because of this, any post–Cold War "advances" in collective security were negated by those very internationalists who were so keen to champion the UN. As [professors] Paul Kennedy and Bruce Russett warned, UN operations such as those to Bosnia and Somalia "far exceed the capabilities of the system as it is now constituted, and they threaten to overwhelm the United Nations and discredit it, perhaps forever, even in the eyes of its warmest supporters." What they did not consider was that some of the UN's "warmest supporters" were those who were most responsible for putting it in desperate straits in the first place.

Patterns of Abuse

Advocates of collective security were almost giddy in the months immediately following the Gulf War. As David Henrickson noted, the end of the Cold War and the Security Council's role in the Gulf War "have produced an unprecedented situation in international society. They have persuaded many observers that we stand today at a critical juncture, one at which the promise of collective security, working through the mechanism of the United Nations, might at last be realized." Think tanks, conferences, workshops, and task-force reports trumpeting a proactive military role for the UN proliferated. In January 1992, the first ever Security Council summit declared that "the world now has the best chance of achieving international peace and security since the foundation of the UN." The heads-of-state asked Secretary-General Boutros-Ghali to prepare a report on steps the UN could take to fulfill their expectations of a more active military role.

In Boutros-Ghali's subsequent *An Agenda for Peace*, he outlined a series of proposals that could take the UN well beyond its traditional military role of classic peacekeeping. The Secretary-General called not only for combat units constituted under the long moribund Article 43 of the UN Charter, but for "peace-enforcement" units "warranted as a provisional measure under Article

40 of the Charter." Although these were largely theoretical and untested ideas, by the time they were published in July 1992, the Security Council had already implemented a similar agenda. A few months prior to *An Agenda for Peace*, large and ambitious UN missions to the former Yugoslavia and Cambodia were already approved and underway.

This initial episode reflected a pattern that would develop over the next several years. The UN, many times reluctantly so, would be thrust into an ambitious and dangerous series of missions and operations by a Security Council that was enthusiastic about new and enlarged mandates for UN peacekeepers—but not so keen on providing the support necessary to make them a success. In 1992, while the Secretary-General was (at the request of the world's most powerful leaders) preparing a draft report on possible new departures in peacekeeping, a series of international crises plunged the organization into what UN official Shashi Tharoor called "a dizzying series of peacekeeping operations that bore little or no resemblance in size, complexity, and function to those that had borne the peacekeeping label in the past."

In the former Yugoslavia, it soon became painfully obvious that despite the deployment of almost 40,000 combat troops, the UN was in over its head. Among American leaders, it was fashionable in both political parties to bemoan the ineffectiveness of the UN peacekeepers. This America was as responsible for what the UN was attempting to do in the former Yugoslavia as any other state or the organization itself. Between September 1991 and January 1996, the Security Council passed 89 resolutions relating to the situation in the former Yugoslavia, of which the United States sponsored one-third. While Russia vetoed one resolution and joined China in abstaining on many others, the United States voted for all 89 to include those twenty resolutions that expanded the mandate or size of the UN peacekeeping mission in the Balkans.

Far from the notion that the UN was pulling the international community into Bosnia, the U.S.-led Security Council was pushing a reluctant UN even further into a series of missions and mandates it could not hope to accomplish. Boutros-Ghali warned the members of the Security Council that "the steady accretion of mandates from the Security Council has transformed the nature of UNPROFOR's [United Nations Protection Force's] mission to Bosnia-Herzegovina and highlighted certain implicit contradictions. The proliferation of resolutions and mandates has complicated the role of the Force." His Undersecretary-General for peacekeeping, Kofi Annan, was more direct. Attempts to further expand the challenging series of missions being given to the UN were "building on sand."

This did not seem to deter the U.S.-led Security Council however, which was happy to expand the mission further while volunteering few additional resources to the force in Bosnia. A June 1993 episode demonstrating this pattern is instructive. Then, the UN field commander estimated he would need some 34,000 more peacekeepers to protect both humanitarian aid convoys and safe areas in Bosnia. The Security Council, having given him these missions in previous resolutions, instead approved a "light option" of 7,600 troops, of whom only 5,000 had deployed to Bosnia some nine months later. Quitting his post in disgust, the Belgian general in command remarked, "I don't read the Security Council resolutions anymore because they don't help me."

The Clinton administration, which had shown unbounded enthusiasm for UN peacekeeping in the first months of the administration, began to sour slightly on its utility by September 1993. By then Ambassador Madeleine Albright's doctrine of "assertive multilateralism" had given way to President Clinton beseeching the UN General Assembly to know "when to say no." But it was the United States and its allies on the Security Council who kept saying yes for the United Nations. Even after that speech, Mrs. Albright voted for all five subsequent resolutions (and sponsored two) that again expanded the size or mandate of the UN peacekeeping mission to the former Yugoslavia. All the while, until the fall of 1995, the U.S. steadfastly resisted participating in the UN mission or intervening itself with military forces through some other forum.

In Somalia, there was an even more direct pattern. There the United States pushed an unwilling UN into a hugely ambitious nation-building mission. In its waning days the Bush administration had put together a U.S.-led coalition that intervened to ameliorate the man-made famine in Somalia. From the very beginning of the mission it had been the intention of the U.S. to turn the operation over to a UN peacekeeping force. Conversely, Boutros Boutros-Ghali, an Egyptian well acquainted with the challenge of nation-building in Somalia, wanted no part of the mission for the UN Ambassador Robert Oakley, the U.S. envoy to Somalia, noted that in a meeting with the Secretary-General and his assistants on 1 December 1992, "the top UN officials rejected the idea that the U.S. initiative should eventually become a UN peacekeeping operation."

The U.S. kept up the pressure on the Secretary-General, who was powerless to resist the idea if it gained momentum in the Security Council. The debate resembled what [Assistant Secretary of State for Africa] Chester Crocker called "bargaining in a bazaar" and "raged out of public view" while the U.S. and the UN negotiated over the follow-[up] mission. For his part, Boutros-Ghali wanted the U.S.-led coalition to accomplish a series of ambitious tasks before the UN would take over. These included the establishment of a reliable cease-fire, the control of all heavy weapons, the disarming of lawless factions, and the establishment of a new Somali police force. For its part, the United States just wanted to leave Somalia as soon as possible. It was now time to put assertive multilateralism to the test. Madeline Albright shrugged off the challenge to the world body and wrote that the difficulties that the UN was bound to encounter in Somalia were "symptomatic of the complexity of mounting international nation-building operations that included a military component."

The debate, with Boutros-Ghali resisting up to the last, effectively ended on 26 March 1993 with the passage of Security Council resolution 814 establishing a new UN operation in Somalia. The resolution authorized, for the first time, Chapter VII enforcement authority for a UN-managed force. More importantly, the resolution greatly expanded the mandate of the UN to well beyond what the American force had accomplished. Former Ambassador T. Frank Crigler called the UN mandate a "bolder and broader operation intended to tackle underlying social, political, and economic problems and to put Somalia back on its feet as a nation." In the meantime, the U.S. withdrew its heavily armed 25,000 troop force and turned the baton over to a lightly armed and still arriving UN force. The transition, set for early May 1993, was so rushed that on the day the UN took command its staff was at only 30 percent of its intended strength. The undermanned and underequipped UN force was left holding a bag not even of its own making.

The travails of the UN mission in Somalia need no further elucidation here. Suffice it to say that the U.S., although no longer a direct player in Somalia, continued to lead the Security Council in piling new mandates on the UN mission there. The most consequential of these was the mandate to apprehend those Somali's responsible for the June 1993 killing of 24 Pakistani peacekeepers. The U.S. further complicated this explosive new mission with an aggressive campaign of disarmament capped by the deployment of a special operations task force that was to lead the manhunt for [Somalian leader] Mohammed Farah Aideed. This task force was not under UN command in any way and when it became engaged in the tragic Mogadishu street battle of 3 October 1993 the UN commanders knew nothing of it until the shooting started. Even MG [Major General] Thomas Montgomery, the American commander and deputy UN commander, was told of the operation only 40 minutes before its launch. A U.S. military report afterward noted that the principal command problems of the UN mission in Somalia were "imposed on the U.S. by itself."

This fact, that the UN was not involved in the deaths of eighteen American soldiers in Mogadishu, was buried by the administration. Even more cynically, several top-level administration officials charged in 1995 with selling the Dayton Peace Accords to a skeptical U.S. public constantly noted that U.S. soldiers in the NATO mission to Bosnia would not be in danger because the UN would not be in command, as it was in Somalia. Few single events have been as damaging to the UN's reputation with the Congress and American public as the continued perception that it was the United Nations that was responsible for the disaster in Somalia. Not only has this myth been left to fester, it was indirectly used, along with the UN's many other U.S.-initiated problems, to call for Boutros Boutros-Ghali's head during the 1996 Presidential campaign. Then, for the first time in several years, the U.S. used its veto to stand alone against the Security Council and bring down the Secretary-General who had resisted the U.S.-led events that so discredited him and his organization.

Conclusion—Friends Like These

After those particular episodes, UN peacekeeping is now happy to be, as a UN official recently told me, in "a bear market." Congress and the administration are happy as well with a low profile for UN military operations—especially as Clinton officials try to get Congress to pay America's share of the unprecedented peacekeeping debt. Fittingly, Madeleine Albright, as Secretary of State, is now chiefly responsible for convincing Congress to pay the bill that she is tacitly accountable for because of her votes during that busy time on the Security Council. Albright also played a central role as the official, more than any other in the Bush and Clinton administrations, who epitomized the keen hopes of liberal internationalists advocating a greater security role for the UN. In early 1993, her speeches were laced with talk of "a renaissance for the United Nations" and ensuring that "the UN is equipped with a robust capacity to plan, organize, lead, and service peacekeeping activities." By 1994, however, after it became obvious that the inherent limitations of a large multinational organization would not allow it effectively to manage complex military operations, Albright stated that "the UN has not yet demonstrated the ability to respond effectively when the risk of combat is high and the level of local cooperation is low." Left unsaid was that the U.S., more than any other member state, was responsible for giving the UN much to do in Somalia and Bosnia and little to do it with. It appeared, as Harvey Sicherman has written, that "the assertive multilateralists of 1992–3 placed more weight upon the UN than it could bear, while ignoring NATO and other regional coalitions."

Regional coalitions or more narrowly focused military alliances were ignored both for reasons of philosophy and political expediency. Philosophically, legitimacy could be gained for collective security in general and the UN in particular by having it directly manage the more dynamic military operations of the post–Cold War era. Thomas Weiss typified this school of thought and wrote, "the UN is the logical convenor of future international military operations. Rhetoric about regional organizations risks slowing down or even making impossible more timely and vigorous action by the UN, the one organization most likely to fulfill adequately the role of regional conflict manager." This appealed in particular to the officials of the Clinton administration who had developed and published many similar thoughts while in academia or the think-tank world.

But for the most part the U.S. promoted unprecedented UN missions to conflicts such as Bosnia and Somalia because they did not want the U.S. or its alliances to be principally responsible for difficult and protracted military operations in areas of limited interest. As Shashi Tharoor wrote, "it is sometimes argued that the peacekeeping deployment to Bosnia-Herzegovina reflected not so much a policy as the absence of policy; that [UN] peacekeeping responds to the need to 'do something' when policy makers are not prepared to expend the political, military, and financial resources required to achieve the outcome that the press and opinion leaders are clamoring for."

The final irony is that the UN's adventurous new role in 1993–1995 and peacekeeping's subsequent demise came about not necessarily by the well intentioned but unsupported design of collective security's most ardent proponents. Instead, it came about by default as these same supporters thrust upon the UN difficult missions they would rather not have addressed more directly. Given the recent and renewed enthusiasm for more missions of the sort that will greatly challenge the UN, the international community would do well to keep this lesson in mind.

POSTSCRIPT

Should the United Nations Be Given Stronger Peacekeeping Capabilities?

The increase in the use of UN peacekeeping forces is partly the result of the changes from the cold war to the post–cold war era, which can be explored in Steven R. Ratner, *The New UN Peacekeeping: Building Peace in Lands of Conflict After the Cold War* (St. Martin's Press, 1996). Within this larger context, the debate over creating a potentially permanent international police force (even army) is heating up in many forums. A good place to begin exploring this topic further is the report issued by Boutros Boutros-Ghali, *An Agenda for Peace: Preventive Diplomacy, Peacemaking, and Peacekeeping* (United Nations, 1992). It is also worthwhile to look into the Internet site for UN peacekeeping, which can be found at http://www.un.org/Depts/dpko/dpko/home.shtml.

Recent events throughout the world have done even more to convince those who advocate a UN standing military force that it is important to create a force that can respond quickly to crises and to have the military power to intervene effectively when necessary. The events in Kosovo, the grisly border war between Ethiopia and Eritrea, and the clashes that occurred in May 1999 between India and Pakistan—both of which have nuclear weapons—are just a few examples of the regular fighting that some hope an enhanced UN force could prevent or stop. Moreover, advocates argue that the fluid, post–cold war international system presents an opportunity to establish such a force that should not be missed. More on this view is available in Lekha Sriram Chancre, Karin Wermester, and Marian de Smet, eds., *From Promise to Practice: Strengthening UN Capacities for the Prevention of Violent Conflict* (Lynne Rienner, 2003).

It is certainly arguable that UN peacekeeping forces have had limited effectiveness. They have often been late to arrive because of the political difficulties of getting a force authorized by the Security Council. UN forces are also lightly armed and often have severe restrictions on their missions. These and other reasons for UN peacekeeping failures (amid successes in other instances) are taken up in Dennis C. Jett, *Why Peacekeeping Fails* (St. Martin's Press, 2000).

Other analysts are skeptical of the possibility or wisdom of a standing UN force and its possible uses. To delve more extensively into Hillen's views, read his *Blue Helmets: The Strategy of UN Military Operations* (Brassey's, 2000). Some of these concerns are based on such narrow factors as cost. The entire UN peacekeeping operation comes approximately to a mere one-tenth of 1 percent of what the world's countries spend on their national military establishments. More substantively, there are worries that a more powerful, proactive UN might undermine the sovereignty of the less developed countries (LDCs), with the UN Security Council serving as a tool of the five big powers that control the council through their veto power. From this perspective, UN intervention carries the danger of neo-colonial control. Other opposition comes from those who believe that a UN force will undermine the national sovereignty of even larger countries. They object, for instance, to the possibility that U.S. troops could be placed under UN command without the authorization of Congress. They worry that there might someday even be an international draft. Such concerns seem far-fetched, but then so once did the very existence of international security forces. For a look to the future, read Olara A. Otunnu and Michael W. Doyle, eds., *Peacemaking and Peacekeeping for the Next Century* (Rowman & Littlefield, 1998).

ISSUE 17

Do International Financial Organizations Require Radical Reform?

YES: Joseph Stiglitz, from "Joseph Stiglitz: The Progressive Interview," interview by Lucy Komisar, *The Progressive* (June 2000)

NO: Kenneth Rogoff, from "An Open Letter to Joseph Stiglitz," International Monetary Fund, http://www.imf.org/external/np/vc/2002/070202.htm (July 2, 2002)

ISSUE SUMMARY

YES: In an interview conducted by Lucy Komisar, Joseph Stiglitz, former chief economist of the World Bank, argues that the policies of the World Bank and the International Monetary Fund (IMF) are driven by the economic model favored by the United States and other powerful and prosperous countries rather than the interests of the poor countries that the World Bank and the IMF are supposed to be helping.

NO: Kenneth Rogoff, economic counselor and director of research for the International Monetary Fund, concedes that the World Bank and the IMF fall short of perfection. He contends, however, that Stiglitz's unbridled criticisms of the two organizations are often factually faulty and that the reforms he favors are unwise.

The World Bank and the International Monetary Fund (IMF) are two of the three key institutions that were created around the end of World War II to try to better regulate the economy. Over time, the World Bank and the IMF have become increasingly concerned with the economic development and monetary stability of the less developed countries (LDCs) of the world.

The formation of the IMF stemmed in part from the belief of many analysts that the Great Depression of the 1930s and World War II were partly caused by inflation, lack of convertibility between currencies, and other monetary problems that characterized the years between 1919 and 1939. To address this problem, 44 nations met at the UN Monetary and Financial Conference at Bretton Woods, New Hampshire, in 1944 to establish a new monetary order. The delegates established the IMF and several other institutions to help promote and regulate the world economy as part of the U.S.-led drive to increase international economic interchange.

The IMF, which began operations in 1947 with 44 member countries, has expanded steadily, and now virtually all countries hold IMF membership. The IMF's primary function is to help maintain exchange-rate stability by making short-term loans to countries with international balance-of-payments problems caused by trade deficits, heavy loan payments, or other factors. The IMF receives its usable funds from hard currency reserves ($265 billion in 2002) that are placed at its disposal by wealthier member-countries and from earnings that it derives from interest on loans made to countries that draw on those reserves.

What the IMF usually does is loan money to a country when overspending or other problems are harming faith in the country's currency. This causes monetary instability because people both outside and inside the country are less willing to accept the country's money. The result is inflation, the inability to import needed goods and services, and other ill effects. To counter this instability, the IMF typically loans a country money to support its currency or to stabilize its financial situation by refinancing its debt.

Although the IMF has many supporters, it also has its detractors. It is possible to divide the controversies over the IMF into two categories: voting and structural adjustment policies.

The first issue centers on the vote distribution of the IMF's board of directors. Voting is based on the level of each member's contribution to the fund's resources. Under this formula, the United States has 17 percent of the votes, the countries of the European Union combined have 30 percent of the votes, and Japan has 6 percent, giving majority control to this handful of countries, which constitute less than 10 percent of the IMF's total membership. The same formula leaves the less developed countries with little or no power in the organization's decision making.

The second criticism of the IMF is that it imposes unfair and unwise economic conditions on countries that use its financial resources. These conditions require recipients to make structural adjustments. Among other things, recipients must move toward a capitalist economy by privatizing state-run enterprises, reduce barriers to trade and to the flow of capital (thus promoting foreign ownership of domestic businesses), reduce domestic social programs to trim budget deficits, and devalue currencies (which increases exports but makes imports more expensive and, thus, increases the cost of living). However reasonable such "strings" might seem on the surface, they evoke sharp criticism by some analysts, such as Joseph Stiglitz in the following selection.

In the second selection, Kenneth Rogoff maintains that it is only sensible to require financial reforms when the existing policies of a country have caused the monetary instability that the IMF is being asked to help remedy. Only when these problems are solved will countries be able to avoid recurring crises and repeatedly asking the IMF to bail them out of their difficulties.

Joseph Stiglitz

An Interview With Joseph Stiglitz

Among the economic policy elite, Joseph Stiglitz is a heretic. The most prestigious critic of the International Monetary Fund (IMF) and the "Washington consensus," Stiglitz has voiced his views in the corridors of power. In 1993, he became a member of the President's Council of Economic Advisers, and later its chairman. In that role, he cautioned against free market "shock therapy" for Russia. Then, in 1997, he became chief economist of the World Bank, where he tried to push the institution in a more progressive direction. He also stepped up his criticisms of IMF and U.S. economic policies toward Russia and East Asia. These criticisms did not endear him to Treasury Secretary Lawrence Summers [1999–2001], who allegedly pushed him out of the World Bank in December.

Stiglitz is a giant among economists; peers assume he will win a Nobel Prize [which he did in 2001]. When he was a junior at Amherst, the economics faculty met and decided that they had nothing more to teach him. One of them was deputized to call MIT to have him admitted to study there. After two years, he went to Oxford as a research fellow, then returned to MIT as assistant professor, and from there he went to Yale. Two years later, the Yale faculty voted to make him a full professor—at twenty-six. He has taught in Nairobi and at Stanford, Oxford, and Princeton. In his field, he helped create a new branch, the economics of information, and he is a leader in the economics of the public sector.

As chairman of the Council of Economic Advisers, Stiglitz was the highest official charged with analyzing the impact of economic policies on Americans. As chief economist of the World Bank, he was the top official responsible for analyzing how economic decisions affected the world's poor. He has been strategically placed to judge the winners and losers of global economic policies.

Stiglitz is an avuncular man of fifty-seven with close-cropped gray hair and a beard. He favors a loose gray V-necked sweater over a blue shirt and black pinstriped trousers. For a Washington official, he is astonishingly amiable and inspires warm affection, almost reverence, among his staff and close associates.

I spoke with Stiglitz several times in April [2000]. He was generous with his time, unpretentious, and genial. He tends to talk not in sound bites or clichés but in professorial explanations. I spoke with him for three hours and read over his recent speeches. Here is an edited transcript of his views.

From Joseph Stiglitz, "Joseph Stiglitz: The Progressive Interview," interview by Lucy Komisar, *The Progressive* (June 2000). Copyright © 2000 by Lucy Komisar. Reprinted by permission.

Q: What was your reaction to the protests in April [2000] against the IMF and World Bank in Washington?

Joseph Stiglitz: I thought they were very effective in conveying the sense of values and concerns that a lot of young people, and people generally in the U.S., have beyond the narrow materialistic issues. They are concerned about poor people in developing countries, about democracy and democratic participation, governance issues, and the environment. There were people advocating protectionist issues, or who were more violent than I would approve of, but the point of marches is to convey a sense of values and concerns. How can one object to Americans caring about issues that go beyond our borders, to caring about poor people?

Q: What do you think of the call to shut down the IMF and the World Bank?

Stiglitz: The world needs an international development agency. I don't think anybody really thinks that one should get rid of the World Bank. Reform is one thing, but getting rid of it I think would be wrong. The IMF is a more complicated issue. I think there is a broad sentiment among both the left and the right that the IMF may be doing more harm than good. On the right, there's the view that it represents a form of corporate welfare that is counter to the IMF's own ideology of markets. But anybody who has watched government from the inside recognizes that governments need institutions, need ways to respond to crises. If the IMF weren't there, it would probably be reinvented. So the issue is fundamentally reform.

Q: The IMF has extraordinary powers to affect countries in times of crisis. Who does it represent? Who controls it?

Stiglitz: Finance ministers and central bank governors have the seats at the table, not labor unions or labor ministers. Finance ministers and central bank governors are linked to financial communities in their countries, so they push policies that reflect the viewpoints and interests of the financial community and barely hear the voices of those who are the first victims of dictated policies.

Q: How does this play out in the IMF's decisions?

Stiglitz: In the midst of the East Asian crisis, there were choices. One choice would have been to encourage countries to implement a bankruptcy law that could have threatened the interests of the lenders. But the mindset of IMF officials was so strong that they acted as if there were no choices.

Even as debate on reforming the international economic architecture has proceeded, the people who would inevitably face many of the costs of the mistaken policy have not been invited to the discussions. Workers' rights should be a central focus of development. But nowhere, in all of these discussions, did issues of workers' rights, including the right to participate in the decisions which would affect their lives in so many ways, get raised. Conditionalities are adopted without social consensus. It's a continuation of the colonial mentality. I often felt myself the lone voice in these discussions suggesting that basic democratic principles be followed. I recommended that not only should workers' voices be heard, but they should actually have a seat at the table. You have the old boys' club discussing how the old boys' club should be reformed.

Q: Who wins when the U.S. and the IMF impose their bailouts on countries in crisis?

Stiglitz: These policies protect foreign creditors. If I came to the problem of what can I do to maintain the Thai economy from the perspective of the chairman of the collection committee of the international creditors, I might mistakenly say the most important thing is to make sure people don't abrogate their debt. Senior people in the IMF actually said that not paying the debt was an abrogation of a contract, whereas anybody who knows about capitalism knows that bankruptcy is an essential part of capitalism.

In East Asia, you had private debtors. The appropriate response when you have private debtors who can't pay is bankruptcy. It is not the nationalization of private debts, which the IMF has facilitated in many countries. Nationalization of private debts undermines prudential lender behavior and is a government intervention in the market. But that's not the view you'd take if you were chairing the creditors.

The IMF insisted that both Russia and Brazil maintain their currency at over-valued levels. Who are you protecting when you try to maintain that exchange rate by having high interest rates? You're protecting domestic and foreign firms that have gambled on the exchange rate. And who is paying the price? The small businesses that did not gamble [and no longer can afford loans], the workers who are going to be put out of jobs.

The lenders in the more advanced countries tend to recover most, if not all, of the amounts lent, with the costs of the bailout (including the costs of economic restructuring) being largely borne by workers. There is a transfer from the workers to the large international banks and other creditors in the U.S., Germany, and Japan.

Q: How else do IMF and World Bank policies affect workers?

Stiglitz: During my three years as chief economist of the World Bank, labor market issues were looked at through the lens of neoclassical economics. "Wage rigidities"—often the fruits of hard-fought bargaining—were thought part of the problem facing many countries. A standard message was to increase labor market flexibility. The not-so-subtle subtext was to lower wages and lay off unneeded workers.

They had a strategy for job destruction. They had no strategy for job creation. Many of the policies the IMF pursued as they were killing off jobs made job creation almost impossible. In the U.S., you couldn't have job creation with interest rates of 30 or 40 percent. They had a philosophy that said job creation was automatic. I wish it were true. Just a short while after hearing, from the same preachers, sermons about how globalization and opening up capital markets would bring them unprecedented growth, workers were asked to listen to sermons about "bearing pain." Wages began falling 20 to 30 percent, and unemployment went up by a factor of two, three, four, or ten.

Q: How would you define "the Washington consensus"?

Stiglitz: It is a set of policies formulated between 15th and 19th streets by the IMF, U.S. Treasury, and World Bank. Countries should focus on stabilization, liberalization, privatization. It's based on a rejection of the state's activist role and the promotion of a minimalist, noninterventionist state. The analysis in the era of [U.S. President Ronald] Reagan and [British Prime Minister Margaret] Thatcher was that government was interfering with the efficiency of the economy through protectionism, government subsidies, and government ownership. Once the government "got out of the way," private markets would allocate resources efficiently and generate robust growth. Development would simply come.

The Washington consensus also considers capital market liberalization essential, and the IMF took it as a central doctrine. Capital market liberalization includes freeing up deposit and lending rates, opening up the market to foreign banks, and removing restrictions on capital account transactions and bank lending. The focus is on deregulation, not on finding the right regulatory structure.

Q: What did the Washington consensus tell the former Communist countries to do?

Stiglitz: Countries were told they had no incentives because of social ownership. The solution was privatization and profit, profit, profit. Privatization would replace inefficient state ownership, and the profit system plus the huge defense cutbacks would let them take existing resources and have a higher GDP [gross domestic product] and an increase in consumption. Worries about distribution and competition—or even concerns about democratic processes being undermined by excessive concentration of wealth—could be addressed later.

Q: How did U.S. Russia policy develop?

Stiglitz: In the early 1990s, there was a debate among economists over shock therapy versus a gradualist strategy for Russia. But Larry Summers [Under Secretary of the Treasury for International Affairs, then Deputy Secretary of the Treasury, (then) Secretary] took control of the economic policy, and there was a lot of discontent with the way he was driving the policy.

The people in Russia who believed in shock therapy were Bolsheviks—a few people at the top that rammed it down everybody's throat. They viewed the democratic process as a real impediment to reform.

The grand larceny that occurred in Russia, the corruption that resulted in nine or ten people getting enormous wealth through loans-for-shares, was condoned because it allowed the reelection of [President Boris] Yeltsin.

Q: What effect did the policies pushed by the United States and the IMF have on the Russian people?

Stiglitz: Both GDP and consumption declined. Living standards collapsed, life spans became shorter, and health worsened. Russia achieved a huge increase in inequality at the same time that it managed to shrink the economy by up to a third. Poverty soared to close to 50 percent from 2 percent in 1989, comparable to that of Latin America—a remarkable achievement in eight years.

Q: How did that happen?

Stiglitz: Put yourself in the shoes of one of these oligarchs who has been given a gift of $10 billion. Russia is in a deep depression. Nobody's investing. There is a widespread political consensus that the way you got your wealth is illegitimate. Through political connections, you got the government to give you a huge oil field. You sell oil. What do you do with the revenue? You have a choice: You can invest it in the booming New York stock market, or you can invest it in Russia, which is in a depression. If you invest it in Russia, you are risking that eventually there will be a new government that says, "Yeltsin was a crook, and you got the money in an illegitimate way." And the IMF invites you to take the money out, because free capital markets are the way of the future. Then, to make your life even easier, in August 1998 the IMF comes in and says we'll give you $5 billion or $6 billion to up the exchange rates so you can get more for your rubles to take over to Cyprus in the next day or so. What would you have done? The incentives worked.

Rather than providing incentives for wealth creation, privatization provided incentives for asset-stripping, with huge movements of capital abroad— $2 billion to $3 billion a month. Policies seemed almost deliberately designed to suppress new enterprise and job creation. The excessive focus on macrostabilization led to interest rates of 20, 30, 40, 250 percent. There is little domestic or foreign investment except in natural resources. How many Americans will start a business if the interest rates are 150 percent?

Q: How did the East Asian crisis happen?

Stiglitz: In East Asia, reckless lending by international banks and other financial institutions combined with reckless borrowing by domestic financial institutions precipitated the crisis, but workers bore the costs. The roots of the crisis in East Asia were in private sector decisions. The biggest problems were the misallocation of investment, most notably to speculative real estate, and risky financing, especially borrowing short-term debt on international markets.

Q: The first country to be hit was Thailand. What happened there?

Stiglitz: The crisis began as a currency crisis with large devaluations in Thailand in July 1997. Pressed by the U.S. Treasury and the IMF on financial and capital market liberalization, Thailand had opened itself to capital flows and foreign banks.

There was a real estate bubble. Bank contraction in Japan led to credit contraction in Thailand, and the real estate bubble burst. When capital stopped flowing in, there was pressure for the exchange rate to fall, the government tried to support it, and in a short time it spent enormous amounts of money and used up its reserves.

The IMF organized a typical rescue package, based on fiscal austerity and high interest. There were massive bankruptcies. The magnitude of unemployment was hard to believe. Real wages went down 20 to 30 percent. The crisis spread to Indonesia, Malaysia, and the Philippines, and eventually Korea. Stock markets lost 40 to 80 percent of their values, and banks failed.

I talked to [IMF Deputy Director] Stan Fischer and others at the beginning of the crisis. They had a peculiar view: "If you're right, we'll correct the policies." But there are important irreversibilities. If you destroy a firm, you can't pull it out of bankruptcy overnight. They ignored that. They ignored that high interest rates wouldn't attract capital and increased the probability of default.

Q: What should the IMF have done in East Asia?

Stiglitz: If you had approached the problem of East Asia from the perspective of what would be best for the global economy, you would say that the countries needed more resources so they wouldn't bring down their neighbors. That was the philosophy of Bretton Woods [the founding conference for the IMF and World Bank in 1944] to stop a global slowdown.

But what did the IMF do? It went to the countries and told them to be more contractionary than they wanted, to increase interest rates enormously. It was just the opposite of the economic analysis that was the basis of the founding of the IMF. Why? In order to make sure that creditors got repaid.

When you're facing the threat of recession, you need to have an expansionary monetary and fiscal policy. Pre-Keynesian, Hooverite views are dead everywhere except on 19th Street in Washington.

Q: Is there a double standard on trade liberalization?

Stiglitz: As people in Seattle were saying, the international institutions go around the world preaching liberalization, and the developing countries see that means open up your markets to our commodities, but we aren't going to open our markets to your commodities. In the nineteenth century, they used gunboats. Now they use economic weapons and arm-twisting.

When you have a march in Seattle, you hear voices saying maybe trade policies haven't been done in the interest of broader constituencies but in the interest of special interests. All of a sudden people get woken up to a different perspective.

Q: What was your reaction to the protests against the WTO [World Trade Organization] in Seattle [in late 1999]?

Stiglitz: The Seattle march was a wake-up call to much of the world that had taken the advantages of globalization for granted, that many people saw globalization through different lenses, saw the process being pursued behind closed doors for the benefit of special interests—banks, businesses, and the rich.

Q: How were decisions on these issues made in the Clinton White House when you were on the Council of Economic Advisers?

Stiglitz: The decision-making process in the White House does not let most issues get up to the President. The Council thought opening up global markets to derivatives that would destabilize other countries wasn't likely to create a lot of jobs in the U.S. and might adversely affect U.S. interests by causing global economic instability.

Larry Summers opposed us.

Q: If there's a disagreement of this sort, doesn't the issue go to the President?

Stiglitz: Treasury made sure there was a consensus not to bring it to the President. I knew the President agreed with me. That was why some of my concerns never went to the President. But it wasn't as if I felt isolated. Treasury is only one view within the Administration, and even some people in the Treasury would agree with me. At the World Bank, Jim Wolfensohn [the president of the Bank] agreed with me on most of the substantive issues. There were other people who agreed with me strongly on the general principle of open discussions.

Q: How did top U.S. officials try to shut you up?

Stiglitz: Stan Fischer wrote an op-ed in the *F.T.* [*Financial Times*] that all he's asking countries to do is to have a balanced budget. Not since Herbert Hoover have we heard something like that! We have fought against a balanced budget amendment to the Constitution on grounds that it would eliminate the ability to use fiscal policy as an anti-cyclical device in a recession.

I wrote an op-ed saying we don't believe in a balanced budget during recession in the U.S., why should we be telling that to other countries? The reaction that I got from the U.S. Treasury and the IMF was enormous. There were phone calls to Jim Wolfensohn, mainly from the Treasury but also from the IMF, not to allow me to talk to the press. Wolfensohn, who believed I was right, tried to manage the difficult process of how do you keep at bay your largest shareholder, how do you maintain relations with your sister institution in a world of clubs in which you don't criticize each other. He told me to tone it down. I had made my point.

Q: I heard that Summers told Wolfensohn that if he didn't fire you, he wouldn't be reappointed.

Stiglitz: I heard the same rumors. I have no way of knowing. In an article in the *Financial Times* when I left, somebody observed that these guys are smart enough that they wouldn't leave fingerprints.

NO ↩

Kenneth Rogoff

An Open Letter to Joseph Stiglitz

Dear Joe:

Like you, I came to my position in Washington from the cloisters of a tenured position at a top-ranking American University. Like you, I came because I care. Unlike you, I am humbled by the World Bank and IMF [International Monetary Fund] staff I meet each day. I meet people who are deeply committed to bringing growth to the developing world and to alleviating poverty. I meet superb professionals who regularly work 80-hour weeks, who endure long separations from their families. Fund staff have been shot at in Bosnia, slaved for weeks without heat in the brutal Tajikistan winter, and have contracted deadly tropical diseases in Africa. These people are bright, energetic, and imaginative. Their dedication humbles me, but in your speeches, in your book [*Globalization and Its Discontents,* 2002], you feel free to carelessly slander them.

Joe, you may not remember this, but in the late 1980s, I once enjoyed the privilege of being in the office next to yours for a semester. We young economists all looked up to you in awe. One of my favorite stories from that era is a lunch with you and our former colleague, Carl Shapiro, at which the two of you started discussing whether Paul Volcker merited your vote for a tenured appointment at Princeton. At one point, you turned to me and said, "Ken, you used to work for Volcker at the Fed. Tell me, is he really smart?" I responded something to the effect of "Well, he *was* arguably the greatest Federal Reserve Chairman of the twentieth century," to which you replied, "But is he smart like *us?*" I wasn't sure how to take it, since you were looking across at Carl, not me, when you said it.

My reason for telling this story is two-fold. First, perhaps the Fund staff who you once blanket-labeled as "third rate"—and I guess you meant to include World Bank staff in this judgment also—will feel better if they know they are in the same company as the great Paul Volcker. Second, it is emblematic of the supreme self-confidence you brought with you to Washington, where you were confronted with policy problems just a little bit more difficult than anything in our mathematical models. This confidence brims over in your new 282 page book. Indeed, I failed to detect a single instance where you, Joe Stiglitz, admit to having been even slightly wrong about a major real world problem. When the

U.S. economy booms in the 1990s, you take some credit. But when anything goes wrong, it is because lesser mortals like Federal Reserve Chairman [Alan] Greenspan or then-Treasury Secretary [Bob] Rubin did not listen to your advice.

Let me make three substantive points. First, there are many ideas and lessons in your book with which we at the Fund would generally agree, though most of it is old hat. For example, we completely agree that there is a need for a dramatic change in how we handle situations where countries go bankrupt. IMF First Deputy Managing Director Anne Krueger—who you paint as a villainess for her 1980s efforts to promote trade liberalization in World Bank policy—has forcefully advocated a far reaching IMF proposal. At our Davos [World Economic Forum] panel in February [2002] you sharply criticized the whole idea. Here, however, you now want to take credit as having been the one to strongly advance it first. Your book is long on innuendo and short on footnotes. Can you document this particular claim?

Second, you put forth a blueprint for how you believe the IMF can radically improve its advice on macroeconomic policy. Your ideas are at best highly controversial, at worst, snake oil. This leads to my third and most important point. In your role as chief economist at the World Bank, you decided to become what you see as a heroic whistleblower, speaking out against macroeconomic policies adopted during the 1990s Asian crisis that you believed to be misguided. You were 100% sure of yourself, 100% sure that your policies were absolutely the right ones. In the middle of a global wave of speculative attacks, that you yourself labeled a crisis of confidence, you fueled the panic by undermining confidence in the very institutions you were working for. Did it ever occur to you for a moment that your actions might have hurt the poor and indigent people in Asia that you care about so deeply? Do you ever lose a night's sleep thinking that just maybe, Alan Greenspan, Larry Summers, Bob Rubin, and Stan Fischer had it right—and that your impulsive actions might have deepened the downturn or delayed—even for a day—the recovery we now see in Asia?

Let's look at Stiglitzian prescriptions for helping a distressed emerging market debtor, the ideas you put forth as superior to existing practice. Governments typically come to the IMF for financial assistance when they are having trouble finding buyers for their debt and when the value of their money is falling. The Stiglitzian prescription is to raise the profile of fiscal deficits, that is, to issue *more* debt and to print *more* money. You seem to believe that if a distressed government issues more currency, its citizens will suddenly think it more valuable. You seem to believe that when investors are no longer willing to hold a government's debt, all that needs to be done is to increase the supply and it will sell like hot cakes. We at the IMF—no, make that we on the Planet Earth—have considerable experience suggesting otherwise. We earthlings have found that when a country in fiscal distress tries to escape by printing more money, inflation rises, often uncontrollably. Uncontrolled inflation strangles growth, hurting the entire populace but, especially the indigent. The laws of economics may be different in your part of the gamma quadrant, but around here we find that when an almost bankrupt government fails to credibly constrain the time profile of its fiscal deficits, things generally get worse instead of better.

Joe, throughout your book, you condemn the IMF because everywhere it seems to be, countries are in trouble. Isn't this a little like observing that where there are epidemics, one tends to find more doctors?

You cloak yourself in the mantle of [the English economist] John Maynard Keynes [1883–1946], saying that the aim of your policies is to maintain full employment. We at the IMF care a lot about employment. But if a government has come to us, it is often precisely because it is in an unsustainable position, and we have to look not just at the next two weeks, but at the next two years and beyond. We certainly believe in the lessons of Keynes, but in a modern, nuanced way. For example, the post-1975 macroeconomics literature—which you say we are tone deaf to—emphasizes the importance of budget constraints across time. It does no good to pile on IMF debt as a very short-run fix if it makes the not-so-distant future drastically worse. By the way, in blatant contradiction to your assertion, IMF programs frequently allow for deficits, indeed they did so in the Asia crisis. If its initial battlefield medicine was wrong, the IMF reacted, learning from its mistakes, quickly reversing course.

No, instead of Keynes, I would cloak your theories in the mantle of Arthur Laffer and other extreme expositors of 1980s Reagan-style supply-side economics. Laffer believed that if the government would only cut tax rates, people would work harder, and total government revenues would rise. The Stiglitz-Laffer theory of crisis management holds that countries need not worry about expanding deficits, as in so doing, they will increase their debt service capacity more than proportionately. George Bush, Sr. once labeled these ideas "voodoo economics." He was right. I will concede, Joe, that real-world policy economics is complicated, and just maybe further research will prove you have a point. But what really puzzles me is how you could be *so* sure that you are 100 percent right, so sure that you were willing to "blow the whistle" in the middle of the crisis, sniping at the paramedics as they tended the wounded. Joe, the academic papers now coming out in top journals are increasingly supporting the interest defense policies of former First Deputy Managing Director Stan Fischer and the IMF that you, from your position at the World Bank, ignominiously sabotaged. Do you ever think that just maybe, Joe Stiglitz might have screwed up? That, just maybe, you were part of the problem and not part of the solution?

You say that the IMF is tone deaf and never listens to its critics. I know that is not true, because in my academic years, I was one of dozens of critics that the IMF bent over backwards to listen to. For example, during the 1980s, I was writing then-heretical papers on the moral hazard problem in IMF/World Bank lending, an issue that was echoed a decade later in the Meltzer report. Did the IMF shut out my views as potentially subversive to its interests? No, the IMF insisted on publishing my work in its flagship research publication *Staff Papers*. Later, in the 1990s, Stan Fischer twice invited me to discuss my views on fixed exchange rates and open capital markets (I warned of severe risks). In the end, Stan and I didn't agree on everything, but I will say that having entered his office 99 percent sure that I was right, I left somewhat humbled by the complexities of price stabilization in high-inflation countries. If only you had crossed over 19th Street from the Bank to the Fund a little more often, Joe, maybe things would have turned out differently.

I don't have time here to do justice to some of your other offbeat policy prescriptions, but let me say this about the transition countries. You accuse the IMF of having "lost Russia." Your analysis of the transition in Russia reads like a paper in which a theorist abstracts from all the major problems, and focuses only on the couple he can handle. You neglect entirely the fact that when the IMF entered Russia, the country was not only in the middle of an economic crisis, it was in the middle of a social and political crisis as well.

Throughout your book, you betray an unrelenting belief in the pervasiveness of market failures, and a staunch conviction that governments can and will make things better. You call us "market fundamentalists." We do not believe that markets are always perfect, as you accuse. But we do believe there are many instances of government failure as well and that, on the whole, government failure is a far bigger problem than market failure in the developing world. Both World Bank President Jim Wolfensohn and IMF Managing Director Horst Köhler have frequently pointed to the fundamental importance of governance and institutions in development. Again, your alternative medicines, involving ever-more government intervention, are highly dubious in many real-world settings.

I haven't had time, Joe, to check all the facts in your book, but I do have some doubts. On page 112, you have Larry Summers (then Deputy U.S. Treasury Secretary) giving a "verbal" tongue lashing to former World Bank Vice-President Jean-Michel Severino. But, Joe, these two have never met. How many conversations do you report that never happened? You give an example where an IMF Staff report was issued prior to the country visit. Joe, this isn't done; I'd like to see your documentation. On page 208, you slander former IMF number two, Stan Fischer, implying that Citibank may have dangled a job offer in front of him in return for his cooperation in debt renegotiations. Joe, Stan Fischer is well known to be a person of unimpeachable integrity. Of all the false inferences and innuendos in this book, this is the most outrageous. I'd suggest you should pull this book off the shelves until this slander is corrected.

Joe, as an academic, you are a towering genius. Like your fellow Nobel Prize winner, John Nash, you have a "beautiful mind." As a policymaker, however, you were just a bit less impressive.

Other than that, I thought it was a pretty good book.

POSTSCRIPT

Do International Financial Organizations Require Radical Reform?

When Stiglitz, a former ranking official in the World Bank, joined the critics of big international financial organizations, it created quite a stir. There are entire Web pages devoted to his views, such as the Global Policy Forum page http://www.globalpolicy.org/socecon/bwi-wto/wbank/stigindx.htm#Key% 20Documents. It is interesting to compare the comments that he made after leaving the World Bank with statements that he made while serving as its senior vice president and chief economist. This commentary is available on the World Bank Group Web site http://www.worldbank.org/knowledge/ chiefecon/stiglitz.htm. Also see Stiglitz's book *Globalization and Its Discontents* (W. W. Norton, 2002). For a less strident view of the World Bank and the IMF, see Michael Fabricius, *Merry Sisterhood or Guarded Watchfulness: The Cooperation Between the IMF and the World Bank* (Institute for International Economics, 2003).

Control is an issue here. Opponents of globalization worry that their political control, including sovereignty for countries, is weakening. A related topic in the case of the IMF involves what some analysts see as racial imbalance, intended or not. The cause is the IMF voting formula, which gives majority power to the small group of countries whose people are predominantly of European heritage while effectively disenfranchising the much larger group of countries peopled largely by Asians, Africans, Latin Americans, and other people of color. For more on the issue of control in an era of globalization, read Strom C. Thacker, "The High Politics of IMF Lending," *World Politics* (October 1999).

The IMF's demand that recipient countries make structural adjustments also faces a rising tide of criticism. None of this particularly persuaded Michael Camdessus, director of the IMF from 1977 to 2000, to relent on this requirement. "The stronger the program [of capitalist reform], the stronger the financing will be," he asserted at one point. His successor, Horst Köhler, former president of the European Bank for Reconstruction and Development, has taken a more moderate approach to conditionality, although critics charge that it is more rhetoric than substance. For a critical look at the IMF, see James R. Vreeland, *The IMF and Economic Development* (Cambridge University Press, 2003).

Finally, it should be noted that yet another charge has been leveled against the IMF, this time from a more conservative perspective. That complaint is that by constantly bailing out financially distressed countries, the IMF too often encourages unsustainable behavior. For this view, see Lawrence J. McQuillan, *The Case Against the International Monetary Fund* (Hoover Institution Press, 1999).

ISSUE 18

Should the United States Ratify the International Criminal Court Treaty?

YES: Lawyers Committee for Human Rights, from Statement Before the Committee on International Relations, U.S. House of Representatives (July 25, 2000)

NO: John R. Bolton, from Statement Before the Committee on International Relations, U.S. House of Representatives (July 25, 2000)

ISSUE SUMMARY

YES: The Lawyers Committee for Human Rights, in a statement submitted to the U.S. Congress, contends that the International Criminal Court (ICC) is an expression, in institutional form, of a global aspiration for justice.

NO: John R. Bolton, senior vice president of the American Enterprise Institute in Washington, D.C., contends that support for an international criminal court is based largely on naive emotion and that adhering to its provisions is not wise.

$\mathbf{H}$istorically, international law has focused primarily on the actions of and relations between states. More recently, the status and actions of individuals have become increasingly subject to international law.

The first significant step in this direction was evident in the Nuremberg and Tokyo war crimes trials after World War II. In these panels, prosecutors and judges from the victorious powers prosecuted and tried German and Japanese military and civilian leaders for waging aggressive war, for war crimes, and for crimes against humanity. Most of the accused were convicted; some were executed. There were no subsequent war crimes tribunals through the 1980s and into the mid-1990s. Then, however, separate international judicial tribunals' processes were established to deal with the Holocaust-like events in Bosnia and the genocidal massacres in Rwanda.

The 11-judge tribunal for the Balkans sits in The Hague, the Netherlands. The 6-judge Rwanda tribunal is located in Arusha, Tanzania. These tribunals have indicted numerous people for war crimes and have convicted and imprisoned a few of them. These actions have been applauded by those who believe that individuals should not escape punishment for crimes against humanity that they commit or order. But advocates of increased and forceful application of international law also feel that ad hoc tribunals are not enough.

Such advocates are convinced that the next step is the establishment of a permanent International Criminal Court (ICC) to prosecute and try individuals for war crimes and other crimes against humanity. The move for an ICC was given particular impetus when President Bill Clinton proposed just such a court in 1995. Just a year later, the United Nations convened a conference to lay out a blueprint for the ICC. Preliminary work led to the convening of a final conference in June 1998 to settle the details of the ICC. Delegates from most of the world's countries met in Rome, where their deliberations were watched and commented on by representatives of 236 nongovernmental organizations (NGOs). The negotiations were far from smooth. A block of about 50 countries informally led by Canada, which came to be known as the "like-minded group," favored establishing a court with broad and independent jurisdiction.

Other countries wanted to narrowly define the court's jurisdiction and to allow it to conduct only prosecutions that were referred to it by the UN Security Council (UNSC). The hesitant countries also wanted the court to be able to prosecute individuals only with the permission of the accused's home government, and they wanted the right to file treaty reservations exempting their citizens from prosecution in some circumstances. Somewhat ironically, given that President Clinton had been the impetus behind the launching of a conference to create the ICC, the United States was one of the principal countries favoring a highly restricted court. U.S. reluctance to support an expansive definition of the ICC's jurisdiction and independence rested on two concerns. One was the fear that U.S. personnel would be especially likely targets of politically motivated prosecutions. The second factor that gave the Clinton administration pause was the requirement that the Senate ratify the treaty. Senate Foreign Relations Committee chairman Jesse Helms has proclaimed that any treaty that gave the UN "a trapping of sovereignty" would be "dead on arrival" in the Senate.

In the following selections, the Lawyers Committee for Human Rights and John R. Bolton present their markedly differing views of the wisdom of founding an ICC. Both analyses agree that, if it works the way that it is intended to, the International Criminal Court will have a profound impact. Where they differ is on whether that impact will be positive or negative.

**Lawyers Committee
for Human Rights**

 **YES**

Statement of the Lawyers Committee
for Human Rights

The United States has compelling reasons to remain open to eventual co-operation with the International Criminal Court (ICC). United States interests may dictate such cooperation, even while the U.S. remains a non-party to the Rome Statute. . . . The following paper describes the U.S. interest in supporting the ICC.

> *I. A strong and independent International Criminal Court serves important na-tional interests of the United States.*

At the end of World War Two, with much of Europe in ashes, some allied leaders urged that the leaders of the defeated Third Reich be summarily executed. The United States disagreed. U.S. leaders insisted that a larger and more valuable contribution to the peace could be made if the Nazis were individually charged and tried for violations of international law. The International Criminal Court is an expression, in institutional form, of an aspiration for justice with which the United States had been deeply identified ever since World War Two. It was created to advance objectives that are totally consistent with the long-term U.S. national interest in a peaceful, stable, democratic and integrated global system. And the Rome Treaty, in its final form, promised to advance that interest in the following ways:

- *First,* the treaty embodies deeply held American values. The establish-ment of the Court responds to the moral imperative of halting crimes that are an offense to our common humanity. The ICC promises to pro-mote respect for human rights; advance the rule of law around the world, both domestically and internationally; reinforce the indepen-dence and effectiveness of national courts; and uphold the principle of equal accountability to international norms.

From U.S. House of Representatives. Committee on International Relations. *The International Crim-inal Court: A Threat to American Military Personnel?* Hearing, July 25, 2000. Washington, D.C.: U.S. Government Printing Office, 2000.

- *Second,* the ICC will help to deter future gross violations. It will not halt them completely, of course. But over time, its proceedings will cause prospective violators to think twice about the likelihood that they will face prosecution. This deterrent effect is already apparent in the former Yugoslavia. Even though leading architects of ethnic cleansing, such as Radovan Karadzic and Ratko Mladic, have not been brought to trial, their indictment has limited their ability to act and has allowed more moderate political forces to emerge, reducing the risk to U.S. and other international peacekeepers still in Bosnia.

- *Third,* through this deterrent effect the ICC will contribute to a more stable and peaceful international order, and thus directly advance U.S. security interests. This is already true of the Yugoslav Tribunal, but it will be much more true of the ICC, because of its broader jurisdiction, its ability to respond to Security Council referrals, and the perception of its impartiality. The court will promote the U.S. interest in the preventing [of] regional conflicts that sap diplomatic energies and drain resources in the form of humanitarian relief and peacekeeping operations. Massive human rights violations almost always have larger ramifications in terms of international security and stability. These include widening armed conflict, refugee flows, international arms and drug trafficking, and other forms of organized crime, all of which involve both direct and indirect costs for the United States.

- *Fourth,* the ICC will reaffirm the importance of international law, including those laws that protect Americans overseas. For many people in the United States, "international law" is seen either as a utopian abstraction, or an unwelcome intrusion into our sovereign affairs. But as Abram Chayes, former Department of State Legal Adviser, remarked shortly before his death [in 2000], there is nothing utopian about international law in today's world. On the contrary, it is a matter of "hardheaded realism." Many nations who voted for the Rome Treaty had similar misgivings about its potential impact on their sovereignty. But they recognized that this kind of trade-off is the necessary price of securing a rule-based international order in the 21st century. France, for example, which participates extensively in international peacekeeping operations, made this calculation, joined the consensus in Rome and . . . ratified the treaty. The United States, likewise, should see the ICC as an integral part of an expanding international legal framework that also includes rules to stimulate and regulate the global economy, protect the environment, control the proliferation of weapons of mass destruction, and curb international criminal activity. The United States has long been a leading exponent, and will be a prime beneficiary, of this growing international system of cooperation.

II. The risks posed by the ICC to U.S. servicemen and officials are negligible in comparison to the benefits of the Court to United States' interests.

In assessing the U.S. government's concerns, it is important to bear in mind some basic threshold considerations about the ICC. Most fundamentally, it will be a court of last resort. It will have a narrow jurisdiction, and is intended to deal with only the most heinous crimes. The ICC will step in only where states are unwilling or unable to dispense justice. Indeed, that is its entire purpose: to ensure that the worst criminals do not go free to create further havoc just because their country of origin does not have a functioning legal system. The Court was designed with situations like Rwanda and Cambodia and Sierra Leone in mind, not to supplant sophisticated legal systems like those of the United States. Furthermore, there are strict guidelines for the selection of ICC judges and prosecutors, as well as a set of internal checks and balances, that meet or exceed the highest existing international standards. The legal professionals who staff the Court will not waste their time in the pursuit of frivolous cases.

Second, the Court will only deal with genocide, war crimes and crimes against humanity, all of which are subject to a jurisdiction narrower than that available to domestic courts under international law. It will not be concerned with allegations of isolated atrocities, but only with the most egregious, planned and large-scale crimes.

Could a member of the U.S. armed forces face credible allegations of crimes of this magnitude? Genocide would seem to be out of the question. War crimes and crimes against humanity are more conceivable. The My Lai massacre in Vietnam revealed the bitter truth that evil knows no nationality: American soldiers can sometimes be capable of serious crimes. If such a crime were committed today, it would appear self-evident that the U.S. military justice system would investigate and prosecute the perpetrators, as it did at My Lai, whether or not an ICC existed. And if it were an isolated act, not committed in pursuit of a systematic plan or policy, it would not meet the threshold for ICC concern in any case.

Benign support by the United States for the ICC as a non-party to the Treaty would reaffirm the standing U.S. commitment to uphold the laws of war and could be offered in the knowledge that the Court would defer to the U.S. military justice system to carry out a good faith investigation in the unlikely event that an alleged crime by an American was brought to its attention. The marginal risk that is involved could then simply be treated as part of the ordinary calculus of conducting military operations, on a par with the risk of incurring casualties or the restraints imposed by the laws of war. The preparation and conduct of military action is all about risk assessment, and the marginal risk of exposure to ICC jurisdiction is far outweighed by the benefits of the Court for U.S. foreign policy.

> *III. The ICC provides an opportunity for the United States to reaffirm its leadership on the issue of international justice, which for so long has been a central goal of U.S. policy.*

We urge the United States to develop a long-term view of the benefits of the ICC. Such an approach would open the door to cooperation with the Court as a non-state party, and eventually to full U.S. participation. This policy shift should be based on the following five premises:

- **The creation of new international institutions requires concessions from all the participants.** As an international agreement, the Rome Statute bears the marks of many concessions to sovereign states—not least the United States. As such, the ICC will have a twofold virtue: it will be imbued with the flexibility of an international institution as well as with the rigor of a domestic criminal court. The risks involved in supporting the present ICC Treaty are more than outweighed by the expansion of an international legal framework that is congenial to U.S. interests and values.

- **The risks of U.S. exposure to ICC jurisdiction are in fact extremely limited, as a result of the extensive safeguards that are built into the Rome Treaty.** Those safeguards are there in large part because the United States insisted on their inclusion. The modest risks that remain can never be fully eliminated without compromising the core principles established at Nuremberg and undermining the basic effectiveness of an institution that can do much to advance U.S. interests. The best way to minimize any residual risk is to remain engaged with others in helping to shape the Court. The risks, in fact, will only be aggravated if the United States decides to withdraw from the ICC process. Joining the ICC, on the other hand, would allow the United States to help nominate, select and dismiss its judges and prosecutors, and so ensure that it operates to the highest standards of professional integrity. More broadly, the ICC's Assembly of States Parties would provide an ideal setting for the United States to demonstrate its leadership in the fight against impunity for the worst criminals.

- **The Pentagon's views, while important, should be balanced among other U.S. policy interests in reference to the ICC.** The U.S. military has an institutional interest in retaining the maximum degree of flexibility in its operational decisions. But this must be put in proper perspective by civilian authorities as they weigh the pros and cons of the ICC. Legislators and others who have so far remained on the sidelines of the ICC debate will have an important part to play in helping the Administration develop a broader approach to the ICC, one that puts long-term stewardship of the national interest into its proper perspective.

- **U.S. leadership requires working in close cooperation with our allies around the world.** It is tempting to believe that U.S. economic and military supremacy is now so absolute that the United States can go it alone and impose its will on the rest of the world. But the evolution of the ICC is a reminder that this kind of unilateralism is not possible in today's more complex world. The United States has tried to impose its will on the ICC negotiations, and it has failed. In its repeated efforts to find a "fix," the United States has succeeded only in painting itself into a corner. Worse, it has disregarded one of the cardinal rules of diplomacy, which is never to commit all your resources to an outcome that is unattainable. Unable to offer credible carrots, decisive sticks, or viable legal arguments, the United States finds itself on what one scholar has called a "lonely legal ledge," able neither to advance nor to retreat. Asking for concessions it cannot win, in a process it can neither leave nor realistically oppose, the United States has so far resisted coming to terms with the limits of its ability to control the ICC process.

- **The costs of opposition to the Court are too high and would significantly damage the U.S. national interest.** Once the ICC is up and running, it seems highly unlikely that the United States would refuse to support the principle of accountability for the worst international crimes simply because the Court was the only viable means of upholding that principle. It is far more likely that a future U.S. administration will see the advantage in supporting the Court, if only as a matter of raw political calculus. Opposition to a functioning Court would undermine faith in a world based on justice and the rule of law and would shake one of the foundation stones on which the legitimacy of U.S. global leadership has rested since World War Two.

For the last half century, U.S. foreign policy has sought to balance military strength with the nurturing of an international system of cooperation based on democracy and the rule of law. It would be a serious mistake to imagine that victory in the Cold War means that the institutional part of this equation can now be abandoned, and that ad hoc applications of force should prevail over the consistent application of law.

NO ⤶

John R. Bolton

Statement of John R. Bolton

Unfortunately, support for the ICC [International Criminal Court] concept is based largely on emotional appeals to an abstract ideal of an international judicial system, unsupported by any meaningful evidence, and running contrary to sound principles of international crisis resolution. Moreover, for some, faith in the ICC rests largely on an unstated agenda of creating ever-more-comprehensive international structures to bind nation states in general, and one nation state in particular. Regrettably, the Clinton Administration's naïve support for the concept of an ICC . . . left the U.S. in a worse position internationally than if we had simply declared our principled opposition in the first place.

Many people have been led astray by analogizing the ICC to the Nuremberg trials, and the mistaken notion that the ICC traces its intellectual lineage from those efforts. However, examining what actually happened at Nuremberg easily disproves this analysis, and demonstrates why the ICC as presently conceived can never perform effectively in the real world. Nuremberg occurred after complete and unambiguous military victories by allies who shared juridical and political norms, and a common vision for reconstructing the defeated Axis powers as democracies. The trials were intended as part of an overall process, at the conclusion of which the defeated states would acknowledge that the trials were prerequisites for their readmission to civilized circles. They were not just political "score settling," or continuing the war by other means. Moreover, the Nuremberg trials were effectively and honorably conducted. Just stating these circumstances shows how different was Nuremberg from so many contemporary circumstances, where not only is the military result ambiguous, but so is the political and where war crimes trials are seen simply as extensions of the military and political struggles under judicial cover.

Many ICC supporters believe simply that if you abhor genocide, war crimes and crimes against humanity, you should support the ICC. This logic is flatly wrong for three compelling reasons.

First, all available historical evidence demonstrates that the Court and the Prosecutor will not achieve their central goal—the deterrence of heinous crimes —because they do not (and should not) have sufficient authority in the real world. Beneath the optimistic rhetoric of the ICC's proponents, there is not a

From U.S. House of Representatives. Committee on International Relations. *The American Service-members' Protection Act of 2000.* Hearing, July 25, 2000. Washington, D.C.: U.S. Government Printing Office, 2000. Notes omitted.

shred of evidence to support their deterrence theories. Instead, it is simply a near-religious article of faith. Rarely, if ever, has so sweeping a proposal for restructuring international life had so little empirical evidence to support it. Once ICC advocate said in Rome that: "the certainty of punishment can be a powerful deterrent." I think that statement is correct, but, unfortunately, it has little or nothing to do with the ICC.

In many respects, the ICC's advocates fundamentally confuse the appropriate role of political and economic power; diplomatic efforts; military force and legal procedures. No one disputes that the barbarous actions under discussion are unacceptable to civilized peoples. The real issue is how and when to deal with these acts, and this is not simply, or even primarily, a legal exercise. The ICC's advocates make a fundamental error by trying to transform matters of international power and force into matters of law. Misunderstanding the appropriate roles of force, diplomacy and power in the world is not just bad analysis, but bad and potentially dangerous policy for the United States.

Recent history is unfortunately rife with cases where strong military force or the threat of force failed to deter aggression or gross abuses of human rights. Why we should believe that bewigged judges in The Hague will prevent what cold steel has failed to prevent remains entirely unexplained. Deterrence ultimately depends on perceived effectiveness, and the ICC is most unlikely to be that. In cases like Rwanda, where the West declined to intervene as crimes against humanity were occurring, why would the mere possibility of distant legal action deter a potential perpetrator? . . .

Moreover, the actual operations of the existing Yugoslav and Rwanda ("ICTR") tribunals have not been free from criticism, criticism that foretells in significant ways how an ICC might actually operate. A UN experts' study (known as the "Ackerman Report," after its chairman) noted considerable room for improvement in the work of the tribunals. . . .

[For example], ICC opponents have warned that it will be subjected to intense political pressures by parties to disputes seeking to use the tribunal to achieve their own non-judicial objectives, such as score-settling and gaining advantage in subsequent phases of the conflict. The Ackerman Report, in discussing the ICTY's quandary about whether to pursue "leadership" cases or low-level suspects, points out precisely how such political pressures work, and their consequences: "[u]navoidable early political pressures on the Office of the Prosecutor to act against perpetrators of war crimes . . . led to the first trials beginning in 1995 against relatively minor figures. And while important developments . . . have resulted from these cases, the cost has been high. Years have elapsed and not all of the cases have been completed." In short, political pressures on the Tribunals, to which they respond, are not phantom threats, but real. . . .

Second, the ICC's advocates mistakenly believe that the international search for "justice" is everywhere and always consistent with the attainable political resolution of serious political and military disputes, whether between or within states, and the reconciliation of hostile neighbors. In the real world, as opposed to theory, justice and reconciliation may be consistent—or they may not be. Our recent experiences in situations as diverse as Bosnia, Rwanda, South

Africa, Cambodia and Iraq argue in favor of a case-by-case approach rather than the artificially imposed uniformity of the ICC.

For example, an important alternative is South Africa's Truth and Reconciliation Commission. After apartheid, the new government faced the difficulty of establishing truly democratic institutions, and dealing with earlier crimes. One option was certainly widespread prosecutions against those who committed human rights abuses. Instead, the new government decided to establish the Commission to deal with prior unlawful acts. Those who had committed human rights abuses may come before the Commission and confess their past misdeeds, and if fully truthful, can, in effect, receive pardons from prosecution.

I do not argue that the South African approach should be followed everywhere, or even necessarily that it is the correct solution for South Africa. But it is certainly a radically different approach from the regime envisioned by the ICC. . . .

Efforts to minimize or override the nation state through "international law" have found further expression in the expansive elaboration of the doctrine of "universal jurisdiction." Until recently an obscure, theoretical creature in the academic domain, the doctrine gained enormous public exposure (although very little scrutiny) during the efforts [of a Spanish judge] to extradite General Augusto Pinochet of Chile from the United Kingdom [to Spain].

Even defining "universal jurisdiction" is not easy because the idea is evolving so rapidly. . . . The idea was first associated with pirates. . . . Because pirates were beyond the control of any state and thus not subject to any existing criminal justice system, the idea developed that it was legitimate for any aggrieved party to deal with them. Such "jurisdiction" could be said to be "universal" because the crime of piracy was of concern to everyone, and because such jurisdiction did not comport with more traditional jurisdictional bases, such as territoriality or nationality. In a sense, the state that prosecuted pirates could be seen as vindicating the common interest of all states. (Slave trading is also frequently considered to be the subject of universal jurisdiction, following similar reasoning.)

. . . [This sense of universal jurisdiction] is a far cry from what "human rights" activists, NGOs [nongovernmental organizations], and academics . . . have in mind today. From a very narrow foundation, theorists have enlarged the concept of universal jurisdiction to cover far more activities, with far less historical or legal support, than arose earlier in the context of piracy. At the same time, they have omitted reference to the use of force, and substituted their preferred criminal prosecution. The proscribed roster of offenses now typically includes genocide, torture, war crimes and crimes against humanity, which are said to vest prosecutorial jurisdiction in all states.

Announcing his decision in the November 25, 1998 decision of *Ex parte Pinochet*, Lord Nicholls described the crimes of which the General stood accused by saying, "International law has made it plain that certain types of conduct . . . are not acceptable conduct on the part of anyone." Although that decision . . . did not actually rest on universal jurisdiction, Lord Nicholls in fact stated the doctrine's essential foundation.

The worst problem with universal jurisdiction is not its diaphanous legal footings but its fundamental inappropriateness in the realm of foreign policy. In effect (and in intention), the NGOs and theoreticians advocating the concept are misapplying legal forms in political or military contexts. What constitutes "crimes against humanity" and whether they should be prosecuted or otherwise handled—and by whom—are not questions to be left to lawyers and judges. To deal with them as such is, ironically, so bloodless as to divorce these crimes from reality. It is not merely naïve, but potentially dangerous, as Pinochet's case demonstrates.

Morally and politically, what Pinochet's regime did or did not do is primarily a question for Chile to resolve. Most assuredly, Pinochet is not, unlike a pirate or a slave trader, beyond the control of any state. Although many people around the world intensely dislike the solution that Chile adopted in order to restore constitutional and democratic rule in 1990, especially the various provisions for amnesty, the terms and implementation of that deal should be left to the Chileans themselves. They (and their democratically elected government) may continue to honor the deal, or they may choose to bring their own judicial proceedings against Pinochet. One may accept or reject the wisdom or morality of either course (and I would argue that they should uphold the deal), but it should be indisputable that the decision is principally theirs to make. The idea that Spain or any other country that subsequently filed extradition requests in the United Kingdom has an interest superior to that of Chile—and can thus effectively overturn the Chilean deal—is untenable. And yet, if the British had ultimately extradited Pinochet to Spain, that is exactly what would have happened. A Spanish magistrate operating completely outside the Chilean system will effectively have imposed his will on the Chilean people. One is sorely tempted to ask: Who elected him? If that is what "universal jurisdiction" means in practice (as opposed to the theoretical world of law reviews), it is hopelessly flawed.*

Spain *does* have a legitimate interest in justice on behalf of Spanish citizens who may have been held hostage, tortured, or murdered by the Pinochet regime. And the Spanish government may take whatever steps it ultimately considers to be in the best interest of Spanish citizens, but its recourse lies with the government of Chile, and certainly not with that of the United Kingdom. . . .

Because of the substantial publicity surrounding the Pinochet matter, we can expect copycat efforts covering a range of other "crimes against humanity" in the near future. But adding purported crimes (shocking though they may be) to the list of what triggers universal jurisdiction does not make the concept any more real. Nor does a flurry of law review articles (and there has been far more than a flurry) make concrete an abstract speculation. In fact, "universal jurisdiction" is conceptually circular: universal jurisdiction covers the most dastardly offenses; accordingly, if the offense is dastardly, there must be universal jurisdiction to prosecute it. Precisely because of this circularity, there is absolutely

---

*Pinochet was ultimately returned to Chile, where he was charged with crimes and awaits trial. —Ed.

no limit to what creative imaginations can enlarge it to cover, and we can be sure that they are already hard at work. . . .

Third, tangible American interests are at risk. I believe that the ICC's most likely future is that it will be weak and ineffective, and eventually ignored, because [it was] naïvely conceived and executed. There is, of course, another possibility: that the Court and the Prosecutor (either as established now, or as potentially enhanced) will be strong and effective. In that case, the U.S. may face a much more serious danger to our interests, if not immediately, then in the long run.

Although everyone commonly refers to the "Court" created at the 1998 Rome Conference, what the Conference actually did was to create not just a Court, but also a powerful and unaccountable piece of an "executive" branch: the Prosecutor. Let there be no mistake: our main concern from the U.S. perspective is not that the Prosecutor will indict the occasional U.S. soldier who violates our own laws and values, and his or her military training and doctrine, by allegedly committing a war crime. Our main concern should be for the President, the Cabinet officers on the National Security Council, and other senior leaders responsible for our defense and foreign policy. They are the real potential targets of the ICC's politically unaccountable Prosecutor.

One problem is the crisis of legitimacy we face now in international organizations dealing with human rights and legal norms. Their record is, to say the least, not encouraging. The International Court of Justice and the UN Human Rights Commission are held in *very* low esteem, and not just in the U.S. ICC supporters deliberately chose to establish it independently of the ICJ to avoid its baggage.

Next is the overwhelming repudiation by the Rome Conference of the American position supporting even a minimal role for the Security Council. Alone among UN governing bodies, the Security Council does enjoy a significant level of legitimacy in America. And yet it was precisely the Council where the U.S. found the greatest resistance to its position. The Council has primacy in the UN for "international peace and security," in all their manifestations, and it is now passing strange that the Council and the ICC are to operate virtually independently of one another. The implicit weakening of the Security Council is a fundamental *new* problem created by the ICC, and an important reason why the ICC should be rejected. The Council now risks both having the ICC interfering in its ongoing work, and even more confusion among the appropriate roles of law, politics and power in settling international disputes.

The ICC has its own problems of legitimacy. Its components do not fit into a coherent international structure that clearly delineates how laws are made, adjudicated and enforced, subject to popular accountability, and structured to protect liberty. Just being "out there" in the international system is unacceptable, and, indeed, almost irrational unless one understands the hidden agenda of many NGOs supporting the ICC. There is real vagueness over the ICC's substantive jurisdiction, although one thing is emphatically clear: this is *not* a court of limited jurisdiction. . . .

Examples of vagueness in key elements of the Statute's text include:

- "Genocide," as defined by the Rome Conference is inconsistent with the Senate reservations attached to the underlying Genocide Convention, and the Rome Statute is not subject to reservations.
- "War crimes" have enormous definitional problems concerning civilian targets. Would the United States, for example, have been guilty of "war crimes" for its WWII bombing campaigns, and use of atomic weapons, under the Rome Statute?
- What does the Statute mean by phrases like "knowledge" of "incidental loss of life or injury to civilians"? "long-term and severe damage to the natural environment"? "clearly excessive" damage?

Apart from problems with existing provisions, and the uncertain development of customary international law, there are many other "crimes" on the waiting list: aggression, terrorism, embargoes (courtesy of Cuba), drug trafficking, etc. The Court's potential jurisdiction is enormous. Article 119 provides: "any dispute concerning the judicial functions of the Court shall be settled by the decision of the Court."

Consider one recent example of the use of force, the NATO air campaign over former Yugoslavia. Although most Americans did not question the international "legality" of NATO's actions, that view was not uniformly held elsewhere. During the NATO air war, Secretary General Kofi Annan expressed the predominant view that "unless the Security Council is restored to its pre-eminent position as the sole source of legitimacy on the use of force, we are on a dangerous path to anarchy." . . .

Implicitly, therefore, in Annan's view, NATO's failure to obtain Council authorization made its actions illegitimate, which is what those pursuing the hidden agenda want to hear: while one cannot stop the United States from using force because it is so big and powerful, one can ensure that it is illegitimate absent Security Council authorization, *and thus a possible target of action by the ICC Prosecutor.* . . .

Many hope to change [U.S. military] behavior as much as the international "rules" themselves, through the threat of prosecution. They seek to constrain military options, and thus lower the potential effectiveness of such actions, or raise the costs to successively more unacceptable levels by increasing the legal risks and liabilities perceived by top American and allied civilian and military planners undertaking military action.

. . . Amnesty [International] asserted . . . that "NATO forces violated the laws of war leading to cases of unlawful killing of civilians." The NGO complained loudly about NATO attacks on a "civilian" television transmitter in Belgrade, even though it served the Milosevic regime's propaganda purposes. Similarly, Human Rights Watch . . . concluded that NATO violated international law, but stopped short of labeling its actions as "war crimes." In another recent report, this NGO announced its opposition to the sale of American air-to-ground missiles to Israel because of the Israeli "war crime" of attacking

Lebanese electrical power stations. Of course, much the same could also be said about American air attacks during the Persian Gulf War, aimed at destroying critical communications and transportation infrastructure inside Iraq, in order to deny it to Saddam's military. If these targets are now "off limits," the American military will be far weaker than it would otherwise be. . . .

What to do next is obviously the critical question. Whether the ICC survives and flourishes depends in large measure on the United States. We should not allow this act of sentimentality masquerading as policy to achieve indirectly what was rejected in Rome. We should oppose any suggestion that we cooperate, help fund, and generally support the work of the Court and Prosecutor. We should isolate and ignore the ICC.

Specifically, I have long proposed for the United States a policy of "Three Noes" toward the ICC: (1) no financial support, directly or indirectly; (2) no collaboration; and (3) no further negotiations with other governments to "improve" the Statute. . . . This approach is likely to maximize the chances that the ICC will wither and collapse, which should be our objective. The ICC is a fundamentally bad idea. It cannot be improved by technical fixes as the years pass. . . . We have alternative approaches and methods consistent with American national interests, as I have previously outlined, and we should follow them.

POSTSCRIPT

Should the United States Ratify the International Criminal Court Treaty?

After the statements by the Lawyers Committee for Human Rights and Bolton were given, the prospects for U.S. adherence to the ICC treaty dimmed even further. In the waning days of his administration, President Clinton directed a State Department representative to sign the ICC treaty on behalf of the United States. That act was mostly symbolic, however, because at the same time Clinton warned that the "United States should have the chance to observe and assess the functioning of the Court, over time, before choosing to become subject to its jurisdiction. Given these concerns, I will not, and do not recommend that my successor submit the Treaty to the Senate for advice and consent until our fundamental concerns are satisfied." In fact, Clinton's cautious advice to the next president was superfluous because the election of George W. Bush in 2000 brought into the Oval Office a president who agreed fully with the opponents of the ICC. In May 2002 the administration announced that it would not even continue to work toward a revision of the treaty that would satisfy Washington. "It's over. We're washing our hands of it," Pierre-Richard Prosper, the State Department's ambassador-at-large for war-crimes issues, said of the ICC.

Despite the U.S. stance, the ICC became a reality on April 12, 2002, when the number of countries ratifying the treaty passed 60, the minimum required for the ICC treaty to take effect. "The long-held dream of a permanent international criminal court will now be realized," Secretary-General Kofi Annan of the United Nations proclaimed. By March 2003 the number of accessions to the ICC treaty stood at 89, with ratification pending in another 50 signatory countries. More on the ICC can be found in William A. Schabas, *An Introduction to the International Criminal Court* (Cambridge University Press, 2001) and Bruce Broomhall, *International Criminal Justice and the International Criminal Court: Between State Consent and the Rule of Law* (Oxford University Press, 2003).

Some of the basic provisions of the ICC are the following:

1. The court's jurisdiction includes genocide and a range of other crimes committed during international and internal wars. Such crimes must be "widespread and systematic" and committed as part of "state, organization, or group policy," not just as individual acts.

2. Except for genocide and complaints brought by the United Nations Security Council (UNSC), the ICC will not be able to prosecute alleged crimes unless either the state of nationality of the accused or the state where the crimes took place has ratified the treaty.

3. Original signatories will have a one-time ability to "opt out" of the court's jurisdiction for war crimes, but not genocide, for a period of seven years.
4. The UNSC can delay one prosecution for one year. The vote to delay will not be subject to veto.
5. The ICC will only be able to try cases when national courts have failed to work.

The irony of the U.S. opposition to the ICC is that the ICC treaty is, in part, an American product. The United States was a major force behind the convening of a conference to draft a treaty, with President Clinton issuing a clarion call during one commencement address for a permanent court that could try and punish those who committed war crimes and other abominations. Once at the conference in Rome, however, the United States retreated from Clinton's rhetorical position and sought to eliminate virtually any possibility that an American civilian or military leader would ever stand before the ICC's bar of justice. For more on the U.S. attitude toward the ICC, see Sarah Sewall and Carl Kaysen, eds., *The United States and the International Criminal Court* (Rowman & Littlefield, 2000).

Technically, U.S. ratification of the ICC treaty and support of the ICC once it begins is not necessary for the court to function. But, in reality, the United States is the world's hegemonic power, and U.S. opposition to the court will almost certainly hinder its operations and could prevent the full establishment of the ICC.

Two Web sites that provide more information on the ICC are the UN site http://www.un.org/law/icc/index.html and the site of the Coalition for the International Criminal Court at http://www.iccnow.org. Finally, Henry Kissinger expresses wariness of the ICC in "The Pitfalls of Universal Jurisdiction," *Foreign Affairs* (July/August 2001).

On the Internet . . .

Centre for Economic and Social Studies on the Environment

The Centre for Economic and Social Studies on the Environment, which is located in Brussels, conducts multidisciplinary research on the qualitative and quantitative evaluation of sustainable development (economic-environmental interactions).

http://www.ulb.ac.be/ceese/

Worldwatch Institute

The Worldwatch Institute, which offers a unique blend of interdisciplinary research, global focus, and accessible writing, is a leading source of information on the interactions among key environmental, social, and economic trends. The institute's work revolves around the transition to an environmentally sustainable and socially just society—and how to achieve it.

http://www.worldwatch.org

The Common-Sense Environmentalist's Suite

The organizations listed on this links page offer research and commentary on environmental topics backed by sound science and cutting-edge economic and legal analyses.

http://www.heartland.org/archives/suites/environment/links.htm

PART 6

The Environment

*W*hen all is said and done, policy is, or at least ought to be, about values. That is, how do we want our world to be? There are choices to make about what to do (and what not to do). It would be easy if these choices were clearly good versus evil. But things are not usually that simple, and the issue in this part shows the disparity of opinions regarding the current state of the environment.

- Do Environmentalists Overstate Their Case?

Do Environmentalists Overstate Their Case?

YES: Bjørn Lomborg, from "Debating 'The Skeptical Environmentalist,'" A Debate Held at the Graduate Center of the City University of New York (April 9, 2002)

NO: Fred Krupp, from "Debating 'The Skeptical Environmentalist,'" A Debate Held at the Graduate Center of the City University of New York (April 9, 2002)

ISSUE SUMMARY

YES: Professor of statistics Bjørn Lomborg argues that it is a myth that the world is in deep trouble on a range of environmental issues and that drastic action must be taken immediately to avoid an ecological catastrophe.

NO: Fred Krupp, executive director of Environmental Defense, asserts that although Lomborg's message is alluring because it says we can relax, the reality is that there are serious problems that, if not addressed, will have a deleterious effect on the global environment.

We live in an era of almost incomprehensible technological boom. In a very short time—less than a long lifetime in many cases—technology has brought some amazing things. If you talked to a 100-year-old person, and there are many, he or she would remember a time before airplanes, before automobiles were common, before air conditioning, before electric refrigerators, and before medicines that could control polio and a host of other deadly diseases were available. A centenarian would also remember when the world's population was 25 percent of what it is today, when uranium was considered to be useless, and when mentioning ozone depletion, acid rain, or global warming would have engendered uncomprehending stares.

There are three points to bear in mind here. One is that technology and economic development are a proverbial two-edged sword. Most people in the economically developed countries (EDCs) and even many people in the less developed countries (LDCs) have benefited mightily from modern technology. For these people, life is longer, easier, and filled with material riches that were the stuff of science fiction not long ago. Yet we are also endangered by the byproducts of progress. There is a burgeoning world population that is now over 6 billion people. Resources are being consumed at an exponential rate. There are many who fear global warming. Acid rain is damaging forests. And extinction claims an alarming array of species of flora and fauna yearly, perhaps daily.

The second notable point is that most of this has occurred very rapidly. Most environmentalists date the escalation of the rate of environmental degradation to the start of the Industrial Revolution in the mid-eighteenth century. Since then, they say, the speed of change has increased steadily, and it is probable that between 80 and 90 percent of all technological advancement has occurred within the last 100 years—that is, the last 2.9 percent of humankind's 3,500 years of recorded history.

The speed of change is important because if there are pressing problems, then they must be addressed quickly. From many people's point of view, the globe cannot stand 100 more years of progress like that of the last century.

In many ways, the issues revolve around whether or not environmental safety requires us to drastically alter some of our consumption patterns; to pay more in taxes and higher prices for technologies that clean the environment; to use more expensive or less satisfactory substitutes for products that threaten the environment or scarce resources; and to alter (some might say lessen) our lifestyles by conserving energy.

Sustainability is one term that is important to this debate. Sustainable development means progress that occurs without further damaging the ecosystem. *Carrying capacity* is another key term. The question is whether or not there is some finite limit to the number of people that the Earth can accommodate. Carrying capacity is about more than just numbers. It also involves how carefully people manage the planet's resources—their lifestyles. If you live to be 100, you may well share the Earth with a world population of 10 billion. Can the world carry 10 billion people if we continue to use resources as rapaciously as we do today? Probably not. Can 10 billion environmentally careful people survive? Maybe.

The third notable point is that individual countries and the global community have collectively begun to try to figure out how to protect the environment while maintaining—indeed increasing and spreading—economic prosperity as well. A significant number of global conferences have been held during the past decade on one or another environmental issue. Data show that there has been progress in some areas but little or no progress in others.

In the following selection, Bjørn Lomborg argues that good progress is being made and will continue to be made and that we should not overreact to predictions of an environmental doomsday. His central assertion is that almost every economic and social change or trend has been positive, as long as the matter is viewed over a reasonably long period of time. In the second selection, Fred Krupp maintains that what is currently being done on many fronts is too little, too late, and he warns of the costs of inaction.

Debating "The Skeptical Environmentalist"

Let me just say there are 2 things that I try to say, and these are like the only two take-home points. One is, we need to remove our myths. Nobody can disagree with that, right. But we need to understand doomsday is actually not now. It's not like we have to act in desperation. This is important, because it means that we can start focusing on making the best possible decisions, or as politicians love to point out, there's only one bag of money but there's lots and lots of good purposes and the basic idea being here if we feel like we've been painted into a corner, we're desperate and we're willing to do pretty much anything, also making really bad decisions. If people come up to you with a gun to your head and say, "Give me your money," you don't stand around and say, "Oh, or would I like to buy a toaster?" You do what he says. That's the basic point. So, what I want to point out is that things are actually getting better—it doesn't mean that there are no problems—and that means that we can also start prioritizing. And that's what I'll get back to at the very end.

Basically, have things been getting better? Yes, on most of the important points we have actually got more leisure time, greater security, fewer accidents, more education, more amenities, higher income, fewer starving, more food, and a healthier and longer life. And this is not only true for the industrialized world, but perhaps more importantly also for the developing world.

The point here is to say, what I'm basically trying to do is take the best and very uncontroversial data we have from for instance the UN organizations—this is not the only thing there is, but at least it should make you somewhat less skeptical of the data that I have produced. What I've tried to show you here is the calories per capita per day in the world, for both the developing and developed world. What you can't see up there in the developed world is that we have enough calories, right? Because our problem is possibly getting too much. But what is really important is to look at the developing world. . . . If you look at the developing world we've gone from in 1961 which is the first data that the UN makes, 1932 calories per person per day in the developing world, that's on average just a little above what it takes to sustain your life. To in 1998, which is the

From Bjørn Lomborg, "Debating 'The Skeptical Environmentalist,'" A Debate Held at the Graduate Center of the City University of New York (April 9, 2002). Copyright © 2002 by Bjørn Lomborg. Reprinted by permission of Environmental Defense. http://www.environmentaldefense.org.

last data that we have, to about 2650 calories, that's an increase of almost 40%—that's a dramatic increase.

And what I'm saying here, and I'm going to say this again and again, this means things have been getting better. I'm not saying that that means things are fine. There's a very, very big difference. Getting better is a scientific discussion that's basically a question of, is this line going up or down?" Whereas, "It's good enough, they don't need any more"? I'm not saying that. That's a political judgement. I would definitely say they need more, so they can decide themselves whether they want to be fat or not. So the idea here is to say, things have actually been getting better. And also note that the UN actually predicts that despite the fact that there are going to be even more people here on Earth, they will continue to get even more food. So we actually end up in 2030 with a situation where they'll have about 3000 calories per person per day or be at the same level as the developed world in 1960. This means it's better and it's getting even better, but it doesn't mean it's fine. I'm just making the scientific point of saying the data actually moves in the right direction.

Now of course you've also heard and I hear this all the time, "There are lies, damn lies, and statistics." But basically you also need to understand it's the only way we can understand our world, is to look the statistics. But of course we shouldn't just take a look at averages, because this could actually hide some very important differences. It's very unlikely it's one person eating all the calories, but it could be the middle class that's eating up this stuff, so on average they have much more food but it still could mean that there's a lot of poor people who don't get fed and possibly even more. But that's not the case. The UN made its first estimate of how many people are starving in the third world in 1970; the answer was back then about 1.2 billion people were starving, or the equivalent of 35 percent of all people in the developing world—35%, that's more than one out of three—today it's down to 18% in 2030; the UN expects it to be down to 6%. This points out both the fact that things are getting better—it's a lot better when it's only 6% starving than 35% starving—but it still means that in 2030, there'll be 400 million people starving needlessly, because we could easily feed them. It's only a question of allowing them the possibility to make enough money so that they could actually feed themselves. So the idea here is to say I'm trying to make the complex—or at least for the press the complex—point of saying, things are getting better but we can still do more.

Let me go through some of the other issue areas. . . .

Basically one of the arguments that I very often hear is people say, "Sure, Bjørn, you may be right when you're talking about money, but that's because you're *only* talking about money. But where does that help us if we're really undermining our future and our kids' future with pollution? Basically, oh sure we'll get more and more money but we're going to cough all the way to the bank." That's an important point. But I try and say, let's look at the most important pollutant, namely air pollution. The US EPA [Environmental Protection Agency] estimates that anywhere from 86 to 96 percent of all social benefits that stem from regulating pollution, any kind of pollution, comes from regulating one single pollutant—particulate pollution. That means that's the most important thing of all to look at when we're talking about pollution. [For the United

Kingdom] we have the data back from 1585 until today. Of course, we love the fact that there've actually been people out measuring it in 1585. We don't. This is based on models, based on very meticulous descriptions of what was imported into London. But basically this is the best data that we have and most of the EPAs and pretty much all over the world accept this data, and then it's correlated with the data we actually have measured from the early '20s on. If you look at smoke, which is by far most important, particulate pollution, you have an increase since 1585 up until about 1890 and from then on a decline down to today; air pollution by the most important factor is down below what it was in 1585. Americans actually believe air pollution is getting worse in their country; it's not true. This is true for all developed countries; things have been getting better at least for the last 30 or 40 years. When we look at the UK data where we actually have data back from 1585, we can actually say, "No, air pollution is an old phenomenon, and it's been getting better for the last 110 years. The air has never been cleaner in London since 1585, in medieval times."

That's an important point because it means we're not painted into a corner, it is not such that air pollution is taking over and breaking down the world. However, that does not mean that we shouldn't do something about it. Actually, it turns out that because particulate pollution is such an important problem, cutting this even further is a very good idea. So not only can we say yes, air pollution has been dropping dramatically, but we should do even more about it. One of the very obvious ways especially in Europe where we have lots of diesel cars, and diesel cars contribute by far the most particulate pollution in the UK, they constitute about 6% of the car pool, but make up 92% of all small particles so yes, we should fit them with filters. It's a very cost-efficient way, and it's probably also one of the best investments that we can make at all, not only in the environmental area but in any area whatsoever. So the idea here is again to say yes, things have been declining, we have not painted ourselves into a corner but we can still do even more. But it's important we do it because it's a good investment, not because we fear the world is coming to an end.

Let me point out one more thing—this is true for the developed world; it's is not true for the developing world. If you live in Beijing or Mexico City things are getting worse, but that's really because they do exactly the same thing as we did. This is one of the World Bank's analyses that—they pretty much show up any way you make it—if you put income out this way and then you put problems with particulate pollution out this way, you basically see first it gets worse and then it gets better. Really, it's no big surprise, it's exactly what we saw with London and it also makes conceptual sense; first you don't have industry, you don't have pollution but you don't have any money either, right? And then you get industrialization and you say cool, I can buy fruit for my kid, give him an education, buy stuff for myself and then never mind our cough. That's what we did. That's what London did. And it's only when you get sufficiently rich that you start saying, "Hmm, it'd be nice to cough a little less," and then we buy some environment. That's what we've got. And it makes good sense. Environment, in this sense, is a luxury good. When you don't know where your next meal comes from you don't care about the environment 10 or 100 years down the line. However if you actually say, "Now I'm sufficiently rich," if we make the developing countries sufficiently rich, they will also start to worry much more about the environment.

So, let me just make two more points here. One is global warming, obviously because it's one of the most important areas of environmental discussion, and then I'm going to finally talk about what are the consequences of us not prioritizing correctly.

First of all, about global warming, what should we do. There's a lot of discussion about global warming—is it happening, all that stuff. Let's just point out global warming is important. I certainly think carbon dioxide does increase warming and I think we need to take our departure as the best possible scientific data we have from the IPCC [Intergovernmental Panel on Climate Change], the UN climate panel, and it doesn't mean it's infallible but it's the best we have. Now, the point here is to say global warming, the total cost will probably be somewhere around 5 to 8 trillion dollars. This is not a trivial amount of money. It means global warming will make great damage to the world. You also need to put it in perspective, that the total worth of the 21st century is about 900 trillion dollars, so we're talking about a 0.5 percent problem. There are not many other problems that reach that scale, but it is not going to drive us to the poor house no matter what we do.

We need to say it's an important problem, but it is not a problem that in any way will damage our future dramatically. It doesn't mean that we shouldn't handle it carefully, because it's an important problem. And so the idea is to say how important, what about the future of carbon dioxide emissions and what should we do. Let me just say very quickly that everyone is worried about should we run out of oil, should we run out of gas or coal or all these things. Sheik Yamani, the guy who founded OPEC [Organization of Petroleum Exporting Countries], loves to say, "The Oil Age isn't going to come to an end because of lack of oil, just as the Stone Age didn't come to the end because of lack of stone." It wasn't like people said, "Oh God, we're out of flint!" Right? They did it because it made good sense, and we're going to do exactly the same with our energy supply; eventually we'll move to renewables [nonfossil energy sources, such as wind and solar power].

Renewables have been dropping in price about 50% per decade over the last 30 years. Even if they continue at a much lower rate, to about 30% per decade, they'll become competitive before mid-century, and that means certainly we will not be using massive amounts of fossil fuel by the end of the century. It means global warming will be a limited problem, that's not the same as saying it won't be a *big* problem, but it will be a limited problem, probably 2 to 3 degrees centigrade temperature increase. Now at that rate, the main problems will occur in the developing countries. Actually the UN IPCC second summary document said, in what was later mangled by a politician, that it is not going to harm the developed world, it *will* harm the developing world, with a median temperature increase of 2 to 3 degrees. And that's an important part. . . . [B]asically, let's just point out it's going to harm by far the most the developing world.

And then we have to ask ourselves: Is it really a good idea what we're talking about doing right now, namely Kyoto [a 1997 treaty aimed at reducing emissions of greenhouse gases]? . . . [F]rom 1990 up through 2100 what's going to happen if we don't do something . . . we're going to have a temperature increase of 2.1 degrees centigrade. If we do Kyoto, it's not like it's going to stop global warming, it's going to simply slow it down slightly . . . —this is totally uncontroversial. All models show this. . . . it will go down to 1.9 degrees, or to put it more clearly, the temperature that we would have had in 2094, we have now postponed until 2100. In other words, we've bought the world six years.

Of course if Kyoto was cheap or something, maybe that would be a good idea, it's something good. But not very good. Basically what we're saying is the fact the guy in Bangladesh, who has to move because his house got flooded in 2100, he only has to move now in 2106. It's a little good but not very much good, right? On the other hand, the cost of Kyoto is going to be anywhere from 150 to 350 billion dollars a year. That's three to seven times the global development aid to the third world. Is that a good investment?

Well actually, all cost/benefit analyses show that it's a very, very bad investment. Just to give you a sense of the cost of this, the cost of Kyoto for one year in 2010, for just that one year cost, we could solve the single biggest problem in the world, once and for all. We could give clean drinking water and sanitation to every single human being on earth once and for all. It would save 2 million lives each year, in fact, half a billion lives each year. We have to ask ourselves, wouldn't that be a better way of helping the developing world? Actually the UN estimates that for 70–80 billion dollars—much less than the cost of Kyoto—we could permanently solve all the basic problems of the developing world: it could include clean drinking water, sanitation, basic health care and education to every single human being on earth. Wouldn't we do better by doing that? And again the idea here is to say, we should not allow ourselves to be painted into a corner to believe we have to do something. If it actually makes sense, we should do not only something that sounds good, but also actually *does* good.

. . . [T]he Harvard Center for Risk Analysis . . . showed all the data publicly available from the US on all legislation which had as its primary purpose to save human life. Notice a lot of environmental legislation does not have as its primary purpose to save human life—if you talk about saving the Bengal tiger for instance, it probably has the opposite effect! The idea here is to say we're talking about all the costs of the legislation that tries to save human life and then they compare what is the median or the typical cost of saving one human life one year. . . . [T]he biggest study we have in the world basically [says] that in the health area it costs 19 thousand dollars to save one human life one year. You can see the other areas—and what we basically have out here is the environment area costing $4.2 million to save one human life a year. And that's when the purpose was to save human life. . . . It's not the same thing as saying there are no good investments to be made in the environmental area. It's simply to say that on average we over-worry about the environmental area—and it does have consequences because if we over-worry about some areas we end up under-worrying about other areas, and that's my last point, that's the reason why we need to focus on what is the real state of the world. Things are actually getting better and better and they're likely to do so in the future. This does not mean that there are no problems and that we don't need to worry, but it means that what we need to understand, the problems are getting smaller, and that means that we have to start focusing and prioritizing correctly, so that we not only make sure that we make a better world—we probably will no matter how stupidly we act—but that we make an even better world for our kids and grandkids and that involves both knowing how the world looks, and how we should prioritize correctly.

Debating "The Skeptical Environmentalist"

I think this debate is timely because now at the beginning of the 21st century scientists have told us that there are some big risks that are posed by climate change, the loss of species and the sorry state of our oceans. It's timely as well because Bjørn [Lomborg] has been on tour saying there really is not much to worry about, a message that's been trumpeted in the media throughout the world. This is important because if Bjørn's right, we can all relax, take a deep breath and reallocate our time and energies to other problems, and the news media or corporate and citizen leaders can also reallocate their time and resources as well.

And it's important though, because if he's wrong, such complacency can have a big cost. . . . I want to stick to the topic of "Is our environmental future secure," and not get into all the graphs and charts that Bjørn has put up. But I do want to point out that others have written articles about those and I commend them to you: *Science, Scientific American,* Union of Concerned Scientists. . . . But basically instead of debating the charts and the graphs, I intend to challenge Bjørn's conclusions. My own answer to this important question, "Is our environmental future secure" is no. *Not yet.* It depends on us, on what we do, on the actions we take. If we make the right choices things can turn out OK, but it could also turn out the other way.

I'll support this answer with three points. First, my involvement with environmental policy over the last 30 years, which teaches me that the progress we have made comes not just from the accumulation of wealth and reaching a certain GDP [gross domestic product] level, but from citizen and government action. Bjørn tells us that progress in the past has come from wealth, and asserts that more wealth will solve any problems in the future. Second, I look to the best scientific assessment and see that yes, important progress has been made. But many important challenges remain that demand urgent attention and good policy choices. Bjørn seems to argue that these problems are largely being solved on their own. Since world scientific bodies and our own National Academy certainly see these problems, I think it's fair, even important to ask why we should rely on Bjørn's views instead. And finally, my third point is that real world experience shows that we can make progress keeping an eye toward keeping cost down, carefully choosing strong scientific standards and incentive-based policies. Given the stakes involved, not to continue to forge ahead on big problems like climate change would be like encouraging my teenage son to go around driving his car without insurance or seat belts. The risks are just too great.

OK. First, my point that environmental progress is not automatic. Others have argued that environmental progress is the inevitable outcome of wealth. Their argument is first we grow, then we clean up. Or as [Bjørn believes], the environment is a luxury good. But this argument neglects to point out the role of citizen and government action in achieving that progress. Take the case . . . of air pollution in London. It didn't just get better; London and the UK passed a series of laws *requiring* that smoke and pollution levels come down. Some of these laws were passed after an infamous inversion that killed many, many folks. Look at the history of air pollution in our own country, or cleaner, high mileage cars. Time and again the prevailing sentiment of industry was against passing a tough Clean Air Act, fighting acid rain controls, and campaigning furiously against cleaner exhaust pipes and higher mileage standards. Opponents claimed each new regulatory proposal would bankrupt them. They haven't.

Regulations to reduce sulfur dioxide pollution largely generated by the burning of coal in power plants were projected by industry to cost as much as $2000 for each ton of sulfur removed from their smokestacks. The nation chose instead a new performance-based approach to cleaning up, giving industry flexibility, but requiring results. And today removing a ton of sulfur costs less than $200, a tenth of the doomsayers' predictions. Reductions of 50% in the annual emissions were required, and so far we've reduced pollution by even more. This happy story was not an automatic happening that was triggered by some level of GDP. It was the result of a campaign waged by many environmental groups (including Environmental Defense) and then of the choices made by our legislators. The victory on sulfur and car exhaust and the Clean Air Act produced much of the air pollution improvements Bjørn points to.

The same citizen and government action was required to take lead out of gasoline, to ban DDT and other pesticides to protect our wetlands, to restore the ozone layer, and to clean up water pollution. Now in addition, other countries really don't need to spend their money cleaning up after the fact when they have the option of doing things right beforehand. China, for example: they don't accept the idea that they need to be consigned to some environmental hell. Their citizens are way down on the ladder in terms of per capita income but they're already demanding a cleanup. And their government is beginning an acid rain control program modeled after our own. Why should they wait?

Second, while I agree that progress has been made on many issues, many important problems remain. It's not true that we're making progress on most of the important problems. Sure, we have made progress on emissions from power plants. But acid rain continues to strip our soils of essential elements and kill our lakes. We need to further cut sulfur emissions. Yes we have cleaned up car exhaust, but with more cars and trucks driving more miles, smog and particulate pollution remain a huge issue both in our country and abroad. Here in this country, 15,000 premature deaths are ascribed by the scientists to our air pollution. And if you look south in Central and South America, 200,000 deaths a year are attributed by the scientists to particulate pollution.

On biodiversity, I think it's fair to say the very web of life is unraveling. For example, let's talk about what happens right before extinction: endangerment. Based on very solid data the World Conservation Union has discovered that of the world's birds, 1 out of every 8 is endangered. The same ratio applies to the world's plant life: 1 out of 8 endangered. These are troubling numbers and they are not the result of natural die-offs. These species are endangered because of what people are doing to their ecosystems here and now. Fortunately, we're talking about endangered species, not extinct species, so there is still time to save some of them. But to deny there's a problem is just plain unscientific.

On the problem of global warming, this needs to be a huge priority because the stakes are so high. It threatens our ecosystem. The parks and reserves that we set aside are at risk, our oceans, our coral reefs. What we know for certain is that the earth is warming, that sea level has been rising partly as a result of this warming, and that human activity has been contributing significantly. Glaciers are melting, the permafrost in Alaska is melting, causing power lines and phone lines to begin to topple. And . . . in New York City we've already seen samplings of the kinds of intense storms and sea level increase that may well inundate some of our own major airports, subways and highways. Our own National Academy has confirmed that this is a very real problem.

Finally, let's get to that question of costs and benefits and buying insurance, with special attention to climate change. Bjørn has [said] that it's not worth the cost to aggressively tackle climate change, but the problem will self-correct as renewables magically become competitive. I disagree. Just looking here at the US, as long as we continue to subsidize the burning of fossil fuels by a 3 to 1 ratio compared to renewables, and as long as we allow the true cost of burning fossil fuels, the true cost to our health, and our environment, to basically frankly be borne by all of us, it's not likely that renewables will be able to play the role that he predicts. Without changes in policy, it just won't happen.

Rather than citing theories about what the cost of reducing emissions will be, the truth is we can rarely know with precision what the costs or benefits of the decision will be, but time and time again we find that the projections of costs have been systematically overestimated. In particular the economic models just cited about climate tell Bjørn that we should only be prepared to spend a small amount on this critical problem. But those models don't include the potentially catastrophic but unknown cost of the collapse of the world's oceans' heat circulation patterns, or the episodic events like the flooding of the Mississippi in 1995, or the intense nor'easter that hit New York in 1992. The models don't account for heat waves that have killed thousands of people. And most importantly, the models assign no monetary value to the loss of natural systems.

As for our experience so far in climate change let's look at the facts. Environmental Defense has been working over the last few years with a number of multinational corporations to reduce emissions. The results of those efforts are now beginning to come in. For example, four years ago, BP [British Petroleum] committed to reduce their worldwide emissions by 10% below 1990 levels by the year 2010. . . . [E]ight years ahead of schedule, they announced that they had already achieved that level of reduction at no net costs whatsoever. Not every experience will turn out this way. But this is a real world example, not some theoretical projection.

[Bjørn] says Kyoto [1997 treaty aimed at reducing emissions of greenhouse gases] will cost too much for too little benefit, but the Kyoto agreement actually is fundamentally structured to minimize cost with flexible mechanisms, incentives for action, and opportunities to invest in reduction strategies like carbon sequestration. And as to the benefits of Kyoto, [Bjørn's evidence] went out to 2100 showing very little benefit. Kyoto only covers reductions mandated up to the year 2012, and everyone in the process anticipates that then subsequent reductions will be required. As for the benefits again, he hasn't taken into account, the models don't take into account the catastrophic storms that climate change can cause, and moreover, the limited economic models don't calculate in what are called co-benefits, the value of protecting forest ecosystems, coral reefs, and these can make a world of difference in the calculations. Simply put, the idea that we should rely on renewables and put the whole future of our world on the cost of renewables coming down stakes a pretty big bet on faith that that will happen. Doesn't it make more sense that we purchase insurance and make sure we put into place policies that will make it happen?

In closing, let me just say my first point, progress has been the result of deliberate decisions often by government spurred on by nongovernmental groups campaigning for action; second, sure, some important problems are getting better, but several important ones are getting worse; and third, while costs are important and they need to be considered, we need to be very wary of the way cost benefit analysis as it's historically been practiced, systematically exaggerating costs and not reliably totaling up the benefits. Time and time again, these projections have been proven wrong. Will our environment be secure? As I indicated I believe the answer is no. Not yet. We can't be sure our environmental future is secure but we can act to secure it. I'm actually quite optimistic about the outcome. But I try to anchor my optimism in realism; that is, in order to solve problems we have to first recognize them, we have to admit they exist, and then we have to understand what has worked before and aggressively apply it to those lessons. Using this approach gives me cause for both hope and concern, and I think that's the combination that can yield the most progress.

POSTSCRIPT

Do Environmentalists Overstate Their Case?

You can't have your cake and eat it too" is a trite phrase. Such bits of folk wisdom, though, often get to be trite because there is a kernel of truth to them that is worth repeating. The environment is akin to our common cake. People have been consuming it gluttonously during the past century, and most experts agree that this cannot go on any longer. The question is whether or not we have to go on a bread-and-water diet. Also, can the world's less developed countries be asked to forgo cake when the developed countries have already consumed so much?

Some, such as Krupp, say we have exceeded the boundaries of responsibility. Lomborg and others are more optimistic. They contend that we have or can develop the technology and the environmental-use policies to continue to develop and to enhance the existence of less developed countries while protecting—even improving—the environment. They may be correct, and it certainly is more comforting to believe Lomborg's optimistic view than to accept Krupp's dire outlook. More on Krupp and his organization, Environmental Defense, is available at the organization's home page at http://www.environmental defense.org/home.cfm. Also see Lomborg's personal Web page at http://www. lomborg.com and his book *The Skeptical Environmentalist: Measuring the Real State of the World* (Cambridge University Press, 2001).

Even if Lomborg is correct, it is important not to ignore the costs of sustainable development. They can be substantial. Because of population and economic development patterns, the less developed countries require particular care and assistance. It is easy to preach about not cutting down Brazilian rain forests or not poaching cheetah skins in Kenya. But what do you say to the poor Brazilian who is trying to scratch out a living by clearing cropland or grazing land? What do you tell the equally poor Kenyan who is trying to earn a few dollars in order to supply food for his family? Questions such as these have brought environmental issues much closer to the forefront of world political concerns.

The point is that environmental protection is not cost free. This is because environmentally safe production, consumption, and waste disposal techniques are frequently much more expensive than current processes. It is also because poorer people will generally do what they must to survive, whether it is environmentally safe or not. Moreover, the less developed countries have precious few financial resources to devote to developing, constructing, and implementing environmentally safe processes. Therefore, if the changes that need to occur are going to be put in place before further massive environmental degradation occurs, there will have to be a massive flow of expensive technology and financial assistance from the developed to the less developed countries.

The debate over the state of the environment is complex. One good place to begin researching it is with the United Nations Environment Programme's report *Global Environment Outlook 3* (Earthscan, 2002). To keep abreast of the current thinking on the environment, read John L. Allen, ed., *Annual Editions: Environment 03/04* (McGraw-Hill/Dushkin, 2003). Finally, to learn more about global environmental politics, consult Lorraine Elliott, *The Global Politics of the Environment* (New York University Press, 2003).

On the Internet . . .   DUSHKIN ONLINE

The Web Site of the George Bush Reelection Campaign

The official site of the Bush campaign contains the candidate's position on both the issues debated in this part.

http://www.georgewbush.com/

The Web Site of the John Kerry for President Campaign

The official site of the Kerry campaign contains the candidate's position on both the issues debated in this part.

http://www.johnkerry.com/

GlobalSecurity.org

The Web site of a military analysis think tank located in Alexandria, Virginia.

http://www.globalsecurity.org/

U.S. Department of State

The trade policy segment of the Web site of the State Department provides an official view of U.S. foreign economic policy.

http://www.state.gov/e/eb/tpp/

PART 7

Bonus Issues

It is said that death and taxes are two things that no one can escape. Much the same can be said about politics. The ever-present flow of events and changes in circumstances continually raise new issues that affect us all. Two such concerns that have come to the fore recently are the national security strategy of President George Bush and whether his administration's trade policy is helping or damaging the U.S. economy. These two issues will play a role in the November 2004 presidential election and will challenge U.S. policy makers in the months and years that follow.

- Does the Bush Administration Have a Sound National Security Strategy?

- Is Current U.S. Trade Policy Harming the American Economy?

ISSUE 20

Does the Bush Administration Have a Sound National Security Strategy?

YES: Donald H. Rumsfeld, from Testimony before the Committee On Armed Services, U.S. House of Representatives (February 4, 2004)

NO: Task Force on a Unified Security Budget for the United States, from "A Unified Security Budget for the United States" (March 2004)

ISSUE SUMMARY

YES: Secretary of Defense Donald H. Rumsfeld argues that national security policy of the Bush administration should be lauded for its remarkable achievements, and that the president has a sound policy.

NO: The Task Force on a Unified Security Budget for the United States, established by the Foreign Policy in Focus Project of the Institute for Policy Studies and by the Center for Defense Information, two private think tanks in Washington, D.C., contends that despite significant increases in the U.S. national security budget, the Bush administration is not spending the money wisely to increase U.S. security.

Prior to World War II, the United States participated in world politics only fitfully. This stemmed in part from a feeling of security based on the fact that the country is protectively flanked by three great bodies of water and has only two relatively weak land neighbors, Canada and Mexico. A great deal of this sense of security was shattered by World War II and its aftermath, and the Cold War confrontation with the Soviet Union and other communist countries. During the war, aircraft carriers, submarines, long-range bombers, and other weapons systems were developed or improved to the point where it became much more possible for an enemy to rapidly attack the U. S. Changes in weaponry further enhanced the capability of potential enemies to launch quick and devastating attacks against the U. S. First the Soviet Union and then other countries developed nuclear weapons with extraordinary killing power. More weapons were produced, and biological weapons were built to add to the deadly nuclear-biological-chemical armaments that became known as weapons of mass destruction (WMD).

The U. S. demobilized most of its military forces and slashed its military spending after World War II, but this was a short-term trend. Soon the onset of the Cold War led to a massive rearmament funded by increases in U.S. national security spending. It rose to over two-thirds of the federal budget and 14 percent of the U.S. gross domestic product (GDP) during the 1950s, and consumed more than half of budget expenditures and 9 percent of the GDP as late as fiscal year (FY) 1961. From that point, with upticks during the Vietnam War years and again during the presidency of Ronald Reagan, defense spending as a percentage of the budget and the GDP generally declined. Spending declined even further after the Soviet Union collapsed in 1991, the Cold War faded and a sense of security grew. The FY2001 budget, the last of President Bill Clinton, allocated 16.4 percent to national security, equal to just 3 percent of the U.S. GDP.

The feeling of improved safety after the Cold War did not last. It was shattered by the terrorist attacks against the U. S. on September 11, 2001. The new president, George W. Bush, also warned the country that other threats existed. Specifically he named Iraq, Iran, and North Korea as part of an axis of evil that was developing WMDs for their own potential use and for international distribution.

The Bush administration responded to the threats it identified in a number of ways. One was by calling for increased national security spending. FY2004 budget allocated $454 billion to national security, a 49% increase over FY2001 expenditures. That expanded the national defense spending to 19.6 percent of the budget and 4 percent of the GDP. Much of this increase went to the military, but part of it was used to augment domestic security functions and agencies ranging from airport security, through the Coast Guard, to the border patrol. A second Bush administration initiative was to consolidate many aspects of domestic security in the newly created Department of Homeland Security. That agency's budget for FY2004 was $31 billion. This amount was about double of what was spent during FY2001 on similar functions. A third part of the Bush administration strategy was to begin to reorient the military to better meet the security challenges of the twenty-first-century world. Over the opposition of the armed services, the White House cancelled such programs as the Army's Crusader, a heavy, self-propelled artillery system that the administration viewed as outmoded by the changing security environment. Finally, the Bush administration, as discussed in Issue 3, took a stand on U.S. security that was more preemptive and unilateral. Exemplifying that strategy was the 2003 war against Iraq, which was launched to counter a perceived threat rather than to meet an attack. This was conducted with only one other country, Great Britain.

The debate in this issue is whether the Bush administration has followed a sound national security policy. In the first reading, U.S. Secretary of Defense Donald H. Rumsfeld says that it has. In addition to projecting a positive image of the administration's utilization of the military to counter threats to the U. S. Rumsfeld outlines many of the completed and ongoing changes in the military's weapons, command structure, and other characteristics made during Bush's first three years in office. The task force report, "A Unified Security Budget for the United States," argues that the changes made under the leadership of Bush and Rumsfeld have been far less than what is needed and that greater security would be provided by a substantial reallocation of funds.

Donald H. Rumsfeld  **YES**

Testimony before the Committee on Armed Services

...**I** am pleased ... to discuss the progress in the global war on terrorism, our transformation efforts, and to discuss the President's 2005 budget request for the Department of Defense.

First, I want to commend the courageous men and women in uniform and the Department civilians who support them. They are remarkable—and what they have accomplished since our country was attacked 28 months ago [on September 11, 2001] is truly impressive. In less than 2½ years, they have:

- Overthrown two terrorist regimes, rescued two nations, and liberated some 50 million people;
- Captured or killed 45 of the 55 most wanted in Iraq—including Iraq's deposed dictator, Saddam Hussein;
- Hunted down thousands of terrorists and regime remnants in Iraq and Afghanistan;
- Captured or killed close to two-thirds of known senior al-Qaeda operatives;
- Disrupted terrorist cells on most continents; and
- Likely prevented a number of planned terrorist attacks.

Our forces are steadfast and determined. We value their service and sacrifice, and the sacrifice of their families, who also serve. . . .

We have a common challenge: to support the troops and to make sure they have what they will need to defend the nation in the years ahead.

We are working to do that in a number of ways:

- By giving them the tools they need to win the global war on terror;
- By transforming for the 21st century, so they will have the training and tools they need to prevail in the next wars our nation may have to fight—wars which could be notably different from today's challenges;
- And by working to ensure that we manage the force properly—so we can continue to attract and retain the best and brightest, and sustain the quality of the all-volunteer force.

From U.S. House of Representatives, Committee on Armed Services. Regarding the President's 2005 Budget Request for the Department of Defense. Hearing February 4, 2004. Washington, D.C.: Government Printing Office, 2004.

Each represents a significant challenge in its own right. Yet we must accomplish all of these critical tasks at once.

When this Administration took office three years ago, the President charged us with a mission—to challenge the status quo, and prepare the Department of Defense to meet the new threats our nation will face as the 21st century unfolds.

We have done a good deal to meet that charge. Consider just some of what has been accomplished:

- We have fashioned a new defense strategy, a new force sizing construct, and a new approach to balancing risks—one that takes into account not just the risks in immediate war plans, but also the risks to people and transformation.
- We have moved from a "threat-based" to a "capabilities-based" approach to defense planning, focusing not only on who might threaten us, or where, or when—but more on *how* we might be threatened, and what portfolio of capabilities we will need to deter and defend against those new threats.
- We have fashioned a new Unified Command Plan, with
  - A new Northern Command, that became fully operational last September, to better defend the homeland;
  - The Joint Forces Command focused on transformation; and
  - A new Strategic Command responsible for early warning of, and defense against, missile attack and the conduct of long-range attacks.
- We have also transformed the Special Operations Command, expanding its capabilities and its missions, so that it can not only support missions directed by the regional combatant commanders, but also plan and execute its own missions in the global war on terror, supported by other combatant commands.
- We have taken critical steps to attract and retain talent in our Armed Forces—including targeted pay raises and quality of life improvements for the troops and their families.
- We have instituted realistic budgeting, so the Department now looks to emergency supplementals for the unknown costs of fighting wars, not to sustain readiness.
- We have reorganized the Department to better focus our space activities.
- Congress has established a new Under Secretary of Defense for Intelligence and an Assistant Secretary of Defense for Homeland Defense.
- We have completed the Nuclear Posture Review, and adopted a new approach to deterrence that will enhance our security, while permitting historic deep reductions in offensive nuclear weapons....
- We have reorganized and revitalized the missile defense research, development and testing program—and are on track to begin deployment of our nation's first rudimentary ballistic missile defenses later this year.
- We have established new strategic relationships, that would have been unimaginable just a decade ago, with nations in Central Asia, the Caucasus, and other critical areas of the world.
- We have transformed the way the Department prepares its war plans [by] ... structuring our plans to be flexible and adaptable to changes in the security environment....

- We made a number of key program decisions that are already having a favorable impact on the capability of the force. Among others:
  - We are converting 4 Trident nuclear SSBN subs into ... subs capable of delivering special forces and cruise missiles into denied areas.
  - The Army ... is replacing the Crusader [armored, mobile artillery system] with a new family of precision artillery that is being developed for the Future Combat System.
  - We have revitalized the B-1 bomber fleet by reducing its size and using the savings to modernize the remaining aircraft with precision weapons and other critical upgrades.
- We have also undertaken a comprehensive review of our global force posture, so we can transform U.S. global capabilities from a structure driven by where the wars of the 20th century ended, to one that positions us to deal with the new threats of the 21st century security environment.
- Using authority granted us last year, we have established a new Joint National Training Capability, that will help us push joint operational concepts throughout the Department, so our forces train and prepare for war the way they will fight it—jointly.
- We have worked with our Allies to bring NATO into the 21st century—standing up a new NATO Response Force that can deploy in days and weeks instead of months or years, and transforming the NATO Command Structure—including the creation of a new NATO command to drive Alliance transformation....

The scope and scale of what has been accomplished is remarkable. It will have an impact on the capability of our Armed Forces for many years to come....

Our challenge is to build on these successes, and continue the transformation efforts that are now underway. In 2004, our objectives are to:

- Successfully prosecute the global war on terror;
- Further strengthen our combined and joint war fighting capabilities;
- Continue transforming the joint force, making it lighter, more agile and more easily deployable, and instilling a culture that rewards innovation and intelligent risk-taking;
- Strengthen our intelligence capabilities, and refocus our intelligence efforts to support the new defense strategy and our contingency plans;
- Reverse the existing WMD [weapon of mass destruction] capabilities of unfriendly states and non-state actors, and stop the global spread of WMD;
- Improve our management of the force;
- Refocus our overseas presence, further strengthen key alliances, and improve our security cooperation with nations that are likely partners in future contingencies;
- Continue improving and refining DoD's [Department of Defense's] role in homeland security and homeland defense; and...

Our task is to prepare now for ... tomorrow's challenges, even as we fight today's war on terror.

Managing the Force

One effect of the global war on terror has been a significant increase in operational tempo, which has resulted in an increased demand on the force. Managing the demand on the force is one of our top priorities. But to do so, we must be clear about the problem—so we can work together to fashion the appropriate solutions.

The increased demand on the force we are experiencing today is likely a "spike," driven by the deployment of nearly 115,000 troops in Iraq. We hope and anticipate that that spike will be temporary. We do not expect to have 115,000 troops permanently deployed in any one campaign....

If the war on terror demands it, we will not hesitate to increase force levels even more using our emergency authorities. And because we are using emergency powers, we have the flexibility to reduce force levels in the period ahead, as the security situation permits, and as our transformation efficiencies bear fruit.

But it should give us pause that even a temporary increase in our force levels was, and remains, necessary. Think about it: At this moment we have a force of 2.6 million people, both active and reserve:

- 1.4 million active forces,
- 876,000 in the Selected Reserve—that is the guard and reserve forces in units;
- And an additional 287,000 in the Individual Ready Reserves.

Yet, despite these large numbers, the deployment of 115,000 troops in Iraq has required that we temporarily increase the size of the force by some 33,000.

That should tell us a great deal about how our forces are organized. It suggests strongly that the real problem is not the size of the force, *per se*, but rather the way the force has been *managed*, and the mix of capabilities at our disposal. And it suggests that our challenge is considerably more complex than simply adding more troops.

General Pete Schoomaker, the Army Chief of Staff, compares the problem to a barrel of rainwater, on which the spigot is placed too high up. When you turn it on, it only draws water off the top, while the water at the bottom can't be used. The answer to this problem is most certainly not a bigger rain barrel; the answer is to move the spigot down, so that more of the water is accessible and can be used. In other words, our challenge today is not simply one of increasing the size of the force. Rather, we must better manage the force we have—to make sure we have enough people in the right skill sets and so that we take full advantage of the skills and talents of everyone who steps forward and volunteers to serve.

Consider another example: I keep hearing people talk about the stress on the Guard and Reserve—that we can't keep calling them up for repeated mobilizations. Well the fact is, since September 11, 2001, we have mobilized roughly 36% of the Selected Reserve—a little over one-third of the available forces—and most of those mobilizations are concentrated in certain skill sets. . . . We have called up 86% of enlisted installation security forces. . . .

But, while certain skills are in demand, only a tiny fraction of the Guard and Reserve—just 7.15 percent—have been called up more than once since 1990. And the vast majority of our Guard and Reserve forces—over 60%—have *not* been mobilized to fight the global war on terror. Indeed, I am told that a full 58% of the current Selected Reserve—or about 500,000 troops—have *not* been involuntarily mobilized in the past 10 years.

What does that tell us? First, it argues that we have too few Guard and Reserve forces with certain skill sets that are high demand—and too many Guard and Reserve with skills that are in little or no demand. Second, it indicates that we need to rebalance the skill sets within the reserve component, and between the active and reserve components, so we have enough of the right kinds of forces available to accomplish our missions. . . .

And we are working to do just that.

Mass vs. Capablity

One thing we have learned in the global war on terror is that, in the 21st century, what is critical to success in military conflict is not necessarily mass as much as it is capability. In Operation Iraqi Freedom, Coalition forces defeated a larger adversary. They did it not by bringing more troops to the fight, which we could have done, but by overmatching the enemy with superior speed, power, precision and agility.

To win the wars of the 21st century, the task is to make certain our forces are arranged in a way to ensure we can defeat any adversary—and conduct all of the operations necessary to achieve our strategic objectives. In looking at our global force posture review, some observers have focused on the number of troops, tanks, or ships that we might add or remove in a given part of the world. I would submit that that may well not be the best measure.

If you have 10 of something—say ships, for the sake of argument—and you reduce the number by five, you end up with 50 percent fewer of them. But if you replace the remaining five ships with ships that have double the capability of those removed, then obviously you have not reduced capability even though the numbers have been reduced.

The same is true as we look at the overall size of the force. What is critical is the capability of the Armed Forces to project power quickly, precisely, and effectively anywhere in the world. For example, today the Navy is reducing force levels. Yet because of the way they are arranging themselves, they will have more combat power available than they did when they had more people.

In Operation Iraqi Freedom, the Navy surged more than half the fleet to the Persian Gulf region for the fight. With the end of major combat operations, instead of keeping two or three carrier strike groups forward deployed, as has been traditional Navy practice, they quickly redeployed all their carrier strike groups to home base. By doing so, they reset their force in a way that will allow them to surge over 50% more combat power on short notice to deal with future contingencies. The result? . . . That capability, coupled with the application of new technologies, gives the Navy growing combat power and greater flexibility to deal with global crises—*all while the Navy is moderately reducing the size of its active force.*

The Army, by contrast, has put forward a plan that, by using emergency powers, will increase the size of its active force by roughly 6% or up to 30,000 troops above authorized end strength. But because of *the way* they will do it, General Schoomaker estimates the Army will be adding not 6%, but up to 30% more combat power. This is possible because, instead of adding more divisions, the Army is moving away from the Napoleonic division structure designed in the 19th century, focusing instead on creating a 21st century "Modular Army" made up of self-contained, more self-sustaining brigades that are available to work for any division commander.

So, for example, in the event of a crisis, the 4th Infantry Division commander could gather two of his own brigades, and combine them with available brigades from, say, the 1st Armored Division and the National Guard, and deploy them together. The result of this approach is jointness *within* the service, as well as *between* the services. And that jointness—combined with other measures—means that 75% of the Army's brigade structure should always be ready in the event of a crisis. . . .

The point is: our focus needs to be on more than just numbers of troops. It should be on finding ways to better manage the forces we have, and by increasing the speed, agility, modularity, capability, and usability of those forces.

DoD Initiatives

Today, using authorities and flexibility Congress has provided, DoD has several dozen initiatives underway to improve management of the force, and increase its capability.

Among other things:

- We are investing in new information age technologies, precision weapons, unmanned air and sea vehicles, and other less manpower-intensive platforms and technologies.
- We are working to increase the jointness of our forces, creating power that exceeds the sum of individual services.
- We are using new flexibility under the Defense Transformation Act to take civilian tasks currently done by uniformed personnel and convert them into civilian jobs—freeing military personnel for military tasks....
- We have begun consultations with allies and friends about ways to transform our global force posture to further increase capability.

We are also working to rebalance the active and reserve components. We are taking skills that are now found almost exclusively in reserve components and moving them into the active force, so that we are not completely reliant on the Guard and Reserve for those needed skills. And in both the active and reserve components, we are moving forces out of low demand specialties, such as heavy artillery, and into high-demand capabilities such as military police, civil affairs, and special operations forces.

Already, in 2003, the services have rebalanced some 10,000 positions within and between the active and reserve components. For example, the

Army is already transforming 18 Reserve field artillery batteries into military police. We intend to expand those efforts this year, with the Services rebalancing an additional 20,000 positions in 2004, and 20,000 more in 2005—for a total of 50,000 rebalanced positions by the end of next year. . . .

. . . On September 11th, war was visited on our country. Our nation was attacked—more than 3,000 innocent men, women, and children were killed in an instant. And at this moment, in caves and underground bunkers half-a-world away, dangerous adversaries are planning new attacks—attacks they hope will be even more deadly than the one on September 11th. . . . We are a nation at war.... Help us to support the Armed Services with the transformational initiatives they now have underway; help us rebalance the active and reserve force, and give the troops more options to contribute along an expanded continuum of service; help us add capability, and transform the force for the future.

2005 Budget

The President's 2005 budget requests the funds to do just that.

The President's first defense budgets were designed while our defense strategy review was still taking place. It was last year's budget—the 2004 request—that was the first to fully reflect the new defense strategies and policies.

One of the key budget reforms we implemented last year is the establishment of a 2-year budgeting process in the Department of Defense—so that the hundreds of people who invest time and energy to rebuild major programs every year can be freed up and not be required to do so on an annual basis, and can focus more effectively on implementation.

The 2005 budget before you is, in a real sense, a request for the second installment of funding for the priorities set out in the President's 2004 request.

We did not rebuild every program. We made changes to just 5% of the Department's planned 2005 budget, and then only on high-interest and must-fix issues—and *then* only when the costs incurred to mitigate risks could be matched by savings elsewhere in the budget.

The President's 2005 budget requests continued investments to support the six transformational goals we identified in our 2001 defense review:

- First, we must be able to defend the U.S. homeland and bases of operation overseas;
- Second, we must be able to project and sustain forces in distant theaters;
- Third, we must be able to deny enemies sanctuary;
- Fourth, we must improve our space capabilities and maintain unhindered access to space;
- Fifth, we must harness our advantages in information technology to link up different kinds of U.S. forces, so they can fight jointly; and
- Sixth, we must be able to protect U.S. information networks from attack—and to disable the information networks of our adversaries.

In all, in 2005, we have requested $29 billion for investments in transforming military capabilities that will support each of these critical objectives.

The President's 2005 budget requests $10.3 billion for missile defense, including:

- $9.2 billion for the Missile Defense Agency—an increase of $1.5 billion above the President's 2004 request; and
- $1 billion for Patriot Advanced Capability-3, the Medium Extended Air Defense System, and other short and medium range capabilities;

The budget also includes $239 million in funding for accelerated development of Cruise Missile Defense, with the goal of fielding an initial capability in 2008;

The 2005 budget request includes critical funds for Army Transformation, including:

- $3.2 billion to support continued development of the Future Combat Systems—an increase of $1.5 billion over the 2004 budget; and
- $1.0 billion to fund continued deployment of the new Stryker Brigade Combat Teams, such as the one now serving in Iraq.

We have also requested additional funds to strengthen intelligence, including critical funds to increase DoD human intelligence (HUMINT) capabilities, persistent surveillance, as well as technical analysis and information sharing to help us better "connect the dots."

To enhance our communications and intelligence activities, we are requesting:

- $408 million to continue development of the Space Based Radar (SBR) which will bring potent and transformational capabilities to joint warfighting—the ability to monitor both fixed and mobile targets, deep behind enemy lines and over denied areas, in any kind of weather. SBR is the only system that can provide such capability.
- $775 million for the Transformational Communications Satellite (TSAT) which will provide the joint warfighter with unprecedented communication capability. To give you an idea of the speed and situational awareness the TSAT will provide, consider: transmitting a Global Hawk image over a current Milstar II, as we do today, takes over 12 minutes—with TSAT it will take less than a second.
- $600 million for the Joint Tactical Radio System, to provide wireless internet capability to enable information exchange among joint warfighters;

The budget also requests $700 million for Joint Unmanned Combat Air Systems (J-UCAS)—a program that consolidates all the various unmanned combat air vehicle programs, and focuses on developing a common operating system.

The budget requests $14.1 billion for major tactical aircraft programs, including:

- $4.6 billion for the restructured Joint Strike Fighter (JSF) program;
- $4.7 billion to continue procurement of the F/A-22;

- 3.1 billion to continue procurement of the F/A-18E/F; and
- $1.7 billion to support development and procurement of 11 V-22 aircraft.

The budget requests funds for Navy fleet transformation, including $1 billion to continue funding the new *CVN-21* aircraft carrier, and $1.6 billion to continue development of a family of 21st century surface combatants including the DDX destroyer, the littoral combat ship, and the CG(X) cruiser.

We have requested $11.1 billion to support procurement of 9 ships in 2005. Fiscal 2005 begins a period of transition and transformation for shipbuilding as the last DDG 51 destroyers are built, and the first DD(X) destroyer and Littoral Combat Ship are procured. This increased commitment is further shown in the average shipbuilding rate for fiscal 2005–2009 of 9.6 ships per year. This will sustain the current force level and significantly add to Navy capabilities.

In all, the President has requested $75 billion for procurement in 2005 and $69 billion for Research, Development, Testing and Evaluation—funds that are vital to our transformation efforts.

Another area critical to transformation is joint training. Last year, Congress approved funding to establish a new Joint National Training Capability (JNTC), an important initiative that will fundamentally change the way our Armed Forces train for 21st century combat.

We saw the power of joint war fighting in Operation Iraqi Freedom. Our challenge is to bring that kind of joint war fighting experience to the rest of the forces, through both live and virtual joint training and exercises. Thanks to the funds authorized in the 2004 budget, the JNTC's initial operating capability is scheduled to come online in October of this year. We have requested $191 million to continue and expand the JNTC in 2005.

With your help, we have put a stop to the past practice of raiding investment accounts to pay for the immediate operation and maintenance needs. The 2005 request continues that practice. We have requested full funding for the military's readiness accounts, providing $140.6 billion for Operation and Maintenance (O&M) including $43 billion for training and operations. These funds are critical to transformation—because they allow us to pay today's urgent bills without robbing the future to do so.

We have also requested funds to support pay and quality of life improvements for the troops—including a 3.5 percent military base pay raise. We have requested funds in the 2005 budget that will also help the Department keep its commitment to eliminate 90% of inadequate military family housing units by 2007, with complete elimination projected for 2009. And we have requested funds to complete the elimination of out-of-pocket housing costs for military personnel living in private housing. Before 2001, the average service member had to absorb over 18 percent of these costs. By the end of FY 2005, it will be zero. These investments are important to the troops, and also to their families, who also serve—and deserve to live in decent and affordable housing. . . .

We are also making progress in getting our facilities replacement and recapitalization rate in proper alignment. When we arrived in 2001, the Department was replacing its buildings at a totally unacceptable average of once every 192 years. Today, we have moved the rate down for the third straight year, though it is still too high—to an average of 107 years. The 2005 budget requests $4.3 billion for facilities recapitalization, keeping us on track toward reaching our target rate of 67 years by 2008. And we have funded 95 percent of facilities maintenance requirements—up from 93 percent in FY 2004. . . .

We also need your continuing support for two initiatives that are critical to 21st century transformation: the Global Posture Review, and the Base Realignment and Closure (BRAC) Commission round scheduled for 2005.

Mr. Chairman, I cannot overemphasize the importance of proceeding with both of these initiatives. We need BRAC to rationalize our infrastructure with the new defense strategy, and to eliminate unneeded bases and facilities that are costing the taxpayers billions of dollars to support.

And we need the global posture review to help us reposition our forces around the world—so they are stationed not simply where the wars of the 20th century ended, but rather are arranged in a way that will allow them to deter, and as necessary, defeat potential adversaries who might threaten our security, or that of our friends and allies, in the 21st century.

These two efforts are inextricably linked.

It is critical that we move forward with both BRAC and the Global Posture Review—so we can rationalize our foreign and domestic force posture. We appreciate Congress' decision to authorize a BRAC round in 2005—and will continue to consult with you as we proceed with the global posture review.

Conclusion

. . . [The] President has asked Congress for a total of $401.7 billion for fiscal year 2005—an increase over last year's budget. Let there be no doubt: it is a large amount of the taxpayer's hard-earned money. Such investments will likely be required for a number of years to come—because our nation is engaged in a struggle that could well go on for a number of years to come.

Our objective is to ensure that our Armed Forces remain the best trained, best equipped fighting force in the world—and that we treat the volunteers who make up the force with respect commensurate with their service, their sacrifice, and their dedication.

Their task is not easy: they must fight and win a global war on terror that is different from any our nation as fought before. And they must do it, while at the same time preparing to fight the wars of 2010 and beyond—wars which may be as different from today's conflict, as the global war on terror is from the conflicts of the 20th century.

So much is at stake.

Opportunity and prosperity are not possible without the security and stability that our Armed Forces provide.

The United States can afford whatever is necessary to provide for the security of our people and stability in the world. We can continue to live as free people because the industriousness and ingenuity of the American people have provided the resources to build the most powerful and capable Armed Forces in human history. . . .

A Unified Security Budget
for the United States

$\mathbf{S}$ecurity took on a new meaning for Americans after Sept. 11, 2001. The worst international terrorist attack in history was also the first to cost numerous lives on the American mainland. Since 9/11, Americans have naturally felt more vulnerable, and have set a higher priority on making America more secure.

In response to this challenge, Congress increased the U.S. military budget for Fiscal Year (FY) 2003 by $49.6 billion [+ $405 billion], which exceeded the total military budget of every other nation on earth. As the budget deficit tops $500 billion, the administration's [FY] 2005 budget projects military spending of $2.2 trillion over the next five years. These figures do not include the cost of actual military operations and occupation in Iraq and Afghanistan, which now exceed $166 billion.

The question is whether all this money is being spent wisely on priorities that will do the most to increase our security. We argue that it is not.

Why? Three reasons. First, the money has been spent on a force structure that does not match today's security threats. Second, a major portion of the force has been committed to the wrong mission. And third, these increases have come at the expense of spending on other tools, in addition to military forces, that we need to make us secure.

- *Mismatched forces.* Our military is still dominated by an obsolete conventional and nuclear structure, designed to counter the least likely threat: a large-scale conventional challenge. As a result, the United States is burdened with a very expensive but misdirected military prepared for large-scale warfare rather than the challenges and operations that American forces now face with increasing strain. The dangers we face today come less from a potential superpower rival and more from failing states that have the potential to destabilize entire regions and to become magnets for transnational terrorist groups.
- *Overstretched forces.* Americans now know, as they were not told going in, that waging war on Iraq was intended as the first phase of a grand strategy to remake the Middle East. It is by no means clear that the U.S. public has either the desire or the means to support such a strat-

From Foreign Policy in Focus, March 2004, pp. 1–20. Copyright © 2004 by Foreign Policy In Focus. Reprinted with permission.

egy. It is, however, now clear that Iraq posed no imminent threat to U.S. security, and had no connection to al Qaeda. The ongoing conflict there is now absorbing troop strength that should be available to counter the real threats to our security.

- *Neglected security tools.* Following the Sept. 11, 2001 attacks, President Bush promised a comprehensive response. It would include the offensive security tools of military force. But it would also include the defensive tools of homeland security, including law enforcement measures to bring terrorists to justice, border and aviation security, physical and cyber protection of critical infrastructure, and public health and safety improvements. It would also include preventive measures, including aid to prevent humanitarian and economic crises, and to prevent the spread of weapons of mass destruction (WMD). In the three years that followed, however, the money flowed overwhelmingly to fund the offensive response. Of the cumulative $240 billion increase for these three kinds of security spending from Sept. 11, 2001 through 2004, four times as much has gone to offense as to defense, and six times as much for offense as for prevention.

The Need for a Unified Security Budget

Part of the problem behind this imbalance in national security funding is that there is no "national security" budget. Spending by numerous different agencies is not brought together in a unified budget category that allows lawmakers to consider all components of security funding as a whole. Hence, the imbalance in resources is obscured, and tradeoffs are not forced between the different programs and tools. Budget presentations and the congressional oversight process could usefully be reorganized to propose, examine, and approve a unified national security budget.

Within such a unified budget it would be possible to reallocate resources, including shifting some from the military tool to the nonmilitary tools of national security, without cutting the overall "national security" budget.

Rebalancing the Security Budget

The Bush administration proposes to spend seven times as much in 2005 for the military portion of the national security budget as for the nonmilitary portion. Its FY 2005 budget requests $430 billion (not including the costs of the wars in Iraq and Afghanistan) for military tools, but only $62 billion for nonmilitary tools, including international security programs and homeland security. When expected costs of Iraq and Afghanistan are added in, the administration allocates twenty times as much for military forces as for international programs ($23 billion) and more than ten times as much for military forces as for homeland security programs ($39 billion).

What follows is an outline of a security budget that corrects these imbalances. It rebalances our military forces to make them more useful for addressing today's threats. It also increases funding for the neglected security tools that will help us to address problems before they become armed conflicts, and to use multilateral approaches to resolve conflicts when they do occur. . . .

The proposals and specific budget recommendations suggested below are meant to be illustrative rather than definitive. It is not a detailed blueprint providing a comprehensive analysis of the details of all the specific programs. Rather it is a broad outline showing the major elements of a unified security budget that incorporates nonmilitary tools into our security strategy and rebalances military forces for today's security challenges.

1. Rebalancing Forces

More and more, the crises of the post-cold war world involve failed states that provide havens for terrorist groups while spreading regional instability. In the last decade, our forces have been deeply engaged in war-fighting and peace-keeping missions to secure order and hope in countries suffering from civil war and collapsed governments. Yet these peace operations have accounted for only about 2% of our defense expenditures over the last decade.

We have had a very mixed record thus far in dealing with such crises, moreover, partly because we have been unprepared for them. Peacekeeping and stability operations are not what America planned to do when we designed our armed forces during the Cold War. We will therefore rebalance our forces to gear a larger proportion of our military toward conducting small- and medium-scale interventions relevant to counterterrorism, and peacekeeping and stability operations.

This realignment will include a greater emphasis on:

- Investing in better strategic airlift capability including improving air-fields abroad and replacing large forward-based troops with more mobile units. . . .
- Strengthened surveillance and reconnaissance systems, and improved communications.
- Increased numbers of special operations units. . . .
- Homeland defenses.
- New specialized units in both the active force and the reserves for peacekeeping, humanitarian relief, and stability operations.
- Retraining much of the National Guard and Reserve forces and some active forces to specialize in homeland defense, counterterrorism, and protection against WMD.

The nature of today's threats also allows us to:

- *Reduce the pace of investment in the next generation of weapons.* The U.S. has a technological edge over all nations, including all of its adversaries. Nonetheless, the U.S. continues rushing expensive new generations of fighters, helicopters, ships, submarines, and tanks into production. Most of these weapons were designed to fight the now-collapsed Soviet Union. . . . New technologies and systems will be developed and tested as prototypes, but they need not be manufac-tured in quantity unless the threat warrants it. It is simply a waste of money and other resources to keep a huge military force on hair-trigger readiness for the conflicts of the last century. In addition, a more restrictive policy of exporting advanced aircraft and other weapons to potentially unstable regions would also help us to safely slow down the pace of developing future weapon systems.

- *Stop deployment of the national missile defense system until the technology is proven and the threat warrants*, while maintaining a robust research program. This would save billions of dollars and insure that America does not close the door on any promising technology. So far, despite spending over $75 billion, we have not found any that . . . works, and we cannot plan our security around doing so. Nor can we risk antagonizing Russia and China and possibly driving them into a military alliance, or alienating our European allies, or sparking a new nuclear arms race in Asia.
- *Reduce our expensive and largely redundant strategic nuclear arsenal* to 1,000 warheads, as a first step to further cuts; take our nuclear forces off hair-trigger alert.
- *Close unnecessary military bases*. While force structures and manpower have been reduced by 37% since the end of the Cold War, bases overseas have been reduced by only 25% and bases in the U.S. by only 20%. There is probably room for even larger reductions since . . . 1988, before the end of the Cold War, an official estimate put excess base capacity at 40%. After the end of the Cold War and the reduction of potential threat, presumably the excess capacity is now even greater. . . .
- *Realign forces* to better prepare them for likely missions, including counterterrorism, peacekeeping, reconstruction, security, and stability operations.

2. Neglected Security Tools

The most important element missing from our current security policy is the community of other nations. . . . Rather than squander our power by single-handedly deploying our forces on missions abroad, we should use it to build stronger and more durable alliances and institutions. A greater emphasis on cooperation will provide a stronger foundation, and more tools, for conflict prevention. It will also discourage the formation of countervailing coalitions, and make sure that if and when diplomacy fails, there is a shared vision on which to launch an enforcement action. And it will allow us to share the human, political, and financial costs of the military burden rather than shouldering them alone. We have only to compare the financial cost of going it alone in Iraq—$128 billion thus far and counting—versus the minority share, about $7 billion, that we paid to wage the Gulf War, to appreciate the virtues of working cooperatively with allies.

We must strengthen those measures that are currently being slighted—diplomacy, arms control treaties, cooperative threat reduction initiatives, and export controls—that work to check state proliferators and terrorist networks. Proposal highlights include:

- Reinvesting in diplomacy. We will refocus resources on diplomacy as preventive action to resolve conflicts before they become violent.

- Reinvigorating the nonproliferation regime. The first line of defense against the spread of WMD is the interlocking set of treaties and institutions that form the global nonproliferation regime. This must include:
  - *Expanding significantly [the support of] . . . initiatives designed to help secure and dismantle the nuclear arsenal of the former Soviet Union,* since this may be the most likely place for terrorists to get their hands on WMD.
  - *Solidifying the norms against proliferation [by strengthening]. . . the* nuclear Nonproliferation Treaty and [by ratifying] the International Atomic Energy Agency (IAEA) Additional Protocol permitting more rigorous inspections . . . [and by ratifying] the Comprehensive Test Ban Treaty, which will create a more powerful nonproliferation tool through its intrusive verification regime.
  - *Working for more effective implementation of the Chemical Weapons Convention,* including an improved inspection system, and resume participation in meetings to develop a biological weapons protocol and strengthen verification and enforcement obligations under the Biological Weapons Convention.
  - *Ratifying the Small Arms Control Pact, the Antipersonnel Landmine Treaty, and the Rome Treaty establishing the International Criminal Court.*
  - *Strengthening existing export control authorities,* focusing especially on regulating truly sensitive exports to hostile and unstable regimes.
- Developing international security forces. . . . We must have effective U.S. military forces acting primarily in conjunction with other nations and international institutions so that burdens and risks are shared and every crisis does not become primarily an American responsibility. . . . The founders of the UN in 1945 foresaw the organization's need to have a permanent standing force at its disposal. The U.S. needs to support the fulfillment of this long-delayed component of the UN charter. An interim step leading toward that goal would be to establish permanent rapid-reaction units drawn from a coalition of those powers able and willing to cooperate, providing the UN with more reliable access to well-trained and equipped international forces in times of crisis.

Proposals for an Alternative National Security Budget

As noted above, the specific program budget levels suggested here are one illustration of how to rebalance the overall national security budget to better address today's threats. They make use of other expert analysis where available, but it is beyond the scope and intent of this report to develop detailed and definitive program analyses here.

Realigning the U.S. Military

The wars in Afghanistan and especially Iraq have reaffirmed that the U.S. military is unmatched in conventional combat. . . . Incremental improvements can be [made,] . . . but the broad capability is already there, and the basic implications of this capability for tactics have been thought through.

The Iraq intervention, however—or rather the political mess left in its wake—has also shown how ill-prepared the military is for missions such as occupation, security and peacekeeping, and how adversaries will learn to avoid our overwhelming strength and attack where we are not so strong. The implication . . . is clear: The priority for our military should not be another generation of expensive aircraft, ships, and missiles designed to combat a superpower, but rather the basic equipment and skills needed to counter adversaries who have less technologically-advanced equipment, but intense commitment to their struggle. . . .

The most obvious candidates for reductions are the weapon platforms—vehicles, aircraft, ships—but there are many other programs designed to develop ever-faster battlefield targeting, communication, and striking hardware that should be lower-priority than programs that address actual, current threats. Additional savings can be achieved in the future by closing unneeded military bases and facilities in the United States and reforming the large but ineffective Defense Department accounting system. Recent estimates of savings in these areas are not available, so the table below does not include them.

Prepare for New Missions   (Improve capabilities for peacekeeping, stability, and counterterrorist missions) Improving U.S. military forces' equipment, doctrine, training, and exercises for peacekeeping, security-building, and similar semi-hostile deployments can raise their readiness for, and success at, such newly-common missions. These types of operations require a small shift in the composition of forces towards more military police, civil affairs, special forces, logistics, engineering, medical and intelligence units, and the addition of regional and foreign language specialists to those units. In order to ensure rapid deployment without maintaining a high-visibility and irritating presence in foreign countries, transportation capabilities also need to be expanded.

Proposed Military Program Changes
Annual change in funding, billions of dollars

Prepare for new missions	+ 5
F/A-22 *Raptor* fighter	– 4.0
Virginia-class submarine	– 2.1
Comanche helicopter	– 1.4
DDX destroyer	– 2.0
Future Combat System	– 0.7
Nuclear warhead maintenance	– 3.2
Nucear weapons	– 1.5
Missile defense	– 8
Army Gaurd divisions	– 4
R & D	– 22
NATO force	– 7
TOTAL	– 51

The occupation of Iraq has illustrated unmet basic equipment needs for security and stability operations. National Guard and Reserve forces in particular may need equipment upgrades. Troops now have to add their own improvised armor protection for Humvee vehicles while awaiting official equipment. At least 10 helicopters have been shot down in Iraq, yet many helicopters lack advanced countermeasures against missiles. A $324 million supplemental request for "urgent" items for Marines deploying to Iraq in 2004 included things like body armor, vehicle protection kits, and communications equipment, generators, shelters, and radios. An estimated $5 billion extra per year could help address all of these needs and preparations for new missions.

F/A-22 Raptor Aircraft (Cancel and buy existing upgraded aircraft) The winner of the prize for single most irrelevant weapon program, the F/A-22 is a fighter aircraft that has long been sold primarily on the promise of being harder to detect on radar than existing aircraft. The Taliban, al Qaeda, Iraqi Baathists, and many other adversaries do not have anti-aircraft radar installations, let alone jet fighters for the F/A-22 to counter. The Air Force, trying to justify a program whose overwhelming purpose (air-to-air combat against high-tech aircraft) has sharply receded, has recently added a whole new mission to try to make it relevant to today's world: bombing. Using the world's most expensive fighter for bombing, however, is not cost-effective. Highly upgraded and effective aircraft such as the [improved] version of the F-16 can be purchased to prevent excessive aging of the aircraft fleet. Savings from canceling the F/A-22 and buying cheaper aircraft would be approximately $4 billion per year. . . .

SSN-774 Virginia-class Submarine (Reduce purchases and stop retiring existing submarines early) This submarine was intended to combat future submarines that the former Soviet Union will never build. It is not clear that a large fleet of nuclear attack submarines are really needed for the few remaining missions of inserting small special forces teams and launching cruise missiles, given the limited occasions for using over-the-beach special forces, alternative delivery means and the high cost of nuclear submarines. Nevertheless, the planned 55-boat fleet can be maintained by halting the practice of retiring highly capable *Los Angeles*-class submarines early, basing submarines closer to their areas of operation, and buying 10 rather than 21 *Virginia*-class submarines. Savings would be $2.1 billion per year.

RAH-66 Comanche Helicopter (Cancel and focus on UAVs) This two decade-old helicopter program has been so poorly developed and managed that its technical problems and cost overruns have forced drastic alteration of the program several times. . . . The remaining primary mission of reconnaissance can be performed by the similarly-equipped but cheaper AH-64 *Apache* attack helicopters, upgraded *Kiowa* helicopters dedicated to reconnaissance, and unmanned aerial vehicles (UAVs) such as the *Predator*. . . . In late February [2004], the administration did wisely cancel this program. Savings will be $1.4 billion per year.

DDX Destroyer   (Replace with smaller ships) The DDX destroyer program, while attempting to incorporate advanced technologies to reduce crew size and operational cost, is still aimed at producing a large, high-end ship, something more attuned to open-ocean warfare against a superpower than support of operations ashore in crowded, dangerous, close-in coastal areas. The DDX would be a substantially larger ship *than any existing U.S. cruisers and destroyers. . . .* Highly capable frigates could similarly provide flexible capability in greater [existing] numbers than the DDX. The Congressional Budget Office has described an option that would buy frigates rather than the DDX. Canceling the 16 DDXs and buying 17 frigates instead would free $2.0 billion a year for other uses.

Future Combat System   (Slow the unrealistic program schedule) The Future Combat System (FCS) is not fully defined yet, but is the Army's broad program for a wide variety of new ground and air vehicles linked together with advanced communications networks into an integrated combat system. Fielding is intended to begin by 2008, a schedule that many experts believe is too aggressive, given the program's ambitious goals. Delaying the planned fielding date by two years would be a more realistic timetable for a technologically risky program that is likely to slip anyway; the delay would save around $700 million a year.

Nuclear Warhead Maintenance   (Reduce rebuilding of nuclear warheads) During the height of the Cold War, the Department of Energy (DoE) spent $3.8 billion per year on its full range of designing, testing, and manufacturing nuclear weapons. Yet the current DoE plan is to spend around $5 billion annually on its stockpile "stewardship" program. The administration does not plan to actually dismantle many of the warheads it is taking off deployed weapons status. In contrast, a program that carefully monitored nuclear warheads and took them out of service as they slowly degraded in reliability, rather than constantly rebuilding them and designing new ones, would cost $1.7 billion per year, saving about $3.2 billion annually.

Nuclear Weapon Delivery Systems   (Reduce strategic nuclear weapon deployment) The U.S. still maintains an excessive nuclear force, given that a large-scale nuclear war with Russia is extremely unlikely. The continuing huge U.S. nuclear arsenal likely hampers U.S. credibility in trying to halt proliferation of other WMD, including chemical and biological weapons more accessible to poorer adversaries. The administration also appears to be willing to use nuclear weapons to attack suspected WMD sites, illustrated by its pursuit of programs such as the nuclear "bunker buster," which undercuts efforts to delegitimize WMD. Funding for the bunker buster, starting out at $50 million, but soon to grow, should be ended. The force of 500 *Minuteman* land-based missiles can be retired, and the fleet of nuclear missile submarines reduced from 14 to 10, fielding 1,000 warheads. Savings would be approximately $1.55 billion a year.

Missile Defense   (Focus on short-range defense and limited national missile defense R&D) The current program allocates too much funding to a program that addresses a low priority threat. Enemy nations could deliver WMD in many cheaper, more reliable, more accurate, more deniable ways than using intercontinental ballistic missiles. A large share of national missile defense funding can be used far more effectively for other tools to reduce or counter the threat of WMD. In addition, a slower pace can allow adequate time for testing and developing a very technologically challenging program. As much as $8 billion a year could be obtained by substantially lowering the priority put on national missile defense, while still providing funding for some R&D and for shorter-range missile defense systems. . . .

Army Guard Divisions   (Reduce the Guard reserve force) Seven of the eight National Guard combat divisions (which do not include 15 "enhanced separate brigades") that were really intended to fight in the Cold War are not adequately trained and ready for quick deployment today. Since they do not have an active role in war plans, the seven divisions can be demobilized while preserving one division and the 15 enhanced brigades, freeing approximately $4 billion for higher priorities. A comprehensive study by the National Defense University finds that there is no shortage of troops oriented towards peace, stability, security, and occupation operations, but that the relevant units are scattered throughout the force. The study suggested designating two new divisions oriented toward "stability and reconstruction" missions, but filling them with existing active and reserve troops.

Weapon and Equipment Research and Development   (Restore a justifiable funding level) The Bush administration used the attacks of 9/11 to justify a rapid increase in military spending. The budget category that received the largest boost from FY 2002 to 2004 was, strangely, R&D, the least urgent category given the commencement of three wars—Afghanistan, Iraq, and the "global war on terrorism." Although defeating terrorists and overthrowing governments that aid them depends largely on having ready, well-trained and well-maintained force now, rather than on developing more high-tech weaponry for the future, the R&D budget has jumped almost $30 billion per year above the level sustained during the latter part of the 1990s.

There is undoubtedly some useful research to be done on new equipment and weapons designed specifically for detecting and attacking terrorists, but these types of products do not generally require the huge levels of funding that items such as aircraft for superpower war require. The R&D budget is now substantially more than was spent in the 1980s at the peak of the Cold War high-tech arms race with the Soviet Union, even taking inflation into account. R&D can safely be restored to $35 billion annually, just above the 1960–89 Cold War average of $34.0 billion (in today's dollars). Counterterrorism operations do not justify a level of R&D spending far in excess of what was spent during the Cold War when the U.S. was in an all-out arms race with the Soviet Union and fighting a major land war in Vietnam. Spending $35 billion annually would amount to a cut of around $22 billion from the $69 billion FY 2005 request, after reductions for specific weapons that are counted separately.

NATO (Make fuller use of NATO military capabilities) The North Atlantic Treaty Organization (NATO) was originally focused exclusively on the Cold War defense of Western Europe against the Soviet Union, but it has now expanded its mission outside of Europe. It is even commanding a peacekeeping force in Afghanistan, a fundamental break with the past. Much of the NATO standing force in Europe is still oriented to the Cold War, and can be demobilized or transformed in order to focus resources and attention on more relevant missions. After shrinking in size, NATO could serve as a useful mechanism for conducting multilateral deployments when an intervention is valid enough to gain international support. . . . If a national security strategy of deliberately and actively using allied forces is followed, this NATO force could allow the U.S. to reduce its ground forces by a division . . . and its air forces by an air wing. This would free up approximately $7 billion. . . .

Funding for the diplomatic, economic, and informational tools of national security, and for mobilizing and strengthening international action to increase global security, is being squeezed by sharply increased military spending. Re-allocating funding to the following programs can help restore the balance.

The International Affairs Budget

The U.S. international affairs budget needs to be viewed as part of the overall national security budget, since building solid international partnerships to address the causes of conflict is cost-effective "preventive medicine" that reduces the need for expensive military responses later. . . . It accounts for only slightly more than 1% of the U.S. discretionary budget. Increases need to be made to both parts of the international affairs budget: to the State Department budget, which includes the cost of U.S.diplomacy and U.S. assessed contributions to international organizations and peacekeeping, and to the foreign operations budget, which includes bilateral development and humanitarian aid. The U.S. is the least generous among all major donor countries in development assistance as a portion of Gross Domestic Product (GDP). The aid budget, in addition to being increased, needs to be redirected to focus most of its resources on countries most in need.

Nonproliferation Programs

A key approach to increasing security is to try to constrain the new opportunities afforded to terrorist groups by the effects of globalization—to prevent them from obtaining particularly powerful weapons, such as nuclear, radiological, chemical, or biological weapons and materials. Nonproliferation programs may significantly raise the barrier to mounting WMD attacks on the United States. The programs include efforts to help secure materials and knowledge around the world, and particularly in Russia, that could be used for WMD attacks if obtained by hostile groups.

An initially-skeptical Bush administration has become a convert to the value of many of these programs. . . . This endorsement has not however been matched by the commitment to financing it. Funding in the 2005 budget request for all non-proliferation programs . . . does slightly exceed . . . $1 billion per year. . . . In 2001 [a] bipartisan . . . commission set what is still the unmet standard for these programs, calling for spending $30 billion over ten years on nuclear weapons and materials in Russia alone. Increasing funding by about $1.5 billion annually would meet that goal.

Diplomatic Operations

In December 2002, eight former national security advisers from both parties argued for a substantial increase in the overall "international affairs" budget, which includes development assistance, security assistance, funding for the Department of State and other U.S. agencies working in foreign affairs, foreign information programs, and international financial programs. On funding for diplomatic operations they noted, "Our diplomats will play a critical role in assembling coalitions that will defeat global terrorist organizations, and they need the tools to do the job. They need secure embassies, capable telecommunications, adequate staffing, and robust public broadcasting facilities to spread America's message of freedom and democracy around the globe." They proposed a 30% overall increase that would restore funding to the peak levels of the Reagan era. Applying that rate to diplomatic operations would raise spending by around $2 billion a year.

Economic Development Assistance

In a 2002 speech, President Bush identified development assistance as a security tool:

> ". . . persistent poverty and oppression can lead to hopelessness and despair. And when governments fail to meet the most basic needs of their people, these failed states can become havens for terror. In Afghanistan, persistent poverty and war and chaos created conditions that allowed a terrorist regime to seize power. And in many other states around the world, poverty prevents governments from controlling their borders, policing their territory, and enforcing their laws. Development provides the resources to build hope and prosperity, and security."

Yet his 2005 budget request cuts nearly $400 million from the seven key humanitarian and development accounts which fund U.S. bilateral and multilateral contributions for humanitarian, health, education and other development programs. The international community agreed in 1970 on a target for official development assistance of 0.7% of national income. For the U.S. that would be $75 billion. Yet in the 2005 budget, proposed U.S. nonmilitary foreign assistance amounts to $13 billion. Five European nations have surpassed the 0.7% goal; four more are past 0.33%. As an interim goal, the U.S. could increase aid by $10 billion.

Increased funding alone is not enough, however. To be effective, these increases must be accompanied by key reforms in U.S. development policy. Reducing animosity around the world toward the U.S. requires redirecting development assistance in the following ways: 1) de-emphasize U.S.strategic advantage in the targeting of aid, and emphasize the poorest of the poor; 2) remove rules requiring aid to flow through U.S. corporations; 3) reduce debt burdens that now have developing countries paying more in debt service than they receive in aid; and 4) advance a trade policy that would level the playing field by eliminating the dumping of U.S. goods on markets in the developing world.

U.S. International Communication

Public diplomacy includes educational and cultural exchanges, academic programs, broadcasting, and language training. The budget for these purposes has been slashed since the 1960s and 1970s. A bipartisan advisory group on public diplomacy formed in June 2003 concluded that this governmental function is seriously underfunded. Doubling the current funding level of approximately $1.2 billion would address the problem. Spending for this purpose must emphasize programs that promote real *dialogue* between Americans and the rest of the world over those that simply seek to *promote* the U.S. around the world. Repairing America's international relations will necessarily involve showing that we know how to listen.

U.S. Contributions to UN and Regional Organization Peacekeeping

U.S. support for peacekeeping consists of assessed contributions to UN operations and voluntary contributions to multilateral operations conducted by sub-regional organizations such as ECOWAS (Economic Community of West African States) and the Organization for Security and Cooperation in Europe (OSCE). The overall responsibilities of international peacekeeping operations have greatly expanded and become much more complex since the end of the Cold War. Yet U.S. funding for peacekeeping operations in recent years has failed to keep pace. The 2005 budget request actually cuts U.S. contributions to UN peacekeeping by $50 million, despite new operations anticipated in the coming year.

Besides chronic underfunding, existing peace operations have to function as ad hoc coalitions without sufficient joint training or fully interoperable weapons systems. A remedy was outlined by the UN Charter: a standing, fully-integrated UN peacekeeping force. Domestic political support for such a force does not currently exist, however. In the absence of such support, the U.S. should undertake the following six interim measures to improve UN and regional peacekeeping capability and support them with a $500 million increase in annual funding.

1. UN headquarters support for peacekeeping should be treated as a core activity of the UN and as such its staff should be funded from the regular UN budget, rather than, as currently, in allocations to a separate peacekeeping budget. This will increase the UN's ability to

plan and manage operations, while reducing U.S. expenses from the current 27% assessment for peacekeeping down to the 22% assessed for the regular budget.

2. At the same time, the current U.S. policy of zero nominal growth in the UN's regular budget should be repealed and replaced with a policy based upon sound fiscal management that would allow for changes in the organization's budget to reflect its evolving responsibilities such as counterterrorism, peace operations and UN reforms.

3. The U.S. should fully support improvements in the UN Stand-by Arrangements System, the voluntary listing of national capacities that the UN can turn to for organized units, personnel, and logistical support for peacekeeping operations and in doing so list at least one brigade-level force as available for rapid deployment for UN peacekeeping operations.

4. Since one of the biggest obstacles to effective deployment of UN operations is logistics and enabling forces, the U.S. should also repeal the legislated limit of $3 million in in-kind military support to any UN-authorized peace operation per year.

5. The U.S. should increase its support for regional training and integration with regional and subregional organizations to enable more effective deployments to potential crisis spots given the range of different national elements operating under UN command.

6. The U.S. should support and develop the UN's capacity for anticipating, planning, and managing operations so that international early warning systems can be developed to provide analysis and intelligence before a crisis occurs.

UN Civilian Police Corps

While the political obstacles to a UN standing military force are daunting, more support exists for a standing UN Civilian Police Corps to restore the rule of law and ensure public safety in post-conflict societies and failed states. Such a force would be designed to address both the short-term need to fill the security gap left by inadequate local capacity, and the long-term goal of rebuilding the indigenous security sector. This is the crucial work that national military forces are neither equipped nor inclined to do. An estimated one-year start-up cost of $700 million would establish a brigade-strength force of 5,000 police officers equipped with light armored transport, protective gear, and weapons. Standing capacity would require a base and an operational headquarters, as well as provisions for a mobile field headquarters. Costs would be substantially lower than those for a military force equipped for robust operations. A U.S. 27% share of a $700 million cost estimate would amount to $189 million.

International Organizations

There is little debate that support for the U.S. around the world has declined drastically since the Sept. 11, 2001 attacks. Nowhere have the costs of U.S. unilateralism been clearer than in Iraq. The urgent task of repairing the tattered relations between the U.S. and the rest of the world argues for a strong, demonstrated recommitment to the fabric of international institutions. The 2005 budget request does include a substantial increase in its largest account for

International Organizations. This increase is misleading, however; most of it is attributable to two factors: a commendable decision to rejoin the United Nations Educational, Scientific and Cultural Organization (UNESCO), and the weakness of the dollar, requiring greater nominal amounts just to keep pace.

One of the most urgent priorities is increased funding for the IAEA. While the United States' Iraq Survey Group, set up by the Bush administration post-Iraq war and led by David Kay, has been spending $100 million a *month* (futilely) seeking WMD in one country, the IAEA is responsible for conducting nuclear inspections around the entire world on a total budget of approximately $268 million a *year.* Curbing nuclear proliferation is rightly one of the administration's highest security priorities; increasing the U.S. contribution to the IAEA by $100 million would be consistent with that goal.

Homeland Security

Although President Bush's FY 2005 budget increases homeland security funding somewhat, certain key priorities are neglected. Department of Homeland Security funding for emergency responders in small- and medium-sized cities, for example, is cut by 46%. Overall federal homeland security-related funding for police drops from $4.9 billion to $3.3 billion. Despite the establishment of a new cabinet department, the U.S. remains woefully vulnerable to terrorist attacks. According to a Brookings Institution study in early 2003, many steps taken already "reflect a response to past tactics of al Qaeda, not an anticipation of possible future innovations in how that organization or other terrorist groups might try to harm Americans." The report called for urgently

> ". . . filling the gaps that remain in the current homeland security effort. These range from creation of a new networked intelligence capability that tries to anticipate and prevent future terrorist actions, to greater protections for private infrastructure like chemical plants and skyscrapers, to a much stronger Coast Guard and Customs service (within DHS)."

A 2003 Council on Foreign Relations Task Force . . . focused specifically on emergency response to a catastrophic attack and found that "[i]f the nation does not take immediate steps to better identify and address the urgent needs of emergency responders, the next terrorist incident could have an even more devastating impact than the Sept. 11 attacks." The Task Force called for increasing spending on police, fire, medical, and other first responders approximately $100 billion over five years, which would also have substantial immediate benefits for day-to-day emergency response unrelated to terrorist attacks. In addition, increasing funding for other homeland security programs can help prevent successful attacks in the first place, such as doubling Coast Guard and Border Patrol programs, and increasing port container inspections tenfold.

Addressing Security Deficits

Proposed Nonmilitary Program Changes

	Increased annual funding, billions of dollars
International Affairs Programs	
Nonproliferation programs	1.5
Diplomatic operations	2
Economic development aid	10
U.S. international communication	1.2
U.S. contributions to UN/regional peace operations	0.5
UN civilian police force	0.2
International organizations	0.1
Homeland Security Programs	
Increase emergency responder preparation	20
Double Coast Guard and Border Patrol programs	11
Increase port container inspection, tenfold	5
TOTAL	52

Conclusion

. . . This proposed security budget will fund a restructured defense policy that provides America with the tools we need to meet the challenges of the new age. Currently we are wasting large sums on the wrong forces for the wrong occasions. It is a mistake to believe that increasing the Pentagon budget alone will guarantee our safety. The strategy outlined by this plan will transform our military into an institution better suited to deal with the new problems of the post-Cold War world and will at the same time leave us with an effective residual capability for conventional military action. It also refocuses resources on diplomacy, humanitarian aid, and the capacity for effective actions to prevent conflicts from turning into wars—and on using multilateral approaches to resolve conflicts when they escalate to war. . . .

POSTSCRIPT

Does the Bush Administration Have a Sound National Security Strategy?

Defense planning is one of the toughest of all policy-making areas because the stakes are so high. To paraphrase President John Kennedy, if you make a mistake in domestic policy it can hurt you, but if you make an error in defense policy it can kill you. For a general discussion of the relationship between military development and capability and world politics, see H. W. Brands, Darren J. Pierson, Reynolds S. Kiefer, eds., *The Use of Force After the Cold War* (Texas A&M University Press, 2000). To put U.S. spending in a global context, see the Military Balance Project report of the Center for Strategic and International Studies (CSIS) `http://www.csis.org/military/index.html`. Details on the current defense budget can be found at the Department of Defense Web site at `http://www.dtic.mil/comptroller/`.

One factor that makes defense planning difficult is arriving at a threat estimate for the present and the future and devising a strategy to the identified dangers. Are there dangers? What are they? How many soldiers, and what types of weapons are needed to provide for an adequate defense? Yes, the Cold War is now long over, but it is also the case that a new, peaceful world order has not arrived. Perils persist and may proliferate in the future. To meet them requires new strategies, organizational structures, and weapons and other changes in the approach to national security. As clear in the previous two articles, there is agreement that the force structure built to meet the Cold War's threats is not appropriate to the new security environment.

During the 2000 presidential campaign, candidate George W. Bush seemed particularly interested in investing in more high-tech weaponry and promised a "new architecture of American defense" including futuristic weapons research, such as building a national missile defense system. Bush accused the Clinton administration of failing to use the U.S. technical advantage to create a more lethal and mobile military. "The [Clinton years] have been wasted in inertia and idle talk," Bush charged. He also called for increased military funding to provide pay raises for troops and to increase the quality of their housing and other support. Where Secretary of Defense Donald Rumsfeld and the Task Force disagree is on how far those changes should go and what percentage of U.S. national security spending should go to preventative diplomacy, foreign aid, support of international peacekeeping, and other alternative approaches rather than to the U.S. military. Two works that will help you better understand many of the issues are Stuart Johnson, Martin Libicki, and Gregory

F. Treverton, eds., *New Challenges, New Tools for Defense Decisiomaking* (Rand 2003) and Doughlas A. Macgregor, *Transformation Under Fire: Revolutionizir How America Fights* (Praeger, 2003). The complexities of defense planning ar further compounded by the significant and rapid changes in both offensiv and defensive technology, a shift that is the "revolution in military affairs (RMA). For more on this, visit the Web site of the Project for Defense Altern: tives at `http://www.comw.org/rma/`.

ISSUE 21

Is Current U.S. Trade Policy Harming the American Economy?

YES: Byron Dorgan, from "Free Trade Imbalances," Remarks in the United States Senate, *Congressional Record* (July 28, 2003)

NO: Robert B. Zoellick, from Testimony Before the Committee on Ways and Means, U.S. House of Representatives (March 11, 2004)

ISSUE SUMMARY

YES: U.S. Senator Byron Dorgan, a Democrat representing North Dakota, contends that under the administration of President George W. Bush the United States has a trade strategy that is in total chaos and that is undermining the economic health of the country.

NO: U.S. Trade Representative Robert B. Zoellick, President Bush's chief trade official, maintains that open international markets create new jobs for Americans and build U.S. economic strength and that a retreat to protectionism would be destructive.

Charges and countercharges of the 2004 presidential campaign began to thunder back and forth between the incumbent candidate for reelection, George W. Bush, and John W. Kerry, the presumptive candidate of the Democratic Party. One issue that Kerry campaigned on in his effort to unseat Bush was jobs—specifically "outsourcing," the practice of U. S. companies hiring workers in foreign countries to perform tasks previously done by Americans. For example, software programmers in India, who earn less than 25% of their American counterparts, now provide about $60 billion annually in services to Microsoft, IBM, and other software giants in the U. S. and elsewhere.

The issue of jobs was especially sensitive because of the lagging U.S. economy during the first three years of the Bush presidency. During this period, the U.S. economy lost 2.2 million jobs, making it likely that Bush would be the first president since Herbert Hoover to have a net decline in jobs during his first term. In the minds of many Americans, the outsourcing issue was just one aspect of the larger concern about job losses as U.S. companies have shifted their production facilities to Asia or other low-wage regions or simply closed down because they were unable to compete with cheaper foreign imports. The depth of concern among Americans was evident in a

March 2004 poll that found that 85% of its respondents said that the issue of keeping American jobs from going overseas would be "very" or "fairly" important in deciding their choice in the November 2004 election.

Seeking to capitalize on this feeling, Senator Kerry in a March 26, 2004 press release charged that "Time after time, [the Bush] administration has put [free trade] ideology first and jobs last," and he pledged to create "10 million new jobs in the next four years," by, among other things, "a plan to replace...incentives to take jobs offshore with new incentives for job creation on our own shores." Less than a week later in Wisconsin, the president warned, "When you hear people talk about, let us reconsider free-trade agreements, what they're really saying is, is that perhaps we ought to wall ourselves off from the rest of the world. I think that would be absolutely wrong for America to be so pessimistic about our ability to compete that we become economic isolationists."

There is no doubt that the U. S. has a massive annual foreign trade deficit in good (tangible items) and services (insurance, tourism, transportation, and other non-investment monetary flows). For 2003, the U.S. exports were $1.019 trillion (goods: $714 billion; services $305 billion), while imports were $1.508 trillion (goods: $1.263 trillion; services $245 billion), for a trade deficit of $489 billion. Notice that the deficit in goods was $549 billion, while the trade in services yielded a surplus of $60 billion. Outsourced jobs would factor into the trade in services as an import.

Also beyond question is the fact that some jobs are lost to trade. For example, the U.S. clothing and textile industries have been devastated, with their combined workforce dropping from 2.1 million in 1980 to 1.2 million in 2000. During the same period, the number of American miners and primary metal workers (those in foundries) also sharply declined from 2.2 million to 1.2 million. Changes in communications have increasingly added white collar and technical positions to the job outflow. Testifying before Congress in 2003, a U.S. Department of Commerce official estimated that "over the next 15 years, 3.3 million U.S. service industrys—including 1 million IT [information technology] service jobs—and $136 billion in wages will 'move offshore'."

Yet trade also creates American jobs. The U. S. is the world's largest exporter, and creating these exports employs some 16 million Americans, about 13% of the total U.S. workforce. Additionally, foreign corporations employ 6 million people in the U. S., or about 4% of the U.S. workforce. Japan's Toyota Motor Corporation alone employs 32,000 Americans manufacturing cars in Kentucky and in other activities, and additionally it buys goods and services from 500 U.S. suppliers, thereby indirectly employing many more U.S. workers. It is also important to note that whether the imports are goods such as toys, or services such as software programming, they often keep prices down for Americans by being cheaper than products and services bearing the label "Made in the U.S.A."

Given that trade both creates and destroys jobs and that there are other costs to either protecting jobs or letting them be lost to foreign competition, it is important not to be swept up in slogans and emotions as you read the following pieces. Also ponder how to protect jobs, if that is what you favor. Remember, U.S. protective measures might spark similar countermeasures by other countries that could cost the jobs in their own right.

Free Trade Imbalances

The Senate will be asked to approve two free-trade agreements with respect to Singapore and Chile. I expect the Senate will approve both trade agreements by very wide margins. (The Senate did so in August 2003.) I intend to oppose both and wanted to explain why. It is not the case that I believe a free-trade agreement with Singapore is inappropriate. It is not the case that I believe a free-trade agreement with Chile is inappropriate. It is the case, however, that this country has a trade regime that is in total chaos and it is a significant mess.

For 20 years, under Republican and Democratic administrations, we have seen our trade deficit ratchet way up. We now have the largest trade deficit in human history that has occurred anywhere on the globe. It has been rising very rapidly. Instead of fixing the problems that exist in international trade and demanding fair trade and demanding from our allies fair trade treatment and doing something to prevent the erosion of American jobs which, incidentally, are now moving overseas at a rapid pace, we have trade negotiators rushing across the world trying to do new agreements.

I say fix the old agreements before we start running around doing new agreements. The reason we are going to consider new agreements today under something called fast track is that Congress decided to handcuff itself and agree to a procedure by which no amendments will be able to be offered to either free-trade agreement [Congress gave the president fast-track trade promotion authority, TPA, in 2002].

Singapore is a tiny nation of 3 million people a half a world away. We already have a very favorable trade relationship with Singapore. It has little manufacturing and little agriculture. It is wide open to imported goods. Singapore is not an example of a trade problem for us. So it does not matter much to me whether we have a free-trade agreement with Singapore.

The trade ambassador [U.S. Trade Representative Robert B. Zoellick] has brought us an 800-page free-trade agreement with Singapore. But demonstrative of the problem we have created for ourselves is a small provision in the free-trade agreement with Singapore that provides an authorization for the opportunity for Singapore to send to our country 5,400 people under a visa program to take jobs in this country.

From the Congressional Record. Hearing on the Free Trade Imbalances held July 28, 2003. Washington, D.C.: Government Printing Office, 2003.

Normally that would be a circumstance that would be dealt with by other committees in Congress, in which we evaluate how many people do we want to come in under a visa to work in this country, but instead this has been negotiated in a foreign-trade agreement negotiation somewhere, perhaps most of it overseas, certainly behind closed doors, inevitably in secret, and they put an immigration provision in this proposal. The immigration provision would allow 5,400 immigrants to come from Singapore to the United States to take jobs in the United States.

Think of this for a second. We have 8 to 10 million people out of work, desperate for jobs, needing to go to work, who cannot find a job in this country. We read a story every day in the major newspapers about someone who has hundreds of resumes out, they spend all day desperately trying to find a job because we have lost $2^{1}/2$ million jobs in the last couple of years.

It is not as if our economy is growing by creating new jobs. To the extent there is any growth at all, it is jobless growth in this country. Some have made the point that, no, there are jobs attached to this growth, it is just that jobs do not exist in the United States. The growth occurs here in terms of profits and economic expansion of sales and profits, but the jobs attached to that growth are in Bangladesh, Indonesia, China, and elsewhere.

So if we have a jobless expansion, which we have, having lost $2^{1}/2$ million jobs in the last couple of years, and we have people desperately searching for jobs, and then we get a free-trade agreement brought to the Senate floor our trade ambassador negotiated with Singapore, and deep in the bowels of that agreement is a provision that says 5,400 people from Singapore will come to this country to take jobs in this country and we ask the question: Why? Why would we do that?

So then the immediate instinct is, if there is a provision in this free-trade agreement with Singapore that is that odious, then let's get rid of it by offering an amendment. Dump it. The problem is, fast track means trade agreements brought to the Senate floor prevent any Member of the Senate from offering any amendment under any circumstance.

This Congress foolishly decided that it would straitjacket itself and whatever is negotiated anywhere by our trade ambassador and brought back in the form of a trade agreement, we will agree that we will be prevented from offering an amendment.

So we will vote on this. The majority of the Senate will vote yes to free trade with Singapore, and yes to 5,400 immigrants from Singapore to come to this country to take American jobs. I am not going to vote for that. Once again, the lesson is, those who believe fast-track trade procedures make sense ought to think again.

Also, this trade agreement with Singapore provides for transshipment. It provides for transshipment of high-tech products from anywhere, China, Burma, Indonesia, if they are transshipped through Singapore to the United States to get the full benefit of the Singapore free-trade agreement.

Singapore is already one of the largest transshipping points in the world. Should we be negotiating trade agreements that encourage transshipment so we do not know the origin of shipments to this country of high-tech products or others? I do not think so.

I understand, interestingly enough, that a bipartisan group of my colleagues will offer a resolution on the immigration piece that is in the free-trade agreement. The resolution is going to be a sense-of-the-Senate amendment. I think I was asked if I put my name on it. I am happy to put my name on it, but it does not mean anything.

It is beating someone over the head with a feather.

It is a sense-of-the-Senate resolution that says: You better watch it; you should not have done this. But it cannot be more than a sense of the Senate because we cannot take out this provision. This provision is stuck in the trade bill and we cannot get it out. This Senate has already agreed we will not allow amendments.

I didn't vote for that; I voted against it. But the majority of this Senate says: Let us line up so we can be subservient to the trade ambassador—whoever it is, Republican or Democrat—and agree whatever they negotiate in secret overseas that affects American jobs, count us out. We will not be able to offer amendments. That is just fine with us.

Apparently, these are colleagues who have forgotten what is written in the Constitution of the United States. The Constitution clearly says that trade is the Senate's responsibility, not anyone else's; not the President but the Senate.

Fast track trade agreements have been disastrous for this country. This chart shows the runaway deficits we have experienced.

It does not matter which administration is in office. A person could be blindfolded and listen and cannot tell if it is a Republican or Democratic administration. They all say the same thing: all we care about is getting another trade agreement. Meanwhile, we had $470 billion in the year 2002 in merchandise trade deficits. Is that alarming to some? One cannot detect it in the Senate. No one seems to care much about it. There are only two or three Members who talk about this, and we are considered the xenophobic isolationist stooges that do not get it.

What I get is this country fought for a century for a series of things that make life better in our country. There are people who died in the streets of America for the right to organize in labor unions. We fought about child labor laws, saying you should not work 12-year-old kids 12 hours a day in a coal mine or manufacturing plant. We fought about prohibiting companies from dumping chemicals into the air and the water. We fought about safe workplaces, believing the American workers have a right to work in safe workplaces. We fought about all those issues for a century.

Now some have decided you can pole-vault over all of that by producing what you want to produce elsewhere, where you do not have to worry about hiring children, where you do not have to worry about clean air and clean water. You do not have to worry about safe workplaces. You could prohibit all workers from organizing any bargaining unit. We have decided that is OK, let companies do that. They pole-vault to China or Indonesia or Bangladesh, produce there but sell here.

The problem is, in the long term, it does not work because the very people who earned the income in the manufacturing plants in this country are the people who were able to purchase the products off the store shelves. Without the incomes from those jobs—and our manufacturing sector is shrinking badly—from that manufacturing sector, who will buy these products?

This morning in the *Wall Street Journal* an article reads, "U.S.-Chinese Trade Becomes a Delicate Issue of Turf." It is talking about the debate within the National Association of Manufacturers between the big manufacturers that are international in scope that want to move their manufacturing to other countries where they can pay pennies on the dollar for labor, and the other businesses, medium and small businesses, that rely on the business from the larger companies to spill over to them. It is a fascinating article. I commend the reading to people who are interested in the subject.

Jim Schollaert, a lobbyist with the American Manufacturing Trade Action Coalition, says simply: The big companies are following a new business model—pay Chinese wages but charge U.S. prices.

That is the question these days for us. Is there a price of admission to the American marketplace? We understand we have a globalization of the international economy, and it will not stop. But have the rules for this new global economy kept pace with globalization itself? The answer, clearly, is no. If a large international company has a choice to decide where it wants to produce, and it flies its jet around the world and looks down at the landscape and sees different kinds of governance, different philosophies, different local politics, and different labor forces and decides to choose where to produce, does it not all too often these days decide to produce where it can hire a 12-year-old, work them 12 hours a day and pay them 12 cents an hour?

You think it does not happen? Of course it does. We can describe it and use names in the Senate, names of workers and names of companies. Not only can they settle on a site in the world where they can put a manufacturing plant, hire kids and adults and pay them pennies on the dollar and pollute the air and water and decide they shall not be allowed to organize as a bargaining unit and they do not have to have safe workplaces in which the workers conduct their daily activities, and then produce there, but they also ship it back to Toledo, Anchorage, Fargo, or Los Angeles and sell it on the store shelves in this country. That is the global marketplace.

Let me talk about a series of specific countries. First, I will talk about China. China has the largest trade deficit with us. It is $103 billion a year. They ship us their trinkets, trousers, shirts, shoes. We are a huge sponge for Chinese production.

One reason we have a very large trade deficit with China, which hurts us and strengthens them, is because the Chinese do not want certain things from us. They are not buying our grain in any significant way. They do not want our wheat. They do not want to buy airplanes. They need airplanes, but do not want to buy our airplanes off the shelf where we manufacture them and send our airplanes to China. They say they want some of our technology, but they want us to build our airplane plant in China and hire Chinese workers. That is the way they would like to buy American airplanes.

The problem is, it does not work that way. That is not what international trade is about. We buy that which we can best use from China, they ought to buy what they can best use from us. That is the doctrine of comparative advantage. It is as old as the study of economics itself.

Our negotiators, our U.S. official negotiators negotiate with other countries and typically underserve American interests.

About $2\frac{1}{2}$ years ago we had a bilateral trade agreement done with China. It was a prelude to China joining the WTO [World Trade Organization]. At the end of the agreement, there was once again celebration by negotiators because negotiators judge their success by whether or not they got a negotiated agreement. It is a terrible agreement, I might say. They decided, for example, that if there is automobile trade between the United States and China in the future, after a long phase-in, the following will exist: China will be allowed a 25-percent tariff on United States automobiles sold in China, and we would have a 2.5-percent tariff on any Chinese automobiles sold in the United States.

Our negotiators went to China and said: All right, we agree if there is automobile trade, vehicle trade between the United States and China. We will agree that you shall have a tariff that is 10 times higher than what we will impose on your products.

Who negotiated this on our behalf? Did they forget who they were working for?

Do you know how many movies we get into China? Before the trade agreement, only 10 imported movies could be shipped to China in a year. Just 10. So after the agreement, we get to ship 20 movies. People say, Look at that; what a great thing that is, to double it to 20. Our expectations on fair trade are pathetic.

The Chinese, by and large, keep their market reasonably closed to us, prevent us from accessing opportunities in their marketplace but expect our marketplace to be wide open to Chinese goods.

We have become a cash cow for the hard currency needs for China, and it is hurting our country. The imbalance in the trade relationship that exists between the United States and China is almost unforgivable. Is anybody doing anything about it? Not a thing. Nothing. Just nothing. All you get, when you talk to the trade ambassador's office, again under Democratic and Republican administrations—all you get from them are a few grunts and groans about we would like to do better and then they rush off and do a new agreement with some other country.

This is what we have with Korea. I mentioned the absurd situation with automobile trade with China. Well, in 2001, 618,000 cars were shipped from Korea to the United States. I believe last year it was 680,000 but use this as a working number; 618,000 cars were shipped from Korea to the United States to U.S. consumers—Hyundais, Daewoos. Probably they are wonderful automobiles. I have not driven one but I am sure they are fine automobiles. They sent us 618,000 into our marketplace. Can anyone guess how many U.S. automobiles were sold in Korea? It was 2,800; 618,000 coming into our marketplace; we got 2,800 into the Korean market. Korea ships us as many cars as they can get into our marketplace and the Korean Government will keep out as many U.S. cars as they can.

A recent example of that is the Dodge Dakota pickup, which showed great promise in the Korean marketplace. The Dodge Dakota pickup, after 2 months, started penetrating the Korean marketplace. The Korean Government cracked down on it, big headlines in the newspapers, and immediately most of the orders were canceled.

My state [North Dakota] produces potatoes in the Red River Valley, great potato country. We produce potatoes and we ship potato flakes to Korea for use in confection food—potato flakes. Do you know what the tariff on potato flakes is to Korea? It is 300 percent. Why do we allow that? I don't know. Our country doesn't seem to be interested in standing up for its economic interests.

Perhaps we should say to the Koreans, these great cars you are shipping into the marketplace, if you don't allow our cars into your marketplace and fair access to your consumers, then you ought to take your cars and sell them in Zaire. Try to sell them in Zaire. If you don't like it, then open your marketplace. Until your marketplace is open, we are not going to absorb more than a half a million of your vehicles. That is simple enough.

But we will not do that because our country is unwilling to stand up for its economic interests. In fact, that which I am presenting today on the floor of the Senate, I can't even present in an op-ed piece in the *Washington Post*. The *Washington Post* wouldn't run an op-ed piece in a million years talking about this because they are for one thing: free trade, free trade, free trade. It is as if they were wearing a robe, standing on a street corner chanting, and they only want one view expressed in their op-ed pages. Those of us who raise questions about the requirement for fair trade to stand up for the interests of American jobs are called protectionists.

My goal is not to put a wall around this country. I want to expand trade. I think expanded trade will be good for everyone, provided the rules are fair. When the rules are not fair, it is time for this country to stand up for itself and stand up for its jobs and stand up for its businesses.

I will give some other examples. I have mentioned Korea and I mentioned China. Now let me discuss Europe. I am using some agricultural examples simply because I come from a farm State. There are so many other examples.

If you take a look at what is happening in beef with Europe, the Europeans do not want U.S. beef in their marketplace because they say it is produced with growth hormones and is therefore harmful to their health. There is no scientific evidence of that. In fact, all the evidence is on the other side. But Europe says, We are not going to allow American beef into the European marketplace. In fact, they portray our beef as two-headed cows, some sort of obscene animal that would be terribly harmful to the marketplace, so they say, Keep it out.

So we go to the World Trade Organization and file a complaint against Europe and we win. It doesn't matter to Europe that we win. They are still not going to allow American beef into Europe. So what do we do? We are going to get tough. This is symbolic of the lack of backbone we have in this country when it comes to trade. How do we get tough? We decide to slap some retaliation on Europe. We hit them with some tariffs on truffles, goose liver, and Roquefort cheese.

God bless us, we are really getting tough with Europe. We are going to sock them around with truffles, goose liver, and Roquefort cheese. So what is Europe's idea to retaliate against us? Tariffs on U.S. steel and textiles.

Can you just see the difference? We simply do not have the backbone, the nerve, or the will to stand up for this country's economic interests.

I am mentioning Europe. There are plenty of problems with Europe in terms of our trade agreements. We continue to see country after country—with respect to Europe, we see the entire continent—with large, abiding, yearly trade deficits that relate to jobs lost in this country.

If we were losing those jobs just because we couldn't compete, that is one thing. That is fine. I wouldn't like it but I would understand it and I would say we better figure out how to compete in the international market-place. But if we are losing those jobs because the basis of competition is fundamentally unfair to America, then I say there is something wrong with the trade agreements.

We connect to other countries in a way that says to other countries: All right. We will trade and this is the circumstance. We will just tie one or two hands behind our back and then we will start. You can hire kids, you can put them in plants that are unsafe, dump your chemicals into the streams and the air, and you can prohibit them from organizing by law. You can do all those things and it is fine. Make your product as cheap as you can make it and ship it to the marketplace in Bismarck, ND, or Boise, ID, or Fairbanks, AK, or Los Angeles, and we would love to purchase that.

How absurd is that? Is there not any basic standard at all? Are the standards we fought for in this country for so long so old-fashioned? Is it not a timeless truth that workers ought to be able to organize, they ought to be able to expect a fair wage, and that you ought not be able to work 12-year-olds 12 hours a day 7 days a week?

If you wonder about that, let me give an example of a story. This story is entitled "Worked Till They Drop." This happens to be about a 19-year-old girl but it is happening way too often in parts of the world where they do not care about the conditions of production that we have cared about for a long while and that we fought over for many decades. This is a story about Li Chunmei, May 13 of last year. She had been on her feet for 16 hours, her coworkers said, "running back and forth inside the Bainan Toy Factory, [in China] carrying toy parts from machine to machine."

Let me read a bit from the piece.

> This was the busy season, before Christmas, when orders peaked from Japan and the United States for the factory's stuffed animals. Long hours were mandatory, and at least 2 months had passed since Li and the other workers had enjoyed even a Sunday off.
>
> Sixteen hours a day, 7 days a week.
>
> Lying on her bed in the night, staring at the bunk above her, the slight 19-year-old complained she felt worn out.

She was massaging her aching legs, coughing, and she told them she was hungry.

The factory food was so bad, she said, she felt as if she had not eaten at all....

"I want to quit," one of her roommates ... remembered her saying. "I want to go home." Her roommates had already fallen asleep when Li started coughing up blood. They found her in the bathroom a few hours later, curled up on the floor....

She was dead.

The exact cause of Li's death remains unknown. But what happened to her last November in this industrial town in southeastern Guangdong province is described by family friends and coworkers as an example of what China's more daring newspapers call *guolaosi*. The phrase means "over-work death," and usually applies to young workers who suddenly collapse and die after working exceedingly long hours, day after day.

This is the sort of thing that is happening in some factories around the world, producing, in this case, stuffed toys. They could have been producing baseball caps. A prominent Ivy League college buys baseball caps from similar factories. They pay $1/5$ cent labor for each cap produced and each cap is sold at $17 on the campus of the Ivy League university. Fair trade?

The question is, What did we fight about all these years? It seems to me we fought about having an economy that gave American businesses a chance to compete fairly and provide good-paying jobs to American workers. On issue after issue in international trade, we have trade agreements being brought to the floor of the Senate that have been negotiated with other countries in a way that is fundamentally incompetent.

One other example I have spent 10 years working on is the aftermatch of a free-trade agreement with Canada. The free-trade agreement with Canada is one I voted against. Incidentally, it was a vote when I was serving in the United States House Ways and Means Committee. It was 34-1. I was the one who voted against it. I was told by my colleagues we really need to make this a unanimous consent vote, that Canada was our good neighbor to the north and we share a common border. I said no. What you are proposing here is wrong. It is going to dramatically injure family farmers in this country.

But the deal was passed under fast track and no one could offer amendments. Oh, we had an assurance in writing from [thru] Trade Ambassador [Clayton] Yeutter [1985–1989] that it would not represent a change or a significant change in the quantity of grain going back and forth across the border. The minute it was passed, we began to see a flood—a virtual avalanche—of Canadian wheat coming into this country sold by the Canadian Wheat Board, a state-sanctioned monopoly that would be illegal in this country. Our farmers were badly undercut by this unfair competition. We haven't been able to do a thing about it—nothing.

I had the GAO [U.S. General Accounting Office] go to the Canadian Wheat Board because we think they are dumping in our marketplace. The Canadian Wheat Board simply thumbed its nose at the General Accounting

Office, saying we don't intend to open our records to you at all. We intend to show you no information.

Year after year, we face this unfair grain trade from Canada. In fact, one day I went to the Canadian border—I have mentioned this many times—with a man named Earl Jensen in a 12-year-old orange truck with a couple hundred bushels of durum wheat. We drove to the Canadian border. All the way to the Canadian border we saw 18-wheelers coming south full of Canadian grain being dumped on our marketplace injuring our farmers. We saw semi load after semi load. I bet we met 20 semi loads of Canadian grain. When we got to the border in the 12-year-old little orange truck, guess what. We were stopped dead in place and we could not get that truck across the border because you couldn't take 200 bushels of durum wheat into Canada. The Canadian market was closed to us, but our market was wide open to unfair Canadian trade in this country. This has gone on for 10 years and we have not been able to do a thing about it.

Today we have a trade ambassador who has been scurrying around the world doing new trade agreements. So we have two new agreements to vote on, one of which has a 5,400 immigrant quota of people coming into our country from Singapore to take American jobs. Everyone knows that is wrong. Everybody in this Chamber knows that is foolish. That is not the way you do immigration policy—behind closed doors in secret on a trade bill. And yet no one in this Chamber will be able to get rid of that provision. That provision will be ratified by this Congress either this afternoon or tomorrow. Not with my vote.

At some point, somehow, somebody will have to wake up on trade. It is not the case that I believe we ought to shut down trade or that we ought to build walls and prevent trade. It is the case that this country needs to have a backbone and some nerve and some will—yes, dealing with China, Japan, Europe, Korea, Canada, and Mexico. And until we get that will and are willing to protect American jobs with the requirement for fair trade, this country is going to continue to lose economic strength.

After the Second World War, for a quarter of a century our trade policy was almost exclusively foreign policy. It wasn't trade. It wasn't economics. It was all foreign policy coming out of the State Department. It didn't matter because we were the biggest, the best, and the strongest country in the world by far and we could tie one hand behind our backs and out-compete anybody under any circumstance. So it was just fine. We could have mushy-headed foreign policy masquerading as trade policy. It didn't matter. We just would win.

But in the second 25 years after the Second World War, we saw the development of some pretty tough and canny competitors—Japan, Europe, now China, and others. Still much of our trade policy is fuzzy-headed foreign policy. Now you tie one hand behind your back with moves that are fairer and this country loses. Again, what do we lose? We lose jobs, economic expansion, opportunity for businesses, opportunity for workers, and some say it doesn't matter; it is just irrelevant.

I do not for the life of me understand that. It makes no sense that this country does not any longer understand that international trade is a significant foundation for this country's economic future. That foundation is either a foundation of cement with strength or quicksand that washes away quickly....

One after another of these trade agreements has traded away this country's economic interests.... It seems to me when something isn't working, you ought to change it. Yet we see no proposal here for change at all. It is just let's have a couple more helpings from the same menu, and the menu isn't working for our country.

There are so many issues related to this. I talked about jobs because, in my judgment, that is central to this. First, you have currency issues and the fact that China, for example, dramatically undervalues its currency against the U.S. dollar. They have a terrific advantage in our marketplace in trade.

There are so many different facets of trade that it is almost hard to describe. You have the political issues. Some countries as a matter of governance decide here is the way we will compete. For example, I have mentioned on a couple of occasions today that some countries will prohibit workers from organizing. We are proud that our country protects those rights. We understand it has strengthened this country and it is good for our country. In fact, the way we have developed a strong middle class in our country is with the development of a manufacturing sector in which workers are organized and have been able through their strength to collect a reasonable share of the national income from manufacturing. But some countries say we will prohibit as a matter of political choice workers from organizing.

Then there are some others who say it doesn't matter that our manufacturing base is eroding; if that is what happens as a result of some natural function of trade, that is all right for our country. Well, it is not all right. There is no country that will long remain a world power—none—without a strong manufacturing base. You cannot be a world economic power without a strong manufacturing base. Those who think this country will remain a strong, vibrant, growing, economic superpower are dead wrong if they allow this manufacturing base to be dissipated. Too many of my colleagues seem to think it is just fine; whatever happens, happens.

It is not fine with me. All you have to do is look at where this country is headed in international trade. Look at what has happened to our manufacturing base. Look at how good jobs have shrunk in this country. I am talking about those people who worked in the coal mines, those who worked in the steel mills, those who worked in our manufacturing plants who used to earn a good wage with good benefits and good job security, and who now discover we are racing toward the bottom to figure out how we can compete with other countries that pay a dime an hour or 20 cents an hour.

How can we compete with other countries that have no laws that prevent them from abusing the environment with chemicals going into the airshed and into the water? If you wonder about that, just travel a bit. Go to those countries—I have—and take a look at what happens. Then ask yourself, Is that the level of competition? Is there an admission price to the American marketplace that says it is almost free? That you don't have to reach any threshold? And any trade—using circumstances I have previously described—is fair trade to which we ought to subject our workers and our employers?

I have explained at great length why I intend to vote no on these two trades agreements. It is not about Chile. It is not about Singapore. It is about a process that is fundamentally bankrupt. It is about trade negotiators who ought to be ashamed of themselves. It is about past trade agreements that are incompetent, whose repercussions we are dealing with today.

I have, from time to time, threatened to offer legislation that would require all U.S. trade negotiators to wear a jersey. When you are representing the United States of America in the Olympics, you wear a jersey that says "USA." It seems to me that perhaps our trade negotiators—more than almost anyone—need to have a jersey to be able to look down at and understand who they represent.

Will Rogers used to say: "The United States of America has never lost a war and never won a conference." He surely must have been thinking about trade negotiators. This country had better develop a backbone and some will and some nerve to stand up for its economy and stand up for its workers and stand up for its employers—no, not in a way that is unfair to any other country but in a way that says to any other country: We are open for business, we are ready for competition, and we will compete anywhere and with anyone in the world, but we, by God, demand that the rules be fair. And if the rules are not fair, then we intend to change them to create rules that are fair to our country.

NO ↩

Robert B. Zoellick

Testimony Before the House Committee on Ways and Means

Consider this statement:

"With America's high standard of living, we cannot successfully compete against foreign producers because of lower foreign wages and a lower cost of production." Perhaps this pessimism sounds familiar. It could very well have come from one of today's opponents of trade, arguing against a modern-day free trade agreement. But in fact these words were written by President Herbert Hoover in 1929, as he successfully urged Congress to pass the disastrous Smoot-Hawley Tariff Act that raised trade barriers, destroyed jobs, and deepened the Great Depression.

Today, as in the 1930s, trade can be a contentious subject. But as we learned 75 years ago, isolating America from the world is not the answer. We need to open markets for American companies to compete in the world economy, so we can create new jobs and build economic strength at home. When we work with the world effectively, America is economically stronger. Ninety-five percent of the world's customers live outside our borders, and we need to open those markets for our manufacturers, our farmers and ranchers, and our service companies. Americans can compete with anybody—and succeed—when we have a fair chance to compete. Our goal is to open new markets and enforce existing agreements so that businesses, workers, and farmers can sell their goods and services around the world and consumers have good choices at lower prices.

Opening foreign markets to U.S. products and services is vital to economic growth, and an expanding economy is the key to better-paying jobs. U.S. exports accounted for about 25 percent of U.S. economic growth during the last decade and supported an estimated 12 million American jobs.

When the world's consumers fly in an airplane, boot up a computer or watch a movie, they are helping to employ Americans. And 6.4 million Americans have jobs working for foreign companies, building cars in Ohio, Kentucky, Tennessee, Alabama and South Carolina—or processing mortgages in Minnesota or engineering software in California.

Although we have opened many markets, too many foreign countries still will not let us compete on an equal footing. They keep our products out,

From U.S. House of Representatives, Testimony Before the Committee on Ways and Means, March 11, 2004. Washington, D.C.: Government Printing Office, 2004.

they illegally copy our technology, and they block us from providing services. We want to make sure our products and services get a fair chance to compete, and to be vigilant and active in enforcing our trade agreements so that American workers have a level playing field.

Recent U.S. trade agreements have cut hidden import taxes and saved every working family in America as much as $2,000 a year, and our newest agreements could add more to these savings. Arguing for trade barriers is like arguing for a tax on single working moms, because that's who pays the most in import taxes as a percentage of household income. Our goal is to cut those hidden import taxes—while other countries cut theirs too—to give working families a boost.

At the same time, we need to help people manage change—particularly when it concerns jobs. Jobs not only provide for our families, they give us hope for a better tomorrow. Losing a job is hard, whether it is because of a recession, changing technology, or competition from another state or overseas. No matter the cause, it is important to help someone who loses a job to get back on his or her feet.

That's why Congress and the President tripled Trade Adjustment Assistance in the Trade Act of 2002. In 2003, this program provided some $1.3 billion in support and retraining, with nearly 200,000 workers eligible for assistance.

That's why the President is focused on helping workers to learn new skills for the jobs of the future. His Jobs for the 21st Century initiative provides over $500 million in new funding for education and job training, including $250 for community colleges to provide workers job training and skill development.

And that's why the private sector has an important role too: Today American companies spend $70 billion a year on worker education and training, and they will need to expand this investment in people for the future.

Some of today's opponents of trade, like those of yesteryear, want to retreat, to cut America off from the world. But we need to remember that what goes around, comes around: If we close America's markets, others will close their markets to America. And the price of closing markets is larger than economic isolationists recognize. Over the last decade, trade helped to raise 140 million people out of poverty, spreading prosperity and peace to parts of the world that have seen too little of both. Americans will not prosper in a world where lives of destitution lead to societies without hope. . . .

Strategic Overview

. . . Through an ambitious trade agenda, the United States is working to secure the benefits of open markets for American families, farmers, workers, consumers and businesses. By pursuing multiple free trade initiatives, we are creating a "competition for liberalization" that provides leverage for openness in all negotiations, establishes models of success that can be used on many fronts, and develops a fresh dynamic that puts America in a leadership role.

This strategy is producing results. . . .

The United States was instrumental in defining and launching a new round of global trade talks at the World Trade Organization (WTO) at Doha in late 2001. That same year we completed the unfinished business of China and Taiwan's entry into the WTO, working from the bilateral trade terms established by President Clinton, so as to establish a legal framework for expanding U.S. exports and integrating China's economy into a system of global rules. Also in 2001, the Administration worked with Congress to pass a Free Trade Agreement (FTA) with Jordan and a basic trade accord with Vietnam. After the 2000 election, President Clinton had announced an interest in FTAs with Singapore and Chile, and this Administration negotiated state-of-the-art accords in 2001–02 and gained Congressional approval in 2003.

A critical aspect of the Trade Act of 2002 was the renewal of the President's trade negotiating authority [Trade Promotion Authority, TPA]. In 2003 and early 2004, the Administration put that authority to good use, promoting global negotiations in the WTO, working toward a Free Trade Area of the Americas (FTAA), completing and winning Congressional approval of free trade agreements with Chile and Singapore, launching bilateral free trade negotiations with 14 more nations (concluding talks with seven of them), announcing its intention to begin free trade negotiations with six additional countries, and putting forward regional trade strategies to deepen U.S. trade and economic relationships in Southeast Asia and the Middle East.

The Trade Act of 2002 also renewed and improved trade preferences covering an estimated $20 billion of business with developing countries in Africa, Latin America, and Asia through the renewal and improvement of the Andean Trade Preference Act, the African Growth and Opportunity Act, and the renewal of benefits under the U.S. Generalized System of Preferences. In addition, the Trade Act of 2002 tripled the level of trade adjustment assistance available to U.S. workers to nearly $6 billion over five years.

USTR [United States Trade Representative, which refers to both the agency and its head], working closely with other federal agencies, works to make sure that our trading partners live up to their commitments. A significant amount of the day-to-day work of USTR is spent pressing foreign officials to abide by their trade obligations. . . .

We resolve most problems without resorting to formal dispute proceedings, which take additional time and involve uncertain outcomes. Most U.S. companies suggest formal dispute proceedings only as a last resort. When we determine it will be the most effective way to settle disputes, we pursue cases under the WTO, NAFTA [North American Free Trade Agreement], or our new FTAs.

In particular, we are devoting more enforcement resources to China. While U.S. exports to China support more jobs for American workers, we face a number of persistent problems that must be resolved. . . . [These include] matters such as Chinese tax policies that disadvantage American exports of products as diverse as semi-conductors and fertilizer; rampant piracy of intellectual property rights; technical commercial standards that are drafted to exclude foreign economic participation—such as on wireless encryption; among other concerns. Ensuring that these trade barriers do not stand is important to achieving the long-term benefits of China's WTO accession package: greater openness, adherence to the rule of law, and the institutionalization of market principles.

We recognize that enforcement of China's commitments requires sticks as well as carrots. . . . These [sticks] include the careful use of the China textile safeguard (which the Administration invoked for three product categories last December); anti-dumping laws; the product-specific safeguards; and WTO dispute settlement, an option that we may need to deploy very soon.

Pressing Forward in the WTO

At key points, the United States has offered crucial leadership to launch, prod, advance and reenergize the Doha Development Agenda, the global trade negotiations at the WTO. At the same time, we have emphasized that in a negotiation with 148 economies seeking consensus, others must also work constructively with us.

After the Doha launch, the United States proposed the elimination of all global tariffs on consumer and industrial goods by 2015, substantial cuts in farm tariffs and trade-distorting subsidies, and broad opening of services markets. . . .

In addition to laying the groundwork for bold market opening, the United States took the lead in resolving the contentious access-to-medicines issue in August 2003. . . . At the Cancun WTO meeting in September, however, some wanted to pocket our offers on agriculture, goods and services without opening their own markets, a position we will not accept. Since Cancun, I believe many countries have concluded the breakdown was a missed opportunity that serves none of our interests. That recognition is a useful starting point for getting the negotiations on track. . . .

By late December [2003], we sensed many WTO members were interested in getting back to the table, probably working from the draft text developed at Cancun. So in January I wrote a letter to all my WTO colleagues putting forward a number of "common sense" suggestions to move the Doha negotiations forward in 2004. I emphasized that the United States did not want 2004 to be a lost year. The letter suggested that progress this year will depend on the willingness of Members to focus on the core agenda of market access for agriculture, manufactured goods, and services.

In agriculture, we believe that WTO Members need to agree to eliminate agricultural export subsidies by a date certain, substantially decrease and harmonize levels of trade-distorting domestic support, and seek a substantial increase in real market access opportunities both in developed and major developing economies. . . .

For manufactured goods, we are proposing that WTO Members pursue an ambitious tariff-cutting formula that includes sufficient flexibility so that the methodology will work for all economies. In addition to the tariff-cutting formula, sectoral zero-tariff initiatives need to be an integral part of the negotiations, perhaps using a "critical mass" approach to define participation—as in the successful Information Technology Agreement. We also underscored the need to develop specific plans to address non-tariff trade barriers effectively in the Doha negotiations.

. . . The important area of services . . . is an increasingly important part of economic development. More open services markets help provide the infrastructure for development. The sector also offers increasing opportunities for developed and developing countries to work together for mutual benefit.

Finally, we are asking that countries not permit the so-called "Singapore Issues" [four issues of concern to wealthier countries: treatment of foreign investors, cartels, financial transparency, and simplifying customs procedures] to be a distraction from our critical work on market access. We need to clear the decks. Based on extensive consultations in Africa and Asia, I believe we can move forward together on trade facilitation, which cuts needless delays and bureaucracy at borders and ports. I have urged my colleagues to drop the other topics. . . .

I believe we are regaining some momentum, although the road ahead is marked by risks. Our ability to make notable progress by this summer depends principally, in my view, on two steps: one, reconciling the conundrum of the "Singapore Issues" by agreeing to focus solely on trade facilitation; and two, by concentrating on the draft agriculture text to see if we can agree on specific frameworks for reform. To secure movement on agriculture, all countries will need to agree to eliminate export subsidies, including the subsidy element of credit, to end State Trading Enterprise monopolies, and discipline food aid in a way that still permits countries to meet vital humanitarian needs.

Advancing Negotiations in the Free Trade Area of the Americas

Since taking office, the Administration has been working to transform years of general talks about a Free Trade Area of the Americas (FTAA) into a real initiative to open markets in the hemisphere, with a focus on first removing the barriers that most affect trade. The FTAA would be the largest free trade zone in the world, covering 800 million people with a combined gross domestic product of over $13 trillion. It would expand U.S. access to Western Hemisphere markets, where tariff barriers are currently much higher than the trade-weighted U.S. average of 2 percent, and where non-tariff barriers are abundant. Studies report that an average family of four would see an income gain, through greater purchasing power and higher income, of more than $800 per year from goods and services liberalization in the FTAA.

At the Summit of the Americas in Quebec City in 2001, the United States started to lead the FTAA into a period of concrete market access negotiations. In February 2003, the Administration put forward—on schedule—its comprehensive and significant market access offers to FTAA partners in the areas of agriculture, industrial goods, services, investment, and government procurement. But others hesitated.

Therefore, in November 2003, at the FTAA Ministerial in Miami co-chaired by the United States and Brazil, we developed a pragmatic approach to match the different circumstances of the 34 nations of the hemisphere—ranging from small Caribbean island states to the United States. We agreed to establish a common set of rights and obligations covering all nine areas under negotiation and that benefits would be commensurate with obligations undertaken. In addition, we agreed that nations that are prepared to go further could do so through plurilateral arrangements in some areas. This higher level of commitment—and benefit—creates incentives for countries to do more, without leaving others behind. The countries most likely to be ambitious are the ones that work with us on our gold-standard bilateral FTAs.

The FTAA will not be an easy negotiation, as this Committee knows. Yet we are committed to working creatively and flexibly with our hemispheric partners to achieve a long-held dream: the free flow of commerce throughout the Americas.

Spanning the Globe with Bilateral Free Trade Agreements

Miami also provided the venue for the announcement of several new U.S. bilateral free trade initiatives, demonstrating how our movement on multiple fronts can support our larger trade goals.

In 2003, the United States signed free trade agreements with Chile and Singapore, and those agreements won strong bipartisan majorities in Congress. These comprehensive, state-of-the-art FTAs set modern rules for 21st Century commerce and broke new ground in areas such as services, e-commerce, intellectual property protection, transparency and anti-corruption measures, and enforcement of environmental and labor laws to help ensure a level playing field for American workers. They also built on the experience of prior free trade agreements and will serve as useful models to advance other U.S. bilateral free trade initiatives in 2004. In Latin America, for example, . . . the Administration concluded a U.S.-Central America Free Trade Agreement (CAFTA) with El Salvador, Guatemala, Honduras, and Nicaragua.

. . . The United States intends to launch new FTA negotiations with Panama, Colombia, and possibly Peru and Ecuador, while continuing preparatory work with Bolivia. Added together, the United States is on track to gain the benefits of free trade with more than two-thirds of the Western Hemisphere through state-of-the-art, comprehensive sub-regional and bilateral FTAs.

. . . We have also concluded a landmark free trade agreement between the United States and Australia. On February 13, . . . Our terms with Australia will eliminate tariffs on more than 99 percent of U.S. manufactured goods exports to Australia on day one. Those exports account for 93 percent of total U.S. sales to Australia's large market, and support 150,000 good-paying American jobs. In creating new export opportunities for America's manufacturers, this deal will help a recovering sector of our economy while also expanding markets for America's services firms, creative artists, and farmers.

With virtually all U.S. manufactured exports going duty-free immediately under this agreement, America's manufacturers estimate they could sell $2 billion more per year to Australia. They predict that U.S. national income would grow by nearly that much as well. Markets for services such as life insurance and express delivery will be opened, too; intellectual property will be better protected; U.S. investments will be facilitated; and American firms will be allowed to compete for Australia's government purchases on a nondiscriminatory basis for the first time. All U.S. farm exports—more than $400 million per year—will go duty-free to Australia, benefiting many sectors such as processed foods, fruits and vegetables, corn oil, and soybean oil.

In Southeast Asia and the Middle East, the President has announced initiatives to offer countries a step-by-step pathway to deeper trade and economic relationships with the United States. The Enterprise for ASEAN [Association of Southeast Asian Nations] Initiative (EAI) and the blueprint for a Middle East Free Trade Area (MEFTA) both start by helping non-member countries to join the WTO, strengthening the global rules-based system. For some countries further along the path toward an open economy, the United States will negotiate Trade and Investment Framework Agreements (TIFAs) and Bilateral Investment Treaties (BITs). These customized arrangements can be employed to resolve trade and investment issues, to improve performance in areas such as intellectual property rights and customs enforcement, and to lay the groundwork for a possible FTA.

President Bush announced the [EAI] in October 2002. Significant progress was made in 2003, and the stage has been set for further achievements in 2004. With the newly enacted Singapore FTA to serve as a guidepost for free trade with ASEAN nations, the President announced that he would begin negotiations for a comprehensive free trade agreement with Thailand. . . . At the Cancun WTO Ministerial last September, Cambodia was offered accession to the World Trade Organization, so it could take another step toward active participation in the global rules-based economy. Spurred by the progress of its neighbors, Vietnam is also working toward WTO membership, building on the foundation of a basic bilateral trade agreement with the United States that was enacted by Congress in 2001. The United States signed a bilateral trade agreement with Laos in 2003, and the Administration continues to support granting Normal Trade Relations (NTR) to Laos. The United States is using TIFAs with the Philippines, Indonesia, and Brunei to solve practical trade problems, build closer bilateral trade ties, and work toward possible FTAs.

The Middle East Free Trade Area initiative, announced by the President in May 2003, offers a similar pathway for the Maghreb, the Gulf states, and the Levant. In addition to helping reforming countries become WTO Members, the initiative will build on the FTAs with Jordan, Israel, and now Morocco; provide assistance to build trade capacity and expand trade so countries can benefit from integration into the global trading system. . . .

The U.S.-Jordan FTA entered into force in December 2001. . . . As a result, trade between the United States and Jordan has nearly tripled in only three years. In 2003, the Administration launched free trade negotiations with Morocco, which we are pleased we completed just last week. Immediately upon the agreements entry into force, 95 percent of bilateral trade in industrial and consumer goods will become duty free, the best day-one tariff elimination in a U.S. free trade agreement with a developing country. Our terms with Morocco provide immediate cuts in Moroccan trade barriers to wheat, corn and soybeans, and new access for U.S. beef and poultry; openings for service providers like audiovisual, telecommunications, distribution, and engineering firms; and new opportunities for manufacturers of construction equipment, chemicals and information technology. In January 2004, the United States [also] began free trade negotiations with Bahrain. Morocco and Bahrain have been leaders in reforming their economies and political systems. Our market opening efforts with these two Arab states are part of the opening act in President Bush's Middle East Initiative, which is aimed at fostering prosperity, encouraging openness, and deepening economic and political reforms throughout the region.

In 2004, the United States will continue its efforts to bring Saudi Arabia into the WTO and will expand its network of TIFAs and BITs throughout the region. The United States now has ten TIFAs in the region, most recently signing agreements with Saudi Arabia, Kuwait, and Yemen. We plan to sign TIFAs with Qatar and the United Arab Emirates soon. As additional countries in the Middle East pursue free trade initiatives with the United States, the Administration will work to integrate these arrangements with the goal of creating a region-wide free trade area by 2013.

In Africa, the African Growth and Opportunity Act (AGOA)—enacted in 2000 and expanded in 2002—has created tangible incentives for commercial and economic reform by providing enhanced access to the U.S. market for products from 37 eligible sub-Saharan nations. Enhancements made in 2002 to the African Growth and Opportunity Act improved access for imports from beneficiary sub-Saharan African countries. . . .

To build on this success . . . the United States launched FTA negotiations with the five countries of the Southern African Customs Union (SACU): Botswana, Lesotho, Namibia, South Africa, and Swaziland. The U.S.-SACU FTA will be a first-of-its-kind agreement with sub-Saharan Africa, building U.S. ties with the region even as it strengthens regional integration among the SACU nations.

The bilateral FTAs we have concluded or are pursuing constitute significant markets for the United States. U.S. goods exports to these countries were $66.6 billion in 2003. This would have made them the third largest U.S. export market behind only Canada and Mexico, and ahead of Japan. The economies of these countries totaled $2.5 trillion in 2002 at purchasing power parity exchange rates, which would rank them as the world's sixth largest economy. And most are developing countries that offer significant growth opportunities in years to come. We are laying free trade foundations for win-win economic ties between America and these partners.

Ensuring a Level Playing Field with China

Since China joined the WTO, it has become America's sixth-largest export market. U.S. exports to China grew 75 percent over the last three years, even as U.S. exports to the rest of the world declined because of slow global growth. China has become a major consumer of U.S. manufactured exports, such as electrical machinery, transportation and telecommunications equipment, numerous components, and chemicals. The market share of U.S. service providers in China has also been increasing rapidly in many sectors. Meanwhile, growth in exports to China of agricultural products has been robust; for example, U.S. exports of soybeans reached an all-time high in 2003 of $2.9 billion and cotton exports were $733 million, up 431 percent over 2002.

In 2003, senior Administration officials met frequently with Chinese counterparts to address shortcomings in China's WTO compliance. We delivered a clear message: China must increase the openness of its market and treat U.S. goods and services fairly if support in the United States for an open market with China is to be sustained.

As a result, China has taken steps to correct systemic problems in its administration of the tariff-rate quota (TRQ) system for bulk agricultural commodities, and relaxed certain market constraints in soybeans and cotton trade, enabling U.S. exporters to achieve record prices and sales. Recent approval of biotech soybeans, cotton and corn—and promised additional approvals—has created greater certainty for U.S. exporters. China has also reduced capitalization requirements for financial services, including opening the motor vehicle financing sector.

China's large installment purchases of billions of dollars of U.S. products—including Boeing 777s and 747s, GE and Pratt & Whitney aircraft engines, Ford and General Motors cars, as well as agricultural products—during recent purchasing missions bode well for 2004. However, we continue to stress the need for structural change that ensures ongoing, open, and fair access—not reliance on one-off sales.

In 2004, the Administration will concentrate on ensuring that: American intellectual property [patents, copyrights] rights [IPRs] are protected; U.S. firms are not subject to discriminatory taxation; market access commitments in areas such as agriculture and financial services are fully met; standards are not used—whether for technology or farm products—to unfairly impede U.S. exports; China's trading regime operates transparently; and promises to grant trading and distribution rights are implemented fully and on time. The Administration will consult closely with Congress and interested U.S. stakeholders in continuing to press China for full WTO compliance, and will not hesitate to take action to enforce trade rules.

China's lax enforcement of intellectual property rights, including counterfeiting, is a fundamental issue. Piracy of movies, music and software is so rampant in China that the practices could subvert the development of knowledge industries and stifle innovation around the world. The scope and magnitude of the problem does not just threaten outsiders, but China's own citizens as well. Counterfeit automobile brakes, electrical switches, medicines and processed foods with pilfered brand names and poor quality control present health and safety risks throughout China. Premier Wen Jiabao has spoken of the importance of IPR and has assigned Vice Premier Wu Yi, a former trade minister who helped defuse the SARS [Severe Acute Respiratory Syndrome] crisis, to chair a working group on IPR enforcement. She will meet with Secretary [of Commerce Donald L.] Evans and me next month as part of our Joint Commission on Commerce and Trade.

In addition, China has adopted discriminatory tax policies—most blatantly on semiconductors—and new wireless encryption standards intended to block U.S. market access. We are pressing China to resolve these disputes promptly.

At the end of this year China and the United States face another challenge. Our Uruguay Round [of WTO negotiators] commitments, ratified by Congress, required us to begin phasing out our textile and apparel quotas in 1995. That process will be completed at year's end. We have urged the Chinese to recognize concerns raised by this important transition. We are committed to using special safeguards, applying unfair trade laws, such as the anti-dumping provisions, and taking action under international trade rules if China falls short in its trade commitments.

Promoting a Cleaner Environment and Better Working Conditions

No country is doing more than the United States to push for strong labor and environmental provisions in international trade agreements. While some other countries talk about labor and the environment in the context of trade, only the United States is actually doing something to integrate these topics as an active part of its trade agenda.

Following the negotiating objectives set forth by Congress in TPA, we are focused on combining effective enforcement with practical cooperation to improve labor and environmental conditions overseas. Our strategy varies depending on the countries we are negotiating with, because conditions vary and one size does not fit all. But in general, we have a ground-breaking, three-part approach:

First, we often find that the issue with working or environmental conditions is not the laws on the books in developing countries, it is with the enforcement of those laws. So our FTAs require that countries effectively enforce their own labor and environmental laws, backed up by enforceable dispute settlement procedures.

Second, we need to understand and address the reasons that laws are not being enforced. Often in poor countries, it is a resource question. Labor Ministries are often poorly funded, and there is a lack of money devoted to enforcement, inspections, and awareness of worker rights. To address this issue, we are pursuing a cooperative approach . . . to focus on real-world problems, such as a lack of trained inspectors at Labor Ministries, the lack of awareness of employees of their rights under existing laws, and the need for education about child labor. . . . [We] also encourage the development of local civil society, through public participation and transparency so that reforms can be sustained by homegrown efforts.

Third, we want to cooperate with countries to improve their laws where there are gaps. Chile, for example, repealed its [dictator Augusto] Pinochet-era [1973–1990] labor laws during the course of negotiating the FTA with the United States because we took a firm but cooperative approach. Just recently, one of my staff returned from Guatemala with news that the government is working hard to reduce its backlog of worker-rights cases in its courts, because they know CAFTA is coming and they want to improve the climate for investment and trade. El Salvador has significantly expanded funding for its Labor Ministry, with monies targeted especially on inspection and enforcement. Morocco enacted a new Labor code that will take effect this year. These are just a few of the many examples where our combination of enforcement standards and cooperation is helping reform these societies.

Of course, free trade also helps developing countries grow, generating the resources for greater protection of workers' rights and the environment. Growing developing countries build a middle class that calls for better environmental and working conditions. Poor people also want better lives for their families. We will not improve their working conditions or environment by making it harder for them to sell the fruit of their labor.

We are putting this multi-faceted approach to trade and development into practice. The Chile and Singapore FTAs create the basis for cooperative projects to promote respect for international core labor standards and to support environmental protection and sound management of natural resources. Both agreements also require that parties effectively enforce their own environmental and labor laws.

The dispute settlement procedures of the new FTAs apply to all obligations of the agreements and set high standards for openness and transparency, such as open public hearings, public release of legal submissions by parties, and the opportunity for interested third parties to submit views. In all cases, the emphasis is on promoting compliance through consultation, joint action plans, and trade-enhancing remedies.

The FTAs with the Central American countries, Morocco, and Australia adopt similar approaches to labor and environmental provisions, but are each tailored to fit individual circumstances. In Central America, for example, the Administration has emphasized trade capacity building projects to enhance the awareness and enforcement of labor laws. We encouraged countries to work with the International Labor Organization (ILO) to identify areas for improvement in labor laws and enforcement. The ILO study found that while the labor laws on the books were generally good, there were some gaps that needed to be addressed, and enforcement needed to be improved. The CAFTA partners are already responding to a number of these recommendations. We are assisting with trade-capacity building and cooperation to help. . . .

Monitoring and Enforcing Trade Agreements

We take pride in the progress we are making to negotiate new commitments to open markets for American products and workers, but the bulk of the work done day-in and day-out at USTR is to ensure that countries live up to their current commitments or to solve problems for American businesses and workers.

Congress created USTR to assure that trade policy—including enforcement—was centrally located within the Executive Branch. We take USTR's enforcement mandate seriously.

The scope of enforcement extends well beyond the number of cases brought before WTO or NAFTA tribunals. On any given day, there is a steady stream of U.S. companies in the Winder Building working with us to figure out how best to press foreign governments to live up to their commitments to open up their markets to U.S. goods and services.

The vast majority of enforcement efforts by USTR are brought to successful resolution without the need to resort to formal litigation. Most U.S. companies urge us to do everything that we can to resolve a problem without bringing a WTO or NAFTA case, given the amount of time such cases take.

In recent years, informal means of resolving trade issues have enabled biotech farm exports and key U.S. financial services to expand their access to the Chinese market. Japan has agreed to lower customs fees by 50 percent as well as increase intellectual property protections. Mexico has implemented rules for pharmaceuticals that respect U.S. patents, and Canada has dropped copyright legislation opposed by U.S. firms that use the internet. We solved pork, poultry, dry bean, and beef issues with Mexico. We increased access for poultry, pork, and beef in Russia. We addressed rice and motorcycle export problems and are improving IPR protection in Taiwan. We headed off Korea's attempt to close the market to Dodge Dakotas based on questionable tax classifications. We encouraged Hong Kong to clean up illegal production of optical discs. The list goes on and on.

But sometimes enforcement can only be achieved through litigation, and we stand prepared to bring WTO and NAFTA cases to secure compliance.

Some of our recent WTO victories include:

An important case against Mexico on telecommunications worth $500 million, according to industry. Under current law, Mexico allows its dominant company, Telmex, the exclusive authority to negotiate, on behalf of all carriers, the rate that U.S. telecom companies must pay to complete their calls in Mexico. These exorbitant rates penalize American and Mexican families seeking to maintain cross-border ties, raise the price of doing business across the border, and burden U.S. telecom firms with unnecessary costs.

In December 2003, the United States won a major case before the WTO holding that Japan's import restrictions on U.S. apples are a violation of Japan's WTO obligations. Japan had argued that the restrictions were needed to protect Japanese plants from disease, but U.S. scientific evidence showed the apples could not transmit the disease. This is a valuable precedent against others that might use Sanitary/Phytosanitary Standards (SPS) to block farm products unfairly.

The United States won an important victory in June 2003 when the WTO rejected India's challenge to U.S. laws on determining the country of origin of textile and apparel products.

We have pending cases against: the European Union's [EU] ban on new imports of genetically-modified foods and against the EU's over-reaching on Geographic Indicators; Mexico's questionable anti-dumping duties on beef and rice; Canada's discriminatory practices affecting wheat; and against Egypt's textile tariffs.

As noted earlier in my testimony, we are focusing more of our enforcement resources on China. While some of China's compliance problems were initially viewed as growing pains as it brought laws and regulations into line with new WTO obligations, China must do more to ensure that it is living up to obligations. Without more progress on matters we have been pressing with China, we will certainly need to avail ourselves of our rights under the WTO.

Of course, our ability to demand that others follow the trade rules is strengthened when we address cases we lose. We very much appreciate the Committee's efforts to repeal the FSC law to end retaliation against U.S. exporters. We also look to work with Congress to remedy other U.S. violations, including the Continued Dumping and Subsidy Offset Act of 2000, the 1916 Act (reflecting early antitrust practice), Section 211 of the Omnibus Appropriations Act of 1998 concerning conditions that permit the banning of trademark enforcement, and the ruling on hot-rolled steel. America should not be a scofflaw of international trade rules.

Conclusion

. . . During 2004, we hope to continue to push forward step-by-step toward the vision set out by President Bush of "a world that trades in freedom." It is a vision of a world in which a working family can save money on everyday household items because trade agreements have cut hidden import taxes. It is a vision of a world in which a Central Valley farmer, a New York financial planner, a Michigan auto worker, a New Orleans longshoreman, an Illinois manufacturer of excavators, or an Iowa pork producer can sell his or her products or services in Costa Rica or Australia or Thailand or Morocco as well as across America. It is a vision of a world in which free trade opens minds as it opens markets, supporting democracy and encouraging tolerance. And it is a vision of a world in which hundreds of millions of people are lifted from poverty through economic growth fueled by trade.

POSTSCRIPT

Is Current U.S. Trade Policy Harming the American Economy?

The debate over the impact of current U.S. trade policy on American jobs and the U.S. economy is in many ways a subset of the larger question debated in Issue 9 over the benefits of free economic interchange. As for the specific issue debated here about of the impact of trade on jobs in the context of the foreign economic policy of the Bush administration, it is a product of the economic doldrums that beset the American economy during Bush's first three years in office and the dynamics of the 2004 presidential campaign.

These factors and the general increase in concern about free economic interchange and globalization are reflected in a cooling of American opinion toward trade and other forms of unfettered economic interchange among countries. A report by University of Maryland's Program on International Policy Attitudes (PIPA) reveals that support for free trade among Americans has dropped off substantially from what it once was and that job losses are a particular concern. That report can be accessed on the PIPA Web site at http://www.pipa.org/ by clicking the hyperlink "international trade" in the right column. Indeed most Americans view job protection as the most important aspect of trade policy. This was clear in a March 2004 poll by the Pew Research Center that found 49 percent of Americans saying the "top priority" of U.S. trade policy should be "protecting the jobs of American workers." This and other related polls are available on the Pew site at http://people-press.org/.

It is safe to say that the popular discontent with free trade and the issue's entry into the electoral arena has alarmed most economists who view the possibilities of renewed protectionism with alarm. For example, one of the Federal Reserve's regional banks warns "In whatever guise, protectionism is pure poison for an economy." The bank's report examines 20 U.S. industries that are being protected by unusually high tariffs and finds that each protected job costs Americans an average of $231,289. Details are available at U.S. Federal Reserve Bank of Dallas, 2002 Annual Report at: http://www.dallasfed.org/fed/annual/2002. Another warning against measures that would restrict trade relating specifically to outsourcing is Daniel W. Drezner, "The Outsourcing Bogeyman," *Foreign Affairs*, May/June 2004, available online at http://www.foreignaffairs.org. For a political view attacking the Kerry position, read Aaron Lukas, "The Party of Protectionism," March 10, 2004, *National Review* online at http://www.nationalreview.com/.

Not all observers agree. Among the sources that argue that trade policy is unsound and that something must be done to stem the flow of jobs is Alan

Tonelson, *The Race to the Bottom: Why a Worldwide Worker Surplus and Uncontrolled Free Trade Are Sinking American Living Standards* (Westview Press, 2002). CNN economic analyst Lou Dobbs also takes this view in "Exporting America: False Choices," March 10, 2004, available at http://money.cnn.com/2004/03/09/commentary/dobbs/dobbs/. For yet another view that calls for change, see U.S. Senator Ernest F. Hollings, "Protectionism Happens to Be Congress's Job," a Washington Post op-ed piece published on March 21, 2004, and available on the senator's Web site at http://hollings.senate.gov/~hollings/opinion/2004321442.html.

Contributors to This Volume

EDITOR

JOHN T. ROURKE, Ph.D., is a professor of political science at the University of Connecticut for campuses in Storrs and Hartford, Connecticut. He has written numerous articles and papers, and he is the author of *Congress and the Presidency in U.S. Foreign Policymaking* (Westview Press, 1985); *The United States, the Soviet Union, and China: Comparative Foreign Policymaking and Implementation* (Brooks/Cole, 1989); and *International Politics on the World Stage*, 8th ed. (McGraw-Hill/Dushkin, 2001). He is also coauthor, with Ralph G. Carter and Mark A. Boyer, of *Making American Foreign Policy*, 2d ed. (Brown & Benchmark, 1996) and editor of *Taking Sides: Clashing Views on Controversial Issues in American Foreign Policy* (McGraw-Hill/Dushkin), now in its second edition. Professor Rourke enjoys teaching introductory political science classes—which he does each semester—and he plays an active role in the university's internship program as well as advising one of its political clubs. In addition, he has served as a staff member of Connecticut's legislature and has been involved in political campaigns at the local, state, and national levels.

STAFF

Larry Loeppke   Managing Editor
Jill Peter   Senior Developmental Editor
Nichole Altman   Developmental Editor
Lori Church   Permissions Coordinator
Beth Kundert   Production Manager
Jane Mohr   Project Manager
Kari Voss Lead Typesetter
Craig Purcell   eContent Coordinator
Maggie Lytle   Cover Designer

AUTHORS

BRUCE BERKOWITZ is a research fellow at the Hoover Institution at Stanford University. His work focuses on defense, intelligence, and technology policy. Berkowitz is currently serving as senior consultant in the Office of the Secretary of Defense. He is the author of several books about national security affairs, including *Best Truth: Intelligence in the Information Age* (Yale University Press, 2000) and *The New Face of War: How War Will Be Fought in the Twenty-First Century* (Free Press, 2003).

P. J. BERLYN is an author of Israelite history and culture and a former associate editor for the *Jewish Bible Quarterly* in Jerusalem. She has also worked for the Council on Foreign Relations in New York, as well as its journal *Foreign Affairs.* She is coauthor, with Shimon Bakon, of *Shani—Her Adventures Beyond the Sambatyon* (En-Gedi Books, 2000).

JOHN R. BOLTON is senior vice president of the American Enterprise Institute. During the presidential administration of George Bush, he served as the assistant secretary of state for international organization affairs. A Yale University–educated lawyer, Bolton has held a variety of posts in both the Reagan and Bush administrations.

JOSÉ BOVÉ, a French farmer and unionist, is an anti-globalization activist. He is best known for leading a group of activists that ransacked a McDonald's outlet in France in 1999 to protest industrialized food.

GEORGE W. BUSH is the 43rd president of the United States. Prior to that, he was the owner of an oil and gas business, managing general partner of the Texas Rangers, and governor of Texas, a position he held for six years. He earned his M.A. in business administration from Harvard Business School in 1975, and he is the author of *A Charge to Keep* (William Morrow, 1999).

MARY CAPRIOLI is an assistant professor of political science at the University of Tennessee. She earned her Ph.D. from the University of Connecticut in 1999.

BILL CHRISTISON served on the analysis side of the CIA for 28 years. From the early 1970s he served as national intelligence officer (at various times) for Southeast Asia, South Asia, and Africa. Before his retirement in 1979, he was director of the CIA's Office of Regional and Political Analysis.

DOCTORS WITHOUT BORDERS/MÉDECINS SANS FRONTIÈRES is an independent humanitarian medical aid agency that is committed to providing medical aid wherever it is needed, regardless of race, religion, politics, or gender, and to raising awareness of the plight of the people that it helps.

WILLIAM EASTERLY is senior adviser in the Development Research Group at the World Bank and the author of *The Elusive Quest for Growth: Economists' Adventures and Misadventures in the Tropics* (MIT Press, 2001).

IVAN ELAND is director of defense policy studies at the Cato Institute. He has also been principal defense analyst at the Congressional Budget Office and an investigator for the U.S. General Accounting Office in national security and intelligence. He has an M.B.A. and a Ph.D. in national security policy from George Washington University, and he is the author of *Putting "Defense" Back Into U.S. Defense Policy: Rethinking U.S. Security in the Post-Cold War World* (Greenwood/Praeger, 2001).

RICHARD D. FISHER, JR., is a senior fellow of the Jamestown Foundation and the managing editor of *China Brief: A Journal of Information and Analysis.* He has also served as senior policy analyst for the Asian Studies Center of the Heritage Foundation.

THOMAS FRIEDMAN is a columnist for the *New York Times* and the author of *The Lexus and the Olive Tree: Understanding Globalization* (Farrar, Straus & Giroux, 1999).

FRANCIS FUKUYAMA, a former deputy director of the U.S. State Department's policy planning staff, is a senior researcher at the RAND Corporation in Santa Monica, California. He is also a fellow of the Johns Hopkins University School for Advanced International Studies' Foreign Policy Institute and director of its telecommunications project. He is the author of *The Great Disruption* (Simon & Schuster, 1999).

ROMILLY GREENHILL is the Jubilee research economist at the New Economics Foundation, a radical think tank that creates practical and enterprising solutions to the social, environmental, and economic challenges facing local, regional, national, and global economies. She also spent two years as an Overseas Development Institute fellow in Uganda.

WILLIAM NORMAN GRIGG is senior editor of *The New American* and host of the John Birch Society's biweekly audio commentary *Review of the News Online.* He is the author of a number of books, including *Global Gun Grab: The United Nations Campaign to Disarm Americans* (John Birch Society, 2001).

JOHN HILLEN is a policy analyst for defense and national security policy at the Heritage Foundation and the author of *Blue Helmets in War and Peace: The Strategy of UN Military Operations* (Brassey's, 1997).

ALAN F. HOLMER is president and chief executive officer of the Pharmaceutical Research and Manufacturers of America.

JOHN C. HULSMAN is a research fellow for European Affairs in the Kathryn and Shelby Cullom Davis Institute for International Policy at the Heritage Foundation. His areas of expertise include the European Union, NATO, U.S.-European trade, and the global war on terrorism. A frequent commentator on all aspects of transatlantic relations, global geopolitics, and international cooperation in fighting terrorism, Hulsman makes regular appearances on major media outlets, such as ABC, Fox News, CNN, MSNBC, and PBS.

A. ELIZABETH JONES is the assistant secretary for the Bureau of European and Eurasian Affairs at the U.S. Department of State. She has also served as ambassador to the Republic of Kazakhstan; deputy director for Lebanon, Jordan, Syria, and Iraq; principal deputy assistant secretary in the Near East Bureau; and as executive assistant to Secretary of State Warren Christopher.

FRED KAPLAN writes the "War Stories" column in *Slate* as well as occasional pieces on music and consumer electronics. He is also a former staff reporter for the *Boston Globe,* for which he served as military correspondent, Moscow bureau chief, and New York bureau chief, and he has written on a variety of subjects for such publications as the *New York Times,* the *Atlantic, The New Yorker,* and *Scientific American.* He earned his Ph.D. in political science from the Massachusetts Institute of Technology, and he is the author of *The Wizards of Armageddon* (Simon & Schuster, 1983).

ROBERT KAPLAN is a correspondent for *The Atlantic Monthly* and the author of *Warrior Politics: Why Leadership Demands a Pagan Ethos* (Random House, 2002).

STEPHEN D. KRASNER is the Graham H. Stuart Professor of Political Science at Stanford University. Krasner earned his Ph.D. from Harvard University. He served as editor for *International Organization* from 1987 to 1992, and he is the author of *Sovereignty: Organized Hypocrisy* (Princeton University Press, 1999).

ANNE O. KRUEGER is the first deputy managing director of the International Monetary Fund. Prior to taking this position, she was the Herald L. and Caroline L. Ritch Professor in Humanities and Sciences in the Department of Economics at Stanford University. She was also director of Stanford's Center for Research on Economic Development and Policy Reform and a senior fellow of the Hoover Institution. From 1982 to 1986 Krueger was the World Bank's vice president for economics and research. She earned her Ph.D. in economics from the University of Wisconsin.

FRED KRUPP is president of the Environmental Defense Fund. He has also been a member of President Bill Clinton's Advisory Committee for Trade Policy and Negotiations, a member of Clinton's Commission on Sustainable Development, and an advisory board member of the Environmental Media Association. Identified as a key figure behind the congressional passage of the 1990 Clean Air Act, Krupp also led the Environmental Defense Fund's delegation for the Buenos Aires climate change negotiations. He holds a J.D. from the University of Michigan.

LAWYERS COMMITTEE FOR HUMAN RIGHTS is a New York–based civil rights advocacy group. The committee seeks to influence the U.S. government to promote the rule of law in both its foreign and domestic policy and presses for greater integration of human rights into the work of the UN and the World Bank. The committee works to protect refugees through the representation of asylum seekers and by challenging legal restrictions on the rights of refugees in the United States and around the world.

BRINK LINDSEY is a senior fellow and director of the Center for Trade Policy Studies at the Cato Institute. An attorney with extensive experience in international trade regulation, he also served as director of regulatory studies at the Cato Institute and senior editor of *Regulation* magazine. He earned his A.B. from Princeton University in 1984 and his J.D. from Harvard Law

School in 1987. He is the author of *Against the Dead Hand: The Uncertain Struggle for Global Capitalism* (John Wiley, 2002).

BJØRN LOMBORG is an associate professor of statistics in the Department of Political Science at the University of Aarhus in Denmark and a frequent participant in topical coverage in the European media. His areas of professional interest include the simulation of strategies in collective action dilemmas, the use of surveys in public administration, and the use of statistics in the environmental arena. In February 2002 Lomborg was named director of Denmark's national Environmental Assessment Institute. He earned his Ph.D. from the University of Copenhagen in 1994.

JOHN MUELLER holds the Woody Hayes Chair of National Security Studies and is a professor of political science at Ohio State University. His interests include international politics, foreign policy, defense policy, public opinion, and democratization. Mueller has been a visiting fellow at the Brookings Institution in Washington, D.C.; the Hoover Institution at Stanford University; and the Nobel Institute in Oslo, Norway. He is the author of *Capitalism, Democracy, and Ralph's Pretty Good Grocery* (Princeton University Press, 1999).

BENJAMIN NETANYAHU is the youngest man ever to become prime minister of Israel. During his tenure, he helped to transform Israel's socialist economy to a modern, high-tech one governed by free-market principles. A former Likud Party chairman, he has also served as deputy chief of mission in the Israeli Embassy in Washington, D.C.; Israel's ambassador to the United Nations; and senior manager of Rim Industries in Jerusalem. He earned his M.S. in management studies from the Massachusetts Institute of Technology, and he is the author or editor of several books, including *A Durable Peace: Israel and Its Place Among the Nations* (Warner Books, 2000).

CHRISTOPHER E. PAINE is a senior research associate in the nuclear program and codirector of the Nuclear Warhead Elimination and Nonproliferation Project at the National Resources Defense Council. He is the author or coauthor of over 60 articles on arms control and defense policy.

MARGOT PATTERSON is a senior writer for *National Catholic Reporter.*

KENNETH ROGOFF is an economic counselor and director of research for the International Monetary Fund. A former professor of economics at Harvard University and professor of economics and international affairs at Princeton University, he has also served as research associate for the National Bureau of Economic Research and coeditor of *NBER Macroeconomics Annual.* Rogoff earned his M.A. in economics from Yale University in 1975 and his Ph.D. from the Massachusetts Institute of Technology in 1980, and he is coeditor of *Workbook for Foundations of International Macroeconomics* (MIT Press, 1998).

LIONEL ROSENBLATT is president emeritus of Refugees International, a Washington-based advocacy organization.

DONALD H. RUMSFELD is the 21st U.S. secretary of defense. A former U.S. Navy pilot, he also served as the 13th secretary of defense, the White

House chief of staff, U.S. ambassador to NATO, U.S. congressman, and chief executive officer of two *Fortune* 500 companies, G. D. Searle & Co. and General Instrument Corporation. In 1977 he was awarded the Presidential Medal of Freedom.

ROSEMARY E. SHINKO teaches in the Department of Political Science at the University of Connecticut, where she is working toward her Ph.D. in international relations and political theory.

JOHN STEINBRUNER is a professor of public policy at the University of Maryland and director of the Center for International and Security Studies at Maryland. He is also a nonresident senior fellow at the Brookings Institution and an academic adviser at the Carnegie Corporation of New York. His work has focused on issues of international security and related problems of international policy.

JOSEPH STIGLITZ is a professor at Columbia's Graduate School of Business and a former chief economist for the World Bank. He has also been chairman of the President's Council of Economic Advisers and vice president of the American Economic Association, for which he is founding editor of *Journal of Economic Perspectives.* Stiglitz was awarded the Nobel Prize in economics in 2001, and he has also received the John Bates Clark Medal. He earned his Ph.D. from the Massachusetts Institute of Technology, and he is the author of *Globalization and Its Discontents* (W. W. Norton, 2002).

LARRY THOMPSON is director of advocacy for Refugees International, a Washington-based advocacy organization.

KIMBERLY WEIR is an assistant professor of political science at Northern Kentucky University. She was director of the GlobalEd High School Simulation Project, which was created to systematically and scientifically evaluate perceived gender differences in leadership and decision-making styles and values and in approaches to technology. She holds a Ph.D. in political science from the University of Connecticut and an M.A. from Villanova University.

Index